SOIL MECHANICS

SOIL MECHANICS

Fourth Edition

R. F. CRAIG
Department of Civil Engineering
University of Dundee

Van Nostrand Reinhold (UK) Co. Ltd

First published 1974
Reprinted 1975, 1976, 1977
Second edition 1978
Reprinted 1979, 1980, 1982
Third edition 1983
Reprinted 1984, 1985, 1986
Fourth edition 1987
Reprinted 1987

Published by Van Nostrand Reinhold (UK) Co. Ltd
Molly Millars Lane, Wokingham, Berkshire, England

Typeset in Times
by Best-set Typesetter Ltd., Hong Kong

Printed and Bound in Great Britain by
T.J. Press (Padstow) Ltd, Padstow, Cornwall

British Library Cataloguing in Publication Data
Craig, R.F.
  Soil mechanics.—4th ed.
  1. Soil mechanics
  I. Title
  624.1′5136   TA710

ISBN 0–278–00019–3

Preface

This book is intended primarily to serve the needs of the undergraduate civil engineering student and aims at the clear explanation, in adequate depth, of the fundamental principles of soil mechanics. The understanding of these principles is considered to be an essential foundation upon which future practical experience in soils engineering can be built. The choice of material involves an element of personal opinion but the contents of this book should cover the requirements of most undergraduate courses to honours level. It is assumed that the student has no prior knowledge of the subject but has a good understanding of basic mechanics. The book includes a comprehensive range of worked examples and problems set for solution by the student to consolidate understanding of the fundamental principles and illustrate their application in simple practical situations. The International System of Units is used throughout the book. A list of references is included at the end of each chapter as an aid to the more advanced study of any particular topic. It is intended also that the book will serve as a useful source of reference for the practising engineer.

In the fourth edition no changes have been made to either the aims or structure of the book. However, several improvements and additions have been made. In particular the sections on the analysis of sheet pile walls have been revised, parts of the chapter on bearing capacity have been up-dated and a new section on the pressuremeter test has been added.

The author wishes to record his thanks to the various publishers, organisations and individuals who gave permission for the use of figures and tables of data, and to acknowledge his dependence on those authors whose works provided sources of material. The author also wishes to express his gratitude to Dr. Ian Christie of the University of Edinburgh for reading the draft manuscript and offering a number of suggestions for improvement. Thanks are also due to Miss Evelyn Clark and Mrs. Sandra Nicoll for their careful work in typing the manuscript.

Material from BS 8004:1986 (Code of Practice for Foundations) and BS 5930:1981 (Code of Practice for Site Investigations) is reproduced by permission of the British Standards Institution, 2 Park Street, London WIA 2BS, from whom complete copies of the codes can be obtained.

Dundee
September 1986

Robert F. Craig

Contents

The unit for stress and pressure used in this book is kN/m^2 (kilonewton per square metre) or, where appropriate, MN/m^2 (meganewton per square metre). In SI the special name for the unit of stress or pressure is the *pascal* (Pa) equal to $1\,N/m^2$ (newton per square metre). Thus:

$1\,kN/m^2 = 1\,kPa$ (kilopascal)

$1\,MN/m^2 = 1\,MPa$ (megapascal)

CHAPTER 1

Basic Characteristics of Soils

1.1 The Nature of Soils

To the civil engineer, soil is any uncemented or weakly cemented accumulation of mineral particles formed by the weathering of rocks, the void space between the particles containing water and/or air. Weak cementation can be due to carbonates or oxides precipitated between the particles or due to organic matter. If the products of weathering remain at their original location they constitute a residual soil. If the products are transported and deposited in a different location they constitute a transported soil, the agents of transportation being gravity, wind, water and glaciers. During transportation the size and shape of particles can undergo change and the particles can be sorted into size ranges.

The destructive process in the formation of soil from rock may be either physical or chemical. The physical process may be erosion by the action of wind, water or glaciers, or disintegration caused by alternate freezing and thawing in cracks in the rock. The resultant soil particles retain the same composition as that of the parent rock. Particles of this type are approximately equidimensional and are described as being of 'bulky' form: the particles may be angular, subangular or rounded. The particles occur in a wide range of sizes, from boulders down to the fine rock flour formed by the grinding action of glaciers. The structural arrangement of bulky particles (Fig. 1.1) is described as *single grain*, each particle being in direct

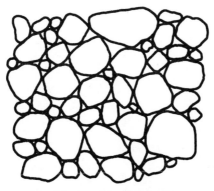

Fig. 1.1   Single grain structure.

contact with adjoining particles without there being any bond or cohesion between them. The structure may be loose, medium dense or dense, depending on the way in which the particles are packed together.

The chemical process results in changes in the mineral form of the parent rock due to the action of water (especially if it contains traces of acid or alkali), oxygen and carbon dioxide. Chemical weathering results in the formation of groups of crystalline particles of colloidal size (<0.002 mm) known as the clay minerals. The clay mineral kaolinite, for example, is formed by the breakdown of felspar by the action of water and carbon dioxide. Most clay mineral particles are of 'plate-like' form having a high specific surface (i.e. a high surface area to mass ratio) with the result that their properties are influenced significantly by surface forces. Long 'needle-shaped' particles can also occur but are comparatively rare.

The basic structural units of most clay minerals consist of a silica tetrahedron and an alumina octahedron (Fig. 1.2a). Silicon and aluminium may be partially replaced by other elements in these units, this being known as isomorphous substitution. The basic units combine to form sheet structures which are represented symbolically in Fig. 1.2b. The various clay minerals are formed by the stacking of combinations of the basic sheet structures with different forms of bonding between the combined sheets. The structures of the principal clay minerals are represented in Fig. 1.3.

Kaolinite consists of a structure based on a single sheet of silica tetrahedrons combined with a single sheet of alumina octahedrons. There is very limited isomorphous substitution. The combined silica-alumina sheets are held together fairly tightly by hydrogen bonding: a kaolinite particle

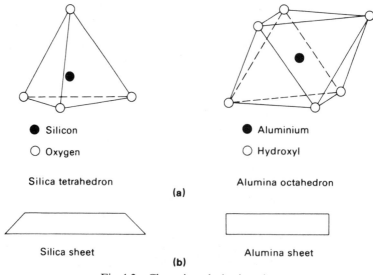

Fig. 1.2   Clay minerals: basic units.

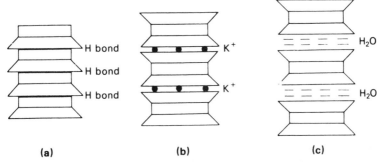

Fig. 1.3　Clay minerals: (a) kaolinite, (b) illite, (c) montmorillonite.

may consist of over one hundred stacks. *Illite* has a basic structure consisting of a sheet of alumina octahedrons between and combined with two sheets of silica tetrahedrons. In the octahedral sheet there is partial substitution of aluminium by magnesium and iron, and in the tetrahedral sheet there is partial substitution of silicon by aluminium. The combined sheets are linked together by fairly weak bonding due to (non-exchangeable) potassium ions held between them. *Montmorillonite* has the same basic structure as illite. In the octahedral sheet there is partial substitution of aluminium by magnesium. The space between the combined sheets is occupied by water molecules and (exchangeable) cations other than potassium. There is a very weak bond between the combined sheets due to these ions. Considerable swelling of montmorillonite can occur due to additional water being adsorbed between the combined sheets.

The surfaces of clay mineral particles carry residual negative charges, mainly as a result of the isomorphous substitution of aluminium or silicon atoms by atoms of lower valency, but also due to disassociation of hydroxyl ions. Unsatisfied charges due to 'broken bonds' at the edges of the particles also occur. The negative charges result in cations present in the water in the void space being attracted to the particles. The cations are not held strongly and, if the nature of the water changes, can be replaced by other cations, a phenomenon referred to as *cation exchange*.

The cations are attracted to a clay mineral particle because of the negative surface charges but at the same time tend to move away from each other because of their thermal energy. The net effect is that the cations form a dispersed layer adjacent to the particle, the cation concentration decreasing with increasing distance from the surface until the concentration becomes equal to that in the 'normal' water in the void space. The term *double layer* describes the negatively charged particle surface and the dispersed layer of cations. For a given particle the thickness of the cation layer depends mainly on the valency and concentration of the cations: an increase in valency (due to cation exchange) or an increase in concentration will result in a decrease in layer thickness. Temperature also affects cation layer thickness, an increase in temperature resulting in a decrease in layer thickness.

Layers of water molecules are held round a clay mineral particle by hydrogen bonding and (because water molecules are dipolar) by attraction to the negatively charged surfaces. In addition the exchangeable cations attract water (i.e. they become hydrated). The particle is thus surrounded by a layer of *adsorbed water*. The water nearest the particle is strongly held and appears to have a high viscosity: the viscosity decreases with increasing distance from the particle surface to that of 'free' water at the boundary of the adsorbed layer. Adsorbed water molecules can move relatively freely parallel to the particle surface but movement perpendicular to the surface is restricted.

Forces of repulsion and attraction act between adjacent clay mineral particles. Repulsion occurs between the like charges of the double layers, the force of repulsion depending on the characteristics of the layers. An increase in cation valency or concentration will result in a decrease in repulsive force, and vice versa. Attraction between particles is due to short-range van der Waals forces; these forces are independent of the double layer characteristics and decrease rapidly with increasing distance between particles. The net interparticle forces influence the structural form assumed by clay mineral particles. If there is net repulsion the particles tend to assume a face-to-face orientation, this being referred to as a *dispersed* structure. If, on the other hand, there is net attraction the orientation of the particles tends to be edge-to-face or edge-to-edge, this being referred to as a *flocculated* structure. These structures, involving interaction between single clay mineral particles, are illustrated in Fig. 1.4a and b.

In natural clays, which normally contain a significant proportion of larger, bulky particles, the structural arrangement can be extremely complex. Interaction between single clay mineral particles is rare, the tendency being for the formation of elementary aggregations of particles (also referred to as domains) with a face-to-face orientation. In turn these elementary aggregations combine to form larger assemblages, the structure of which is influenced by the depositional environment. Two possible forms

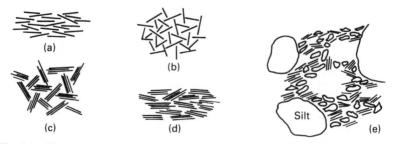

Fig. 1.4   Clay structures: (a) dispersed, (b) flocculated, (c) bookhouse, (d) turbo-stratic, (e) example of a natural clay.

of particle assemblage, known as the bookhouse and turbostratic structures, are illustrated in Fig. 1.4c and d. Assemblages can also occur in the form of connectors or a matrix between larger particles. An example of the structure of a natural clay is shown in Fig. 1.4e.

Particle sizes in soils can vary from over 100 mm to less than 0·001 mm. In British Standards the size ranges detailed in Fig. 1.5 are specified. In Fig. 1.5 the terms 'clay', 'silt', etc. are used to describe only the *sizes* of particles between specified limits. However, the same terms are also used to describe particular *types* of soil. For example clay is a type of soil possessing cohesion and plasticity which normally consists of particles in both the *clay size* and *silt size* ranges. Most types of soil consist of a graded mixture of particles from two or more size ranges. All clay size particles are not necessarily clay mineral particles: the finest rock flour particles may be of clay size. If clay mineral particles are present they usually exert a considerable influence on the properties of a soil, an influence out of all proportion to their percentage by weight in the soil.

In general terms, a soil is considered to be *cohesive* if the particles adhere after wetting and subsequent drying and if significant force is then required to crumble the soil: this does not include soils whose particles adhere when wet due to surface tension.

Soils whose properties are influenced mainly by clay and silt size particles are referred to as *fine* or *fine-grained* soils. Those whose properties are influenced mainly by sand and gravel size particles are referred to as *coarse* or *coarse-grained* soils.

1.2 Particle Size Analysis

The particle size analysis of a soil sample involves determining the percentage by weight of particles within the different size ranges. The particle size distribution of a coarse-grained soil can be determined by the method of *sieving*. The soil sample is passed through a series of standard test sieves having successively smaller mesh sizes. The weight of soil retained in each sieve is determined and the cumulative percentage by weight passing each sieve is calculated. If fine-grained particles are present in the soil, the sample should be treated with a deflocculating agent and washed through the sieves.

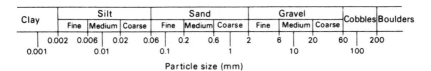

Fig. 1.5 Particle size ranges.

The particle size distribution of a fine-grained soil or the fine-grained fraction of a coarse-grained soil can be determined by the method of *sedimentation*. The method is based on Stokes' law which governs the velocity at which spherical particles settle in a suspension: the larger the particles the greater is the settling velocity and vice versa. The law does not apply to particles smaller than 0·0002 mm, the settlement of which is influenced by Brownian movement. The size of a particle is given as the diameter of a sphere which would settle at the same velocity as the particle. Initially the soil sample is pretreated with hydrogen peroxide to remove any organic material. The sample is then made up as a suspension in distilled water to which a deflocculating agent has been added to ensure that all particles settle individually. The suspension is placed in a sedimentation tube. From Stokes' law it is possible to calculate the time, t, for particles of a certain 'size', D (the equivalent settling diameter), to settle a specified depth in the suspension. If, after the calculated time t, a sample of the suspension is drawn off with a pipette at the specified depth below the surface, the sample will contain only particles smaller than the size D at a concentration unchanged from that at the start of sedimentation. If pipette samples are taken at the specified depth at times corresponding to other chosen particle sizes the particle size distribution can be determined from the weights of the residues. An alternative procedure to pipette sampling is the measurement of the specific gravity of the suspension by means of a special hydrometer, the specific gravity depending on the weight of soil particles in the suspension at the time of measurement. Full details of the determination of particle size distribution by both the sieving and sedimentation methods are given in BS 1377 [1.2].

The particle size distribution of a soil is presented as a curve on a semi-logarithmic plot, the ordinates being the percentage by weight of particles smaller than the size given by the abscissa. The flatter the distribution curve the larger the range of particle sizes in the soil; the steeper the curve the smaller the size range. A coarse-grained soil is described as *well graded* if there is no excess of particles in any size range and if no intermediate sizes are lacking. In general a well graded soil is represented by a smooth, concave distribution curve. A coarse-grained soil is described as *poorly graded* (a) if a high proportion of the particles have sizes within narrow limits (a *uniform* soil) or (b) if particles of both large and small sizes are present but with a relatively low proportion of particles of intermediate size (a *gap-graded* soil). Particle size is represented on a logarithmic scale so that two soils having the same degree of uniformity are represented by curves of the same shape regardless of their positions on the particle size distribution plot. Examples of particle size distribution curves appear in Fig. 1.8. The particle size corresponding to any specified value on the 'percentage smaller' scale can be read from the particle size distribution curve. The size such that 10% of the particles are smaller than that size is denoted by D_{10}. Other sizes such as D_{30} and D_{60} can be defined in a similar way. The size

D_{10} is defined as the *effective size*. The general slope and shape of the distribution curve can be described by means of the *coefficient of uniformity* (C_U) and the *coefficient of curvature* (C_C), defined as follows:

$$C_U = \frac{D_{60}}{D_{10}}$$ (1.1)

$$C_C = \frac{D_{30}^2}{D_{60}D_{10}}$$ (1.2)

The higher the value of the coefficient of uniformity the larger the range of particle sizes in the soil. A well graded soil has a coefficient of curvature between 1 and 3.

1.3 Plasticity of Fine-Grained Soils

Plasticity is an important characteristic in the case of fine-grained soils, the term plasticity describing the ability of a soil to undergo unrecoverable deformation at constant volume without cracking or crumbling. Plasticity is due to the presence of clay minerals or organic material. The physical state of a fine-grained soil at a particular water content (defined as the ratio of the mass of water in the soil to the mass of solid particles) is known as its consistency. Depending on its water content a soil may exist in the liquid, plastic, semi-solid or solid state. The water contents at which the transitions between states occur vary from soil to soil. The consistency depends on the interaction between the clay mineral particles. Any decrease in water content results in a decrease in cation layer thickness and an increase in the net attractive forces between particles. For a soil to exist in the plastic state the magnitudes of the net interparticle forces must be such that the particles are free to slide relative to each other, with cohesion between them being maintained. A decrease in water content also results in a reduction in the volume of a soil in the liquid, plastic or semi-solid state.

Most fine-grained soils exist naturally in the plastic state. The upper and lower limits of the range of water content over which a soil exhibits plastic behaviour are defined as the *liquid limit* (*LL* or w_L) and the *plastic limit* (*PL* or w_P) respectively. The water content range itself is defined as the *plasticity index* (*PI* or I_P), i.e.:

$$I_P = w_L - w_P$$

However the transitions between the different states are gradual and the liquid and plastic limits must be defined arbitrarily. The natural water content (*w*) of a soil relative to the liquid and plastic limits can be represented by means of the *liquidity index* (*LI* or I_L), where:

$$I_L = \frac{w - w_P}{I_P}$$

The degree of plasticity of the clay size fraction of a soil is expressed by the ratio of the plasticity index to the percentage of clay size particles in the soil: this ratio is called the *activity*.

The transition between the semi-solid and solid states occurs at the *shrinkage limit*, defined as the water content at which the volume of the soil reaches its lowest value as it dries out.

The liquid and plastic limits are determined by means of arbitrary test procedures, fully detailed in BS 1377. The soil sample is dried sufficiently to enable it to be crumbled and broken up, using a mortar and a rubber pestle, without crushing individual particles: only material passing a 425 μm BS sieve is used in the tests.

The apparatus for the liquid limit test consists of a penetrometer fitted with a 30° cone of stainless steel, 35 mm long: the cone and the sliding shaft to which it is attached have a mass of 80 g. The test soil is mixed with distilled water to form a thick homogeneous paste and stored for 24 h. Some of the paste is then placed in a cylindrical metal cup, 55 mm internal diameter by 40 mm deep, and levelled off at the rim of the cup to give a smooth surface. The cone is lowered so that it just touches the surface of the soil in the cup, the cone being locked in its support at this stage. The cone is then released for a period of 5 s and its depth of penetration into the soil is measured. A little more of the soil paste is added to the cup and the test is repeated until a consistent value of penetration has been obtained. (The average of two values within 0·5 mm or of three values within 1·0 mm is taken.) The entire test procedure is repeated at least four times using the same soil sample but increasing the water content each time by adding distilled water. The penetration values should cover the range of approximately 15 mm to 25 mm, the tests proceeding from the drier to the wetter state of the soil. Cone penetration is plotted against water content and the best straight line fitting the plotted points is drawn. An example appears in Fig. 1.9. The liquid limit is defined as the percentage water content (to the nearest integer) corresponding to a cone penetration of 20 mm.

In an alternative test for liquid limit the apparatus consists of a flat metal cup, mounted on an edge pivot: the cup rests initially on a hard rubber base. A mechanism enables the cup to be lifted to a height of 10 mm and dropped onto the base. Some of the soil paste is placed in the cup, levelled off horizontally and divided by cutting a groove, on the diameter through the pivot of the cup, using a standard grooving tool. The two halves of the soil gradually flow together as the cup is repeatedly dropped onto the base at a rate of two drops per second. The number of drops, or blows, required to close the bottom of the groove over a distance of 13 mm is recorded. Repeat determinations should be made until two successive determinations give the same number of blows. The water content of the soil in the cup is then determined. This test is also repeated at least four times, the water content of the soil paste being increased for each test: the number of blows should be within the limits of 50 and 10. Water content is plotted against the logarithm of the number of blows and the best straight line fitting the plotted points is drawn. For this test the liquid limit is defined as the water content at which 25 blows are required to close the bottom of the groove over a distance of 13 mm. Also given in BS 1377 are details of the determination of liquid limit based on a single test, provided the number of blows is between 35 and 15.

For the determination of the plastic limit the test soil is mixed with distilled water until it becomes sufficiently plastic to be moulded into a ball. Part of the soil sample (approximately 2·5 g) is formed into a thread, approximately 6 mm in diameter between the first finger and thumb of each hand. The thread is then placed on a glass plate and rolled with the tips of the fingers of one hand until its diameter is reduced to approximately 3 mm: the rolling pressure must be uniform throughout the test. The thread is then remoulded between the fingers (the water content being reduced by the heat of the fingers) and the procedure is repeated until the thread of soil shears both longitudinally and transversely when it has been rolled to a diameter of 3 mm. The procedure is repeated using three more parts of the sample and the percentage water content of all the crumbled soil is determined as a whole. This water content (to the nearest integer) is defined as the plastic limit of the soil. The entire test is repeated using four other sub-samples and the average taken of the two values of plastic limit: the tests must be repeated if the two values differ by more than 0·5%.

If the plastic limit cannot be determined for the given soil or if the plastic limit is equal to or greater than the liquid limit, the soil is reported as being non-plastic (*NP*).

1.4 Soil Description and Classification

It is essential that a standard language should exist for the description of soils. A comprehensive description should include the characteristics of both the soil material and the in-situ soil mass. Material characteristics can be determined from disturbed samples of the soil, i.e. samples having the same particle size distribution as the in-situ soil but in which the in-situ structure has not been preserved. The principal material characteristics are particle size distribution (or grading) and plasticity, from which the soil name can be deduced. Particle size distribution and plasticity properties can be determined either by standard laboratory tests or by simple visual and manual procedures. Secondary material characteristics are the colour of the soil and the shape, texture and composition of the particles. Mass characteristics should ideally be determined in the field but in many cases they can be detected in undistorted samples, i.e. samples in which the in-situ soil structure has been essentially preserved. A description of mass characteristics should include an assessment of in-situ firmness or strength, and details of any bedding, discontinuities and weathering. The arrangement of minor geological details, referred to as the soil macro-fabric, should be carefully described as this can influence the engineering behaviour of the in-situ soil to a considerable extent. Examples of macro-fabric features are thin layers of fine sand and silt in a clay; silt-filled fissures in clay; small lenses of clay in a sand; organic inclusions and root holes. The name of the geological formation, if definitely known, should be included in the description; in addition the type of deposit may be stated (e.g. till, alluvium, river terrace) as this can indicate, in a general way, the likely behaviour of the soil.

It is important to distinguish between soil description and soil classification. Soil description includes details of both material and mass characteristics, therefore it is unlikely that any two soils will have identical descriptions. In soil classification, on the other hand, a soil is allocated to one of a limited number of groups on the basis of material characteristics only (viz. particle size distribution and plasticity): soil classification is thus independent of the in-situ condition of the soil mass. If the soil is to be employed in its undisturbed condition, for example to support a foundation, a full soil description will be adequate and the addition of the soil classification is discretionary. However, classification is particularly useful if the soil in question is to be used as a construction material, for example in an embankment or a fill.

Rapid Assessment Procedures

Soil description and classification both require a knowledge of grading and plasticity. These can be determined by the full laboratory procedure using standard tests, as described in Sections 1.2 and 1.3, in which values defining the particle size distribution and the liquid and plastic limits are obtained for the soil in question. Alternatively, grading and plasticity can be assessed using a rapid procedure which involves personal judgements based on the appearance and feel of the soil. The rapid procedure can be used in the field and in other situations where the use of the laboratory procedure is not possible or not justified. In the rapid procedure the following indicators should be used.

Particles of 0·06 mm, the lower size limit for coarse soils, are just visible to the naked eye and feel harsh but not gritty when rubbed between the fingers; finer material feels smooth to the touch. The size boundary between sand and gravel is 2 mm and this represents about the largest size of particles which will hold together by capillary attraction when moist. A purely visual judgement must be made as to whether the sample is well graded or poorly graded, this being more difficult for sands than for gravels.

If a predominantly coarse soil contains a significant proportion of fine material it is important to know whether the fines are essentially plastic or non-plastic (i.e. whether the fines are predominantly clay or silt respectively). This can be judged by the extent to which the soil exhibits cohesion and plasticity. A small quantity of the soil, with the largest particles removed, should be moulded together in the hands, adding water if necessary. Cohesion is indicated if the soil, at an appropriate water content, can be moulded into a relatively firm mass. Plasticity is indicated if the soil can be deformed without cracking or crumbling, i.e. without losing cohesion. If cohesion and plasticity are pronounced then the fines are plastic. If cohesion and plasticity are absent or only weakly indicated then the fines are essentially non-plastic.

The plasticity of fine soils can be assessed by means of the dry strength, toughness and dilatancy tests described below. Any coarse particles are first removed then a small sample of the soil is moulded in the hand to a consistency judged to be just above the plastic limit; water is added or the soil is allowed to dry as necessary. The procedures are then as follows.

Dry Strength Test. A pat of soil about 6 mm thick is allowed to dry completely either naturally or in an oven. The strength of the dry soil is then assessed by breaking and crumbling between the fingers. Inorganic clays have relatively high dry strength, the greater the strength the higher the liquid limit. Inorganic silts of low liquid limit have little or no dry strength, crumbling easily between the fingers.

Toughness Test. A small piece of soil is rolled out into a thread on a flat surface or on the palm of the hand, moulded together and rolled out again until it has dried sufficiently to break into lumps at a diameter of around 3 mm. In this condition, inorganic clays of high liquid limit are fairly stiff and tough; those of low liquid limit are softer and crumble more easily. Inorganic silts produce a weak and often soft thread which may be difficult to form and readily breaks and crumbles.

Dilatancy Test. A pat of soil, with sufficient water added to make it soft but not sticky, is placed in the open (horizontal) palm of the hand. The side of the hand is then strcuk against the other hand several times. Dilatancy is indicated by the appearance of a shiny film of water on the surface of the pat: if the pat is then squeezed or pressed with the fingers the surface becomes dull as the pat stiffens and eventually crumbles. These reactions are pronounced only for predominantly silt size material and for very fine sands. Plastic clays give no reaction.

Organic soils contain a significant proportion of dispersed vegetable matter which usually produces a distinctive odour and often a dark brown, dark grey or bluish grey colour. Peats consist predominantly of plant remains, usually dark brown or black in colour and with a distinctive odour.

Soil Description Details

A detailed guide to soil description is given in BS 5930 [1.3]. According to this standard the basic soil types are boulders, cobbles, gravel, sand, silt and clay, defined in terms of the particle size ranges shown in Fig. 1.5: added to these are organic clay, silt or sand, and peat. These names are always written in capital letters in a soil description. Mixtures of the basic soil types are referred to as composite types.

A soil is of basic type sand or gravel (these being termed coarse soils) if, after the removal of any cobbles or boulders, over 65% of the material is of sand and gravel sizes. A soil is of basic type silt or clay (termed fine soils) if, after the removal of any cobbles or boulders, over 35% of the material is of silt and clay sizes. Sand and gravel may each be subdivided into coarse, medium and fine fractions, as defined in Fig. 1.5. Sand and gravel can be described as well graded, poorly graded, uniform or gap-graded as defined in section 1.2. In the case of gravels, particle shape (angular, subangular, subrounded, rounded, flat, elongate) and texture (rough, smooth, polished) can be described if necessary. Particle composition can also be described. Gravel particles are usually rock fragments (e.g. sandstone); sand particles usually consist of individual mineral grains (e.g. quartz).

Composite soil types are named in Table 1.1, the predominant component being written in capitals. Deposits containing over 50% of boulders and cobbles are referred to as very coarse and normally can be described only in excavations or exposures. Mixtures of very coarse material with finer soils can be described by combining the descriptions of the two components, e.g. COBBLES with some FINER MATERIAL (sand); gravelly SAND with occasional cobbly BOULDERS.

The firmness or strength of the in-situ soil can be assessed by means of the tests described in Table 1.2.

Table 1.1

Slightly sandy GRAVEL	up to 5% sand
Sandy GRAVEL	5%–20% sand
Very sandy GRAVEL	over 20% sand
GRAVEL/SAND	about equal proportions
Very gravelly SAND	over 20% gravel
Gravelly SAND	5%–20% gravel
Slightly gravelly SAND	up to 5% gravel
Slightly silty SAND (or GRAVEL)	up to 5% silt
Silty SAND (or GRAVEL)	5%–15% silt
Very silty SAND (or GRAVEL)	15%–35% silt
Slightly clayey SAND (or GRAVEL)	up to 5% clay
Clayey SAND (or GRAVEL)	5%–15% clay
Very clayey SAND (or GRAVEL)	15%–35% clay
Sandy SILT (or CLAY)	35%–65% sand
Gravelly SILT (or CLAY)	35%–65% gravel

Table 1.2

Soil type	Term	Field test
Sands, gravels	Loose	Can be excavated with a spade; 50 mm wooden peg can be easily driven
	Dense	Requires a pick for excavation; 50 mm wooden peg is hard to drive
	Slightly cemented	Visual examination; pick removes soil in lumps which can be abraded
Silts	Soft or loose	Easily moulded or crushed in the fingers
	Firm or dense	Can be moulded or crushed by strong pressure in the fingers
Clays	Very soft	Exudes between the fingers when squeezed in the hand
	Soft	Moulded by light finger pressure
	Firm	Can be moulded by strong finger pressure
	Stiff	Cannot be moulded by the fingers; can be indented by the thumb
	Very stiff	Can be indented by the thumb nail
Organic, peats	Firm	Fibres already compressed together
	Spongy	Very compressible and open structure
	Plastic	Can be moulded in the hand and smears the fingers

The terms in Table 1.3 are used to describe the structure of soil deposits. Some examples of soil descriptions are:

Dense, reddish-brown, subangular, well graded, gravelly SAND

Firm, grey, laminated CLAY of low plasticity with occasional silt partings 0·5–2·0 mm

Dense, brown, heterogeneous, well graded, very silty SAND and GRAVEL with some COBBLES: Till

Stiff, brown, closely fissured CLAY of high plasticity: London Clay

Spongy, dark brown, fibrous PEAT

Table 1.3

Homogeneous	Deposit consists essentially of one soil type
Interstratified	Alternating layers of varying types or with bands or lenses of other materials (An interval scale for bedding spacing or layer thickness can be used)
Heterogeneous	A mixture of soil types
Weathered	Coarse particles may be weakened and may show concentric layering Fine soils usually have crumb or columnar structure
Fissured (clays)	Breaks into polyhedral fragments along fissures (Interval scale for spacing of discontinuities may be used)
Intact (clays)	No fissures
Fibrous (peats)	Plant remains are recognizable and retain some strength
Amorphous (peats)	Recognizable plant remains are absent

The British Soil Classification System

The British Soil Classification System is shown in detail in Table 1.4. Reference should also be made to the plasticity chart (Fig. 1.6). The soil groups in the classification are denoted by group symbols composed of main and qualifying descriptive letters having the meanings given in Table 1.5.

The axes of the plasticity chart are plasticity index and liquid limit; thus if these parameters have been determined in the laboratory the plasticity characteristics of a fine soil can be represented by a point on the chart. Classification letters are allotted to the soil according to the zone within which the point lies. The plasticity chart is divided into five ranges of liquid limit; the four highest ranges (I, H, V and E) can be combined as the upper plasticity range (U) if closer designation is not required or if the rapid procedure has been used.

The letter describing the dominant size fraction is placed first in the group symbol. If a soil has a significant content of organic matter the suffix O is added as the last letter of the group symbol. A group symbol may consist of two or more letters, for example:

SW:	well graded SAND
SCL:	very clayey SAND (clay of low plasticity)

Table 1.4 British Soil Classification System for Engineering Purposes

Soil groups			Subgroups and laboratory identification			
GRAVEL and SAND may be qualified sandy GRAVEL and gravelly SAND, etc., where appropriate			Group symbol	Subgroup symbol	Fines (% less than 0.06 mm)	Liquid limit
COARSE SOILS less than 35% of the material is finer than 0·06 mm	GRAVELS More than 50% of coarse material is of gravel size (coarser than 2 mm)	Slightly silty or clayey GRAVEL	GW G GP	GW GPu GPg	0 to 5	
		Silty GRAVEL	G–M G–F	GWM GPM	5 to 15	
		Clayey GRAVEL	G–C	GWC GPC		
		Very silty GRAVEL	GM GF	GML, etc	15 to 35	
		Very clayey GRAVEL	GC	GCL GCI GCH GCV GCE		
	SANDS More than 50% of coarse material is of sand size (finer than 2 mm)	Slightly silty or clayey SAND	SW S SP	SW SPu SPg	0 to 5	
		Silty SAND	S–M S–F	SWM SPM	5 to 15	
		Clayey SAND	S–C	SWC SPC		
		Very silty SAND	SM SF	SML, etc	15 to 35	
		Very clayey SAND	SC	SCL SCI SCH SCV SCE		
FINE SOILS more than 35% of the material is finer than 0·06 mm	Gravelly or sandy SILTS and CLAYS 35% to 65% fines	Gravelly SILT	MG FG	MLG, etc		
		Gravelly CLAY	CG	CLG CIG CHG CVG CEG		< 35 35 to 50 50 to 70 70 to 90 > 90
		Sandy SILT	MS FS	MLS, etc		
		Sandy CLAY	CS	CLS, etc		
	SILTS AND CLAYS 65% to 100% fines	SILT (M-SOIL)	M F	ML, etc		
		CLAY	C	CL CI CH CV CE		< 35 35 to 50 50 to 70 70 to 90 > 90
ORGANIC SOILS		Descriptive letter 'O' suffixed to any group or subgroup symbol				
PEAT		Pt				

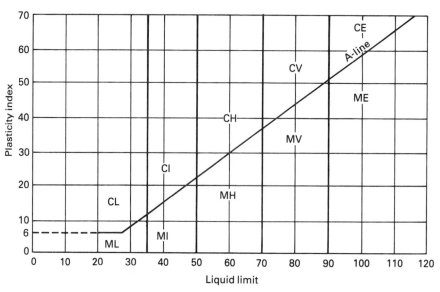

Fig. 1.6 Plasticity chart: British system. (BS 5930: 1981). Reproduced by permission of BSI. Complete copies can be obtained from them at Linford Wood, Milton Keynes, MK14 6LE.

CIS: sandy CLAY of intermediate plasticity
MHSO: organic sandy SILT of high plasticity.

The name of the soil group or subgroup should always be given, as above, in addition to the symbol, the extent of subdivision depending on the particular situation. If the rapid procedure has been used to assess grading and plasticity the group symbol should be enclosed in brackets to indicate the lower degree of accuracy associated with this procedure.

The term FINE SOIL or FINES (F) is used when it is not required, or not possible, to differentiate between SILT (M) and CLAY (C). SILT (M) plots below the *A*-line and CLAY (C) above the *A*-line on the plasticity chart, i.e. silts exhibit plastic properties over a lower range of water content than clays having the same liquid limit. SILT or CLAY is qualified as gravelly if more than 50% of the coarse fraction is of gravel size and as sandy if more than 50% of the coarse fraction is of sand size. The alternative term M-SOIL is introduced to describe material which, regardless of its particle size distribution, plots below the *A*-line on the plasticity chart: the use of this term avoids confusion with soils of predominantly silt size (but with a significant proportion of clay-size particles) which plot *above* the *A*-line. Fine soils containing significant amounts of organic matter usually have high to extremely high liquid limits and plot below the *A*-line as organic silt. Peats usually have very high or extremely high liquid limits.

Any cobbles or boulders (particles retained on a 63 mm BS sieve) are removed from the soil before the classification tests are carried out but their percentages in the total sample should be determined or estimated.

Table 1.5

Main terms		Qualifying terms	
GRAVEL	G	Well graded	W
SAND	S	Poorly graded	P
		Uniform	Pu
		Gap-graded	Pg
FINE SOIL, FINES	F	Of low plasticity (LL < 35)	L
SILT (M-SOIL)	M	Of intermediate plasticity (LL 35–50)	I
CLAY	C	Of high plasticity (LL 50–70)	H
		Of very high plasticity (LL 70–90)	V
		Of extremely high plasticity	
		(LL > 90)	E
		Of upper plasticity range (LL > 35)	U
		Organic (may be a suffix to any group)	O
PEAT	Pt		

Mixtures of soil and cobbles or boulders can be indicated by using the letters Cb (COBBLES) or B (BOULDERS) joined by a + sign to the group symbol for the soil, the dominant component being given first, for example:

GW + Cb: well graded GRAVEL with COBBLES
B + CL: BOULDERS with CLAY of low plasticity

The Unified Soil Classification System

In the Unified Soil Classification System, developed in the United States, the group symbols consist of a primary and a secondary descriptive letter. The letters and their meaning are given in Table 1.6. The Unified system, including the laboratory classification criteria, is detailed in Table 1.7 and the associated plasticity chart is shown in Fig. 1.7. Classification may be based on either laboratory or field test procedures. Soils exhibiting the

Table 1.6

Primary letter		Secondary letter	
G:	Gravel	W:	Well graded
S:	Sand	P:	Poorly graded
M:	Silt	M:	With non-plastic fines
C:	Clay	C:	With plastic fines
O:	Organic soil	L:	Of low plasticity (LL < 50)
Pt:	Peat	H:	Of high plasticity (LL > 50)

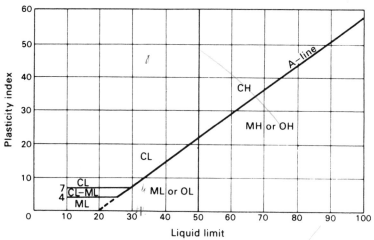

Fig. 1.7   Plasticity chart: Unified system. (Reproduced from Wagner, A. A. (1957) *Proceedings of the Fourth International Conference on Soil Mechanics and Foundation Engineering*, by permission of Butterworth & Co.)

characteristics of two groups should be given a boundary classification denoted by dual symbols connected by a hyphen.

Example 1.1

The results of particle size analyses of four soils A, B, C and D are shown in Table 1.8. The results of limit tests on soil D are:

Liquid limit:

Cone penetration (mm)	15·5	18·0	19·4	22·2	24·9
Water content (%)	39·3	40·8	42·1	44·6	45·6

Plastic limit:

Water content (%)	23·9	24·3

The fine fraction of soil C has a liquid limit of 26 and a plasticity index of 9. (a) Determine the coefficients of uniformity and curvature for soils A, B and C. (b) Classify the four soils according to both the British and Unified systems.

The particle size distribution curves are plotted in Fig. 1.8. For soils A, B and C the sizes D_{10}, D_{30} and D_{60} are read from the curves and the values of C_U and C_C are calculated:

Soil	D_{10}	D_{30}	D_{60}	C_U	C_C
A	0·47	3·5	16	34	1·6
B	0·23	0·30	0·41	1·8	0·95
C	0·003	0·042	2·4	800	0·25

Table 1.7 Unified Soil Classification System
Based on Wagner, A. A. (1957) *Proceedings of the Fourth International Conference SMFE, London*, Vol. 1. Reproduced by permission of Butterworth & Co.

Description		Soil description	Group Symbols	Laboratory criteria			Notes
				Fines (%)	Grading	Plasticity	
Coarse grained (more than 50% larger than 63 μm BS or No. 200 US sieve size)	Gravels (more than 50% of coarse fraction of gravel size)	Well graded gravels, sandy gravels, with little or no fines	GW	0–5	$C_U > 4$ $1 < C_C < 3$		Dual symbols if 5-12% fines.
		Poorly graded gravels, sandy gravels, with little or no fines	GP	0–5	Not satisfying GW requirements		Dual symbols if above A-line and $4 < PI < 7$
		Silty gravels, silty sandy gravels	GM	> 12		Below *A*-line or $PI < 4$	
		Clayey gravels, clayey sandy gravels	GC	> 12		Above *A*-line and $PI > 7$	
	Sands (more than 50% of coarse fraction of sand size)	Well graded sands, gravelly sands, with little or no fines	SW	0–5	$C_U > 6$ $1 < C_C < 3$		
		Poorly graded sands, gravelly sands, with little or no fines	SP	0–5	Not satisfying SW requirements		
		Silty sands	SM	> 12		Below *A*-line or $PI < 4$	
		Clayey sands	SC	> 12		Above *A*-line and $PI > 7$	

Table 1.7   (continued)

Fine grained (more than 50% smaller than 63 μm BS or No. 200 US sieve size)	Silts and clays (liquid limit less than 50)	Inorganic silts, silty or clayey fine sands, with slight plasticity	ML	Use plasticity chart
		Inorganic clays, silty clays, sandy clays of low plasticity	CL	Use plasticity chart
		Organic silts and organic silty clays of low plasticity	OL	Use plasticity chart
	Silts and clays (liquid limit greater than 50)	Inorganic silts of high plasticity	MH	Use plasticity chart
		Inorganic clays of high plasticity	CH	Use plasticity chart
		Organic clays of high plasticity	OH	Use Plasticity chart
Highly organic soils		Peat and other highly organic soils	Pt	

Table 1.8

BS sieve	Particle size*	Percentage smaller			
		Soil A	Soil B	Soil C	Soil D
63 mm		100		100	
20 mm		64		76	
6·3 mm		39	100	65	
2 mm		24	98	59	
600 μm		12	90	54	
212 μm		5	9	47	100
63 μm		0	3	34	95
	0·020 mm			23	69
	0·006 mm			14	46
	0·002 mm			7	31

* From sedimentation test.

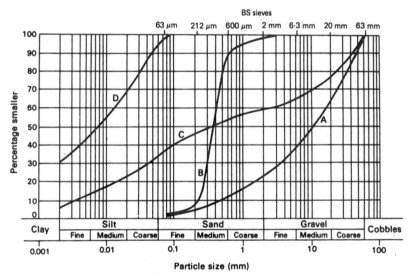

Fig. 1.8 Particle size distribution curves (Example 1.1).

For soil *D* the liquid limit is obtained from Fig. 1.9, in which cone penetration is plotted against water content. The percentage water content, to the nearest integer, corresponding to a penetration of 20 mm is the liquid limit and is 42. The plastic limit is the average of the two percentage water contents, again to the nearest integer, i.e. 24. The plasticity index is the difference between the liquid and plastic limits, namely 18.

Soil A consists of 100% coarse material (76% gravel size; 24% sand size) and is classified as GW: well graded, very sandy GRAVEL.

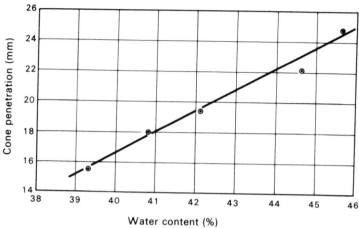

Fig. 1.9   Determination of liquid limit.

Soil B consists of 97% coarse material (95% sand size; 2% gravel size) and 3% fines. It is classified as SPu: uniform, slightly silty, medium SAND.

Soil C comprises 66% coarse material (41% gravel size; 25% sand size) and 34% fines ($LL = 26$, $PI = 9$, plotting in the CL zone on the plasticity chart). The classification is GCL: very clayey GRAVEL (clay of low plasticity). This is a till, a glacial deposit having a large range of particle sizes.

Soil D contains 95% fine material: the liquid limit is 42 and the plasticity index is 18, plotting just above the A-line in the CI zone on the plasticity chart. The classification is thus CI: CLAY of intermediate plasticity.

According to the Unified system the four soils would be classified as GW, SP, GC and CL respectively.

1.5   Phase Relationships

Soils can be of either two-phase or three-phase composition. In a completely dry soil there are two phases, namely the solid soil particles and pore air. A fully saturated soil is also two-phase, being composed of solid soil particles and pore water. A partially saturated soil is three-phase, being composed of solid soil particles, pore water and pore air. The components of a soil can be represented by a phase diagram as shown in Fig. 1.10a. The following relationships are defined with reference to Fig. 1.10a.

The *water content* (w), or *moisture content* (m), is the ratio of the mass of water to the mass of solids in the soil, i.e.:

$$w = \frac{M_w}{M_s}$$

(1.3)

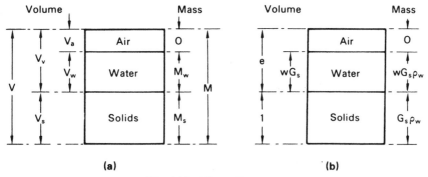

Fig. 1.10 Phase diagrams.

The water content is determined by weighing a sample of the soil then drying the sample in an oven at a temperature of 105–110 °C and reweighing. Drying should continue until the differences between successive weighings at four-hourly intervals are not greater than 0·1% of the original mass of the sample. A drying period of 24 h is normally adequate for most soils. (See BS 1377.)

The *degree of saturation* (S_r) is the ratio of the volume of water to the total volume of void space, i.e.:

$$S_r = \frac{V_w}{V_v} \tag{1.4}$$

The degree of saturation can range between the limits of zero for a completely dry soil and 1 (or 100%) for a fully saturated soil.

The *void ratio* (e) is the ratio of the volume of voids to the volume of solids, i.e.:

$$e = \frac{V_v}{V_s} \tag{1.5}$$

The *porosity* (n) is the ratio of the volume of voids to the total volume of the soil, i.e.:

$$n = \frac{V_v}{V} \tag{1.6}$$

The void ratio and the porosity are inter-related as follows:

$$e = \frac{n}{1 - n} \tag{1.7}$$

$$n = \frac{e}{1 + e} \tag{1.8}$$

The *specific volume* (*v*) is the total volume of soil which contains unit volume of solids, i.e.:

$$v = 1 + e \tag{1.9}$$

The *air content* (*A*) is the ratio of the volume of air to the total volume of the soil, i.e.:

$$A = \frac{V_a}{V} \tag{1.10}$$

The *bulk density* (*ρ*) of a soil is the ratio of the total mass to the total volume, i.e.:

$$\rho = \frac{M}{V} \tag{1.11}$$

Convenient units for density are kg/m^3 or Mg/m^3. The density of water ($1000\ kg/m^3$ or $1\cdot00\ Mg/m^3$) is denoted by ρ_w.

The specific gravity of the *solid soil particles* (G_s) is given by:

$$G_s = \frac{M_s}{V_s \rho_w} \tag{1.12}$$

Procedures for determining the value of G_s are detailed in BS 1377.

From the definition of void ratio, if the volume of solids is 1 unit then the volume of voids is *e* units. The mass of solids is then $G_s \rho_w$ and, from the definition of water content, the mass of water is $wG_s \rho_w$. The volume of water is thus wG_s. These volumes and masses are represented in Fig. 1.10b. The following relationships can now be obtained.

The degree of saturation can be expressed as

$$S_r = \frac{wG_s}{e} \tag{1.13}$$

In the case of a fully saturated soil, $S_r = 1$, hence:

$$e = wG_s \tag{1.14}$$

The air content can be expressed as:

$$A = \frac{e - wG_s}{1 + e} \tag{1.15}$$

or, from Equations 1.13 and 1.8:

$$A = n(1 - S_r) \tag{1.16}$$

The bulk density of a soil can be expressed as:

$$\rho = \frac{G_s(1 + w)}{1 + e} \rho_w \tag{1.17}$$

or, from Equation 1.13:

$$\rho = \frac{G_s + S_r e}{1 + e} \rho_w \tag{1.18}$$

For a fully saturated soil ($S_r = 1$):

$$\rho_{sat} = \frac{G_s + e}{1 + e} \rho_w \tag{1.19}$$

For a completely dry soil ($S_r = 0$):

$$\rho_d = \frac{G_s}{1 + e} \rho_w \tag{1.20}$$

The *unit weight* (γ) of a soil is the ratio of the total weight (a force) to the total volume, i.e.:

$$\gamma = \frac{W}{V} = \frac{Mg}{V}$$

Equations similar to 1.17 to 1.20 apply in the case of unit weights, for example:

$$\gamma = \frac{G_s(1 + w)}{1 + e} \gamma_w \tag{1.17a}$$

$$\gamma = \frac{G_s + S_r e}{1 + e} \gamma_w \tag{1.18a}$$

where γ_w is the unit weight of water. Convenient units are kN/m^3, the unit weight of water being $9 \cdot 8 \, kN/m^3$.

When a soil in-situ is fully saturated the solid soil particles (volume 1 unit, weight $G_s \gamma_w$) are subjected to upthrust (γ_w). Hence the *buoyant unit weight* (γ') is given by:

$$\gamma' = \frac{G_s \gamma_w - \gamma_w}{1 + e} = \frac{G_s - 1}{1 + e} \gamma_w \tag{1.21}$$

i.e.:

$$\gamma' = \gamma_{sat} - \gamma_w \tag{1.22}$$

In the case of sands the *relative density* (D_r) is used to express the relationship between the actual void ratio (e) and the limiting values e_{max} and e_{min}. The relative density is defined as:

$$D_r = \frac{e_{max} - e}{e_{max} - e_{min}} \tag{1.23}$$

Thus the relative density of a soil in its densest possible state ($e = e_{min}$) is 1 (or 100%) and the relative density in its loosest possible state ($e = e_{max}$) is 0.

The minimum void ratio of a sand (or a gravel with a maximum particle size of 20 mm) can be determined by compacting an oven-dry sample in three layers in a standard mould, each layer being compacted by means of a vibrating hammer for a period of $1\frac{1}{2}$ min. The density is obtained from the mass and volume of the compacted sand and the void ratio calculated using Equation 1.20. In an alternative procedure the sand is compacted under water, also using a vibrating hammer.

The maximum void ratio of a sand can be determined by pouring an oven-dry sample through a filter funnel (with opening large enough to allow the sand to pass) into a standard mould, the bottom of the funnel being maintained 5 mm above the surface of the sand in the mould. In an alternative method a 1 kg sample is placed in a 2 l measuring cyclinder (75 mm diameter). A stopper, or the palm of the hand, is placed over the top of the cylinder which is shaken a few times then turned upside down and quickly turned back again: this procedure should ensure that the sand is in its loosest state. The volume of sand is then read and the void ratio calculated. In a third method the sand is placed under water in the funnel and poured into a cylinder filled with water. No reliable procedure has been devised to determine the maximum void ratio of gravels.

Example 1.2

In its natural condition a soil sample has a mass of 2290 g and a volume of $1 \cdot 15 \times 10^{-3} \, \text{m}^3$. After being completely dried in an oven the mass of the sample is 2035 g. The value of G_s for the soil is $2 \cdot 68$. Determine the bulk density, unit weight, water content, void ratio, porosity, degree of saturation and air content.

$$\text{Bulk density, } \rho = \frac{M}{V} = \frac{2 \cdot 290}{1 \cdot 15 \times 10^{-3}} = 1990 \, \text{kg/m}^3 \, (1.99 \, \text{Mg/m}^3)$$

$$\text{Unit weight, } \gamma = \frac{Mg}{V} = 1990 \times 9 \cdot 8 = 19{,}500 \, \text{N/m}^3$$
$$= 19 \cdot 5 \, \text{kN/m}^3$$

$$\text{Water content, } w = \frac{M_w}{M_s} = \frac{2290 - 2035}{2035} = 0 \cdot 125 \text{ or } 12 \cdot 5\%$$

From Equation 1.17,

$$\text{Void ratio, } e = G_s(1 + w)\frac{\rho_w}{\rho} - 1$$

$$= \left(2 \cdot 68 \times 1 \cdot 125 \times \frac{1000}{1990}\right) - 1$$
$$= 1 \cdot 52 - 1$$
$$= 0 \cdot 52$$

Porosity, $n = \dfrac{e}{1+e} = \dfrac{0\cdot52}{1\cdot52} = 0\cdot34$ or 34%

Degree of saturation, $S_r = \dfrac{wG_s}{e} = \dfrac{0\cdot125 \times 2\cdot68}{0\cdot52} = 0\cdot645$ or 64·5%

Air content, $A = n(1 - S_r) = 0\cdot34 \times 0\cdot355$
$$= 0\cdot121 \text{ or } 12\cdot1\%$$

1.6 Soil Compaction

Compaction is the process of increasing the density of a soil by packing the particles closer together with a reduction in the volume of *air*: there is no significant change in the volume of water in the soil. In the construction of fills and embankments, loose soil is placed in layers ranging between 75 mm and 450 mm in thickness, each layer being compacted to a specified standard by means of rollers, vibrators or rammers. In general the higher the degree of compaction the higher will be the shear strength and the lower will be the compressibility of the soil.

The degree of compaction of a soil is measured in terms of dry density, i.e. the mass of solids only per unit volume of soil. If the bulk density of the soil is ρ and the water content w, then from Equations 1.17 and 1.20 it is apparent that the dry density is given by:

$$\rho_d = \frac{\rho}{1+w} \tag{1.24}$$

The dry density of a given soil after compaction depends on the water content and the energy supplied by the compaction equipment (referred to as the *compactive effort*).

The compaction characteristics of a soil can be assessed by means of standard laboratory tests. The soil is compacted in a cylindrical mould using a standard compactive effort. In BS 1377 three compaction procedures are detailed. In the *Proctor* test the volume of the mould is 1000 cm^3 and the soil (with all particles larger than 20 mm removed) is compacted by a rammer consisting of a 2·5 kg mass falling freely through 300 mm: the soil is compacted in three equal layers, each layer receiving 27 blows with the rammer. In the *modified AASHTO* test the mould is the same as is used in the above test but the rammer consists of a 4·5 kg mass falling 450 mm: the soil (with all particles larger than 20 mm removed) is compacted in five layers, each layer receiving 27 blows with the rammer. In the *vibrating hammer* test approximately 2360 cm^3 of soil (with all particles larger than 37·5 mm removed) is compacted in three layers, in a mould 152 mm in diameter, under a circular tamper fitted in the vibrating hammer, each layer being compacted for a period of 60 s.

After compaction using one of the three standard methods, the bulk density and water content of the soil are determined and the dry density calculated. For a given soil the process is repeated at least five times, the water content of the sample being increased each time. Dry density is plotted against water content and a curve of the form shown in Fig. 1.11 is obtained. This curve shows that for a particular method of compaction (i.e. a particular compactive effort) there is a particular value of water content, known as the *optimum water content* (w_{opt}) at which a maximum value of dry density is obtained. At low values of water content most soils tend to be stiff and are difficult to compact. As the water content is increased the soil becomes more workable, facilitating compaction and resulting in higher dry densities. At high water contents, however, the dry density decreases with increasing water content, an increasing proportion of the soil volume being occupied by water.

If all the air in a soil could be expelled by compaction the soil would be in a state of full saturation and the dry density would be the maximum possible value for the given water content. However, this degree of compaction is unattainable in practice. The maximum possible value of dry density is referred to as the 'zero air voids' dry density or the saturation dry density and can be calculated from the expression:

$$\rho_d = \frac{G_s}{1 + wG_s} \rho_w \tag{1.25}$$

In general the dry density after compaction at water content w to an air content A can be calculated from the following expression, derived from Equations 1.15 and 1.20:

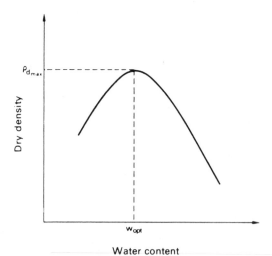

Fig. 1.11   Dry density-water content relationship.

$$\rho_d = \frac{G_s(1 - A)}{1 + wG_s} \rho_w \tag{1.26}$$

The calculated relationship between zero air voids dry density and water content (for $G_s = 2\cdot65$) is shown in Fig. 1.12; the curve is referred to as the zero air voids line or the saturation line. The experimental dry density/water content curve for a particular compactive effort must lie completely to the left of the zero air voids line. The curves relating dry density at air contents of 5% and 10% with water content are also shown in Fig. 1.12, the values of dry density being calculated from Equation 1.26. These curves enable the air content at any point on the experimental dry density/water content curve to be determined by inspection.

For a particular soil, different dry density/water content curves are obtained for different compactive efforts. Curves representing the results of

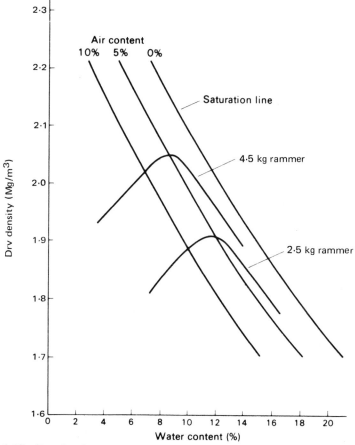

Fig. 1.12 Dry density-water content curves for different compactive efforts.

tests using the 2·5 kg and 4·5 kg rammers are shown in Fig. 1.12. The curve
for the 4·5 kg test is situated above and to the left of the curve for the 2·5 kg
test. Thus a higher compactive effort results in a higher value of maximum
dry density and a lower value of optimum water content: however, the
values of air content at maximum dry density are approximately equal.

The dry density/water content curves for a range of soil types using the
same compactive effort (the BS 2·5 kg rammer) are shown in Fig. 1.13. In
general coarse-grained soils can be compacted to higher dry densities than
fine-grained soils.

The following types of compaction equipment are used in the field.

Smooth-Wheeled Rollers. These consist of hollow steel drums, the mass of
which can be increased by water or sand ballast. They are suitable for most
types of soil except uniform sands and silty sands and provided a mixing or

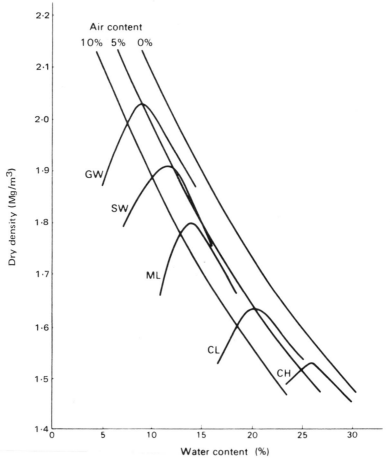

Fig. 1.13   Dry density-water content curves for a range of soil types.

kneading action is not required. A smooth surface is produced on the compacted layer, encouraging the run-off of any rainfall but resulting in relatively poor bonding between successive layers: the fill as a whole will therefore tend to be laminated. Smooth-wheeled rollers, and the other types of roller described below, can be either towed or self-propelled.

Pneumatic-tyred Rollers. This equipment is suitable for a wide range of coarse and fine soils but not uniformly graded material. Wheels are mounted close together on two axles, the rear set overlapping the lines of the front set to ensure complete coverage of the soil surface. The tyres are relatively wide with a flat tread so that the soil is not displaced laterally. This type of roller is also available with a special axle which allows the wheels to wobble, thus preventing the bridging over of low spots. Pneumatic-tyred rollers impart a kneading action to the soil. The finished surface is relatively smooth, resulting in a low degree of bonding between layers. If good bonding is essential, the compacted surface must be scarified between layers. Increased compactive effort can be obtained by increasing the tyre inflation pressure or, less effectively, by adding kentledge to the body of the roller.

Sheepsfoot Rollers. This type of roller consists of hollow steel drums with numerous tapered or club-shaped feet projecting from their surfaces. The mass of the drums can be increased by ballasting. The arrangement of the feet can vary but they are usually from 200–250 mm in length with an end area of 40–65 cm². The feet thus impart a relatively high pressure over a small area. Initially, when the layer of soil is loose, the drums are in contact with the soil surface. Subsequently, as the projecting feet compact below the surface and the soil becomes sufficiently dense to support the high contact pressure, the drums rise above the soil. Sheepsfoot rollers are most suitable for fine soils, both plastic and non-plastic, especially at water contents dry of optimum. They are also suitable for coarse soils with more than 20% of fines. The action of the feet causes significant mixing of the soil, improving its degree of homogeneity, and will break up lumps of stiff material. Due to the penetration of the feet, excellent bonding is produced between successive soil layers, an important requirement for water-retaining earthworks. *Tamping* rollers are similar to sheepsfoot rollers but the feet have a larger end area, usually over 100 cm², and the total area of the feet exceeds 15% of the surface area of the drums.

Grid Rollers. These rollers have a surface consisting of a network of steel bars forming a grid with square holes. Kentledge can be added to the body of the roller. Grid rollers provide high contact pressure but little kneading action and are suitable for most coarse soils.

Vibratory Rollers. These are smooth-wheeled rollers fitted with a power-driven vibration mechanism. They are used for most soil types and are more efficient if the water content of the soil is slightly wet of optimum. They are particularly effective for coarse soils with little or no fines. The mass of the roller and the frequency of vibration must be matched to the soil type and layer thickness. The lower the speed of the roller the fewer the number of passes required.

Vibrating Plates. This equipment, which is suitable for most soil types, consists of a steel plate with upturned edges, or a curved plate, on which a vibrator is mounted. The unit, under manual guidance, propells itself slowly over the surface of the soil.

Power Rammers. Manually controlled power rammers, generally petrol driven, are used for the compaction of small areas where access is difficult or where the use of larger equipment would not be justified. They are also used extensively for the compaction of backfill in trenches. They do not operate effectively on uniformly graded soils.

A minimum number of passes must be made with the chosen compaction equipment to produce the required value of dry density. This number, which depends on the type and mass of the equipment and on the thickness of the soil layer, is usually within the range 3 to 12. Above a certain number of passes no significant increase in dry density is obtained. In general the thicker the soil layer the heavier the equipment required to produce an adequate degree of compaction.

The results of laboratory compaction tests are not directly applicable to field compaction because the compactive efforts in the laboratory tests are different, and are applied in a different way, from those produced by the field equipment. Further, the laboratory tests are carried out only on material smaller than either 20 mm or 37·5 mm size. However, the maximum dry densities obtained in the laboratory using the 2·5 kg and 4·5 kg rammers cover the range of dry density normally produced by field compaction equipment. Laboratory test results provide only a rough guide to the water content at which the maximum dry density will be obtained in the field. The main value of laboratory tests is in the classification and selection of soils for use in earthworks.

The required standard of field compaction can be specified in terms of a minimum percentage of the maximum dry density obtained in one of the standard laboratory tests. For example, it might be stated in a specification that the dry density must not be less than 95% of the maximum dry density in the BS Proctor (2.5 kg rammer) test. In addition water content limits should be specified, compaction being allowed to proceed only if the natural water content of the soil is within these limits. Relatively small variations in particle size distribution can significantly affect the values of maximum dry density and optimum water content. Specification in terms of a percentage of the maximum dry density in a standard test, rather than a stated value of dry density, is therefore preferable because it ensures that a compactive effort comparable to that of the laboratory standard is applied in the field. An alternative method of controlling field compaction is to specify a maximum air content with an associated maximum water content. Water content limits for fine soils are usually stated as a certain percentage above or below the plastic limit and for coarse soils as a certain percentage above or below the optimum water content determined in a standard laboratory test.

Field density tests can be carried out, if considered necessary, to verify the standard of compaction in earthworks, dry density or air content being calculated from measured values of bulk density and water content. A number of methods of measuring bulk density in the field are detailed in BS 1377.

For major projects field compaction trials should be carried out to determine the most suitable type of compaction equipment, the required number of passes and the optimum layer thickness for the soil in question. The required degree of compaction should be obtained using the most economic procedure. In the UK the performance of the various types of compaction equipment on different types of soils has been comprehensively investigated by the Transport and Road Research Laboratory [1.9; 1.10].

Problems

1.1 The results of particle size analyses and, where appropriate, limit tests on samples of four soils are given in Table 1.9. Classify each soil according to the British and Unified systems. only

1.2 A soil has a bulk density of $1.91 \, \text{Mg/m}^3$ and a water content of 9.5%. The value of G_s is 2.70. Calculate the void ratio and degree of saturation of the soil. What would be the values of density and water content if the soil were fully saturated at the same void ratio?

1.3 Calculate the dry unit weight, the saturated unit weight and the buoyant unit weight of a soil having a void ratio of 0.70 and a value of G_s of 2.72. Calculate also the unit weight and water content at a degree of saturation of 75%.

1.4 A soil specimen is 38 mm in diameter and 76 mm long and in its natural condition weighs $168.0 \, \text{g}$. When dried completely in an oven the specimen weighs $130.5 \, \text{g}$. The value of G_s is 2.73. What is the degree of saturation of the specimen?

1.5 Soil has been compacted in an embankment at a bulk density of $2.15 \, \text{Mg/m}^3$ and a water content of 12%. The value of G_s is 2.65. Calculate the dry density, void ratio, degree of saturation and air content. Would it be possible to compact the above soil at a water content of 13.5% to a dry density of $2.00 \, \text{Mg/m}^3$?

1.6 The following results were obtained from a standard compaction test on a soil:

Table 1.9

| BS sieve | Particle size | Percentage smaller | | | |
		Soil E	Soil F	Soil G	Soil H
63 mm					
20 mm		100			
6·3 mm		94	100	—	
2 mm		69	98		
600 μm		32	88	100	
212 μm		13	67	95	100
63 μm		2	37	73	99
	0·020 mm		22	46	88
	0·006 mm		11	25	71
	0·002 mm		4	13	58
Liquid limit			Non-plastic	32	78
Plastic limit				24	31

Mass (g)		2010	2092	2114	2100	2055
Water content (%)		12·8	14·5	15·6	16·8	19·2

The value of G_s is 2·67. Plot the dry density/water content curve and give the optimum water content and maximum dry density. Plot also the curves of zero, 5% and 10% air content and give the value of air content at maximum dry density. The volume of the mould is 1000 cm³.

References

1.1 American Society for Testing and Materials: *Annual Book of ASTM Standards*, Part 19.

1.2 British Standard 1377 (1975): *Methods of Test for Soils for Civil Engineering Purposes*, British Standards Institution, London.

1.3 British Standard 5930 (1981): *Code of Practice for Site Investigations*, British Standards Institution, London.

1.4 British Standard 6031 (1981): *Code of Practice for Earthworks*, British Standards Institution, London.

1.5 Collins, K. and McGown, A. (1974): 'The Form and Function of Microfabric Features in a Variety of Natural Soils', *Geotechnique*, Vol. 24, No. 2.

1.6 Department of Transport (1976): 'Earthworks' in *Specification for Road and Bridge Works*, HMSO, London.

1.7 Grim, R. E. (1953): *Clay Mineralogy*, McGraw-Hill, New York.

1.8 Kolbuszewski, J. J. (1948): 'An Experimental Study of the Maximum and Minimum Porosities of Sands', *Proceedings 2nd International Conference SMFE, Rotterdam*, Vol. 1.

1.9 Lewis, W. A. (1954): *Further Studies in the Compaction of Soil and the Performance of Compaction Plant*, Road Research Technical Paper No. 33, HMSO, London.

1.10 Lewis, W. A. (1960): 'Full Scale Compaction Studies at the British Road Research Laboratory', *US Highway Research Board Bulletin*, No. 254.

1.11 Rowe, P. W. (1972): 'The Relevance of Soil Fabric to Site Investigation Practice', *Geotechnique*, Vol. 22, No. 2.

1.12 Wagner, A. A. (1957): 'The Use of the Unified Soil Classification System by the Bureau of Reclamation', *Proceedings 4th International Conference SMFE, London*, Vol. 1, Butterworths.

CHAPTER 2

Seepage

2.1 Soil Water

All soils are *permeable* materials, water being free to flow through the interconnected pores between the solid particles. The pressure of the pore water is measured relative to atmospheric pressure and the level at which the pressure is atmospheric (i.e. zero) is defined as the *water table* (WT) or the *phreatic surface*. Below the water table the soil is assumed to be fully saturated although it is likely that, due to the presence of small volumes of entrapped air, the degree of saturation will be marginally below 100%. The level of the water table changes according to climatic conditions but the level can change also as a consequence of constructional operations. A *perched* water table can occur locally, contained by soil of low permeability, above the normal water table level. *Artesian* conditions can exist if an inclined soil layer of high permeability is confined locally by an overlying layer of low permeability: the pressure in the artesian layer is governed not by the local water table level but by a higher water table level at a distant location where the layer is unconfined.

Below the water table the pore water may be static, the hydrostatic pressure depending on the depth below the water table, or may be seeping through the soil under hydraulic gradient: this chapter is concerned with the second case. Bernoulli's theorem applies to the pore water but seepage velocities in soils are normally so small that velocity head can be neglected. Thus:

$$h = \frac{u}{\gamma_w} + z \qquad\qquad (2.1)$$

where h = total head, u = pore water pressure, γ_w = unit weight of water (9·8 kN/m^3), and z = elevation head above a chosen datum.

Above the water table, water can be held at negative pressure by capillary tension: the smaller the size of the pores the higher the water can rise above the water table. The capillary rise tends to be irregular due to the random pore sizes occurring in a soil. The soil can be almost completely saturated in the lower part of the capillary zone but in general the degree of saturation decreases with height. When water percolates through the soil from the surface towards the water table some of this water can be held by surface tension around the points of contact between particles. The negative pressure of water held above the water table results in attractive forces between the particles: this attraction is referred to as soil suction and is a function of pore size and water content.

2.2 Permeability

In one dimension, water flows through a fully saturated soil in accordance with Darcy's empirical law:

$$q = Aki \qquad (2.2)$$

or

$$v = \frac{q}{A} = ki$$

where q = volume of water flowing per unit time, A = cross-sectional area of soil corresponding to the flow q, k = coefficient of permeability, i = hydraulic gradient, and v = discharge velocity. The units of the coefficient of permeability are those of velocity (m/s).

The coefficient of permeability depends primarily on the average size of the pores, which in turn is related to the distribution of particle sizes, particle shape and soil structure. In general, the smaller the particles the smaller is the average size of the pores and the lower is the coefficient of permeability. The presence of a small percentage of fines in a coarse-grained soil results in a value of k significantly lower than the value for the same soil without fines. For a given soil the coefficient of permeability is a function of void ratio. If a soil deposit is stratified the permeability for flow parallel to the direction of stratification is higher than that for flow perpendicular to the direction of stratification. The presence of fissures in a clay results in a much higher value of permeability compared with that of the unfissured material.

The coefficient of permeability also varies with temperature, upon which the viscosity of the water depends. If the value of k measured at 20 °C is taken as 100% then the values at 10 °C and 0 °C are 77% and 56% respectively. The coefficient of permeability can also be represented by the equation:

$$k = \frac{\gamma_w}{\eta} K$$

where γ_w is the unit weight of water, η is the viscosity of water and K (units m^2) is an absolute coefficient depending only on the characteristics of the soil skeleton.

The values of k for different types of soil are typically within the ranges shown in Table 2.1. For sands, Hazen showed that the approximate value of k is given by:

Table 2.1   Coefficient of Permeability (m/s) (BS 8004:1986)

1	10^{-1}	10^{-2}	10^{-3}	10^{-4}	10^{-5}	10^{-6}	10^{-7}	10^{-8}	10^{-9}	10^{-10}

Clean gravels	Clean sands and sand-gravel mixtures	Very fine sands, silts and clay-silt laminate	Unfissured clays and clay-silts (> 20% clay)
	Desiccated and fissured clays		

$$k = 10^{-2}D_{10}^2 \quad \text{(m/s)} \tag{2.3}$$

where D_{10} is the effective size in mm.

On the microscopic scale the water seeping through a soil follows a very tortuous path between the solid particles but macroscopically the flow path (in one dimension) can be considered as a straight line. The average velocity at which the water flows through the soil pores is obtained by dividing the volume of water flowing per unit time by the average area of voids (A_v) on a cross-section normal to the macroscopic direction of flow: this velocity is called the seepage velocity (v'). Thus:

$$v' = \frac{q}{A_v}$$

The porosity of a soil is defined in terms of volume:

$$n = \frac{V_v}{V}$$

However, on average, the porosity can also be expressed as:

$$n = \frac{A_v}{A}$$

Hence:

$$v' = \frac{q}{nA} = \frac{v}{n}$$

or

$$v' = \frac{ki}{n} \tag{2.4}$$

Determination of Coefficient of Permeability

Laboratory Methods. The coefficient of permeability for coarse-grained soils can be determined by means of the *constant head* permeability test (Fig. 2.1a). The soil specimen, at the appropriate density, is contained in a perspex cylinder of cross-sectional area A: the specimen rests on a coarse filter or a wire mesh. A steady vertical flow of water, under a constant total head, is maintained through the soil and the volume of water flowing per unit time (q) is measured. Tappings from the side of the cylinder enable the hydraulic gradient (h/l) to be measured. Then from Darcy's law:

$$k = \frac{ql}{Ah}$$

A series of tests should be run, each at a different rate of flow. Prior to running the test a vacuum is applied to the specimen to ensure that the degree of saturation under flow will be close to 100%. If a high degree of saturation is to be maintained the water used in the test should be de-aired.

For fine-grained soils the *falling-head* test (Fig. 2.1b) should be used. In the case of fine-grained soils, undisturbed specimens are normally tested and the containing cylinder in the test may be the sampling tube itself. The length of the specimen is l and the cross-sectional area A. A coarse filter is placed at each end of the specimen and a standpipe of internal area a is connected to the top of the cylinder. The water drains into a reservoir of constant level. The standpipe is filled with water and a measurement is made of the time (t_1) for the water level (relative to the water level in the reservoir) to fall from h_0 to h_1. At any intermediate time t the water level in the standpipe is given by h and its rate of change by $-dh/dt$. At time t the difference in total head between the top and bottom of the specimen is h. Then applying Darcy's law:

$$-a\frac{dh}{dt} = Ak\frac{h}{l}$$

$$\therefore \quad -a\int_{h_0}^{h_1} \frac{dh}{h} = \frac{Ak}{l}\int_0^{t_1} dt$$

$$\therefore \quad k = \frac{al}{At_1}\ln\frac{h_0}{h_1} = 2\cdot3\frac{al}{At_1}\log\frac{h_0}{h_1}$$

Again, precautions must be taken to ensure that the degree of saturation remains close to 100%. A series of tests should be run using different values of h_0 and h_1 and/or standpipes of different diameters.

The coefficient of permeability of fine grained soils can also be determined indirectly from the results of consolidation tests (see Chapter 7).

In-Situ Methods. The reliability of laboratory permeability determinations depends on the extent to which the test specimens are representative of the in-situ soil mass as a whole. For important projects the in-situ determination of permeability may be justified.

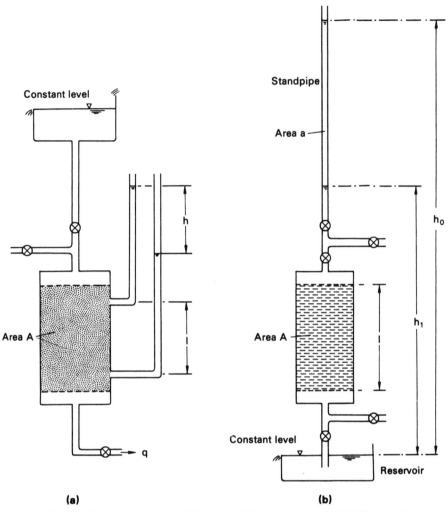

Fig. 2.1   Laboratory permeability tests: (a) constant head, (b) falling head.

One in-situ method is the *well pumping test* which is suitable for use in homogeneous coarse-grained soil strata. The method involves continuous pumping from a well which penetrates to the bottom of the stratum and observing the water levels in a number of adjacent boreholes. Pumping is continued until steady seepage conditions become established. Seepage takes place radially towards the well and the boreholes are located, therefore, on a number of radial lines from the centre of the well: at least two boreholes are required on each line. A section through the well and two boreholes is shown in Fig. 2.2. Drawdown of the water table takes place as the result of pumping and when the steady state is established the water

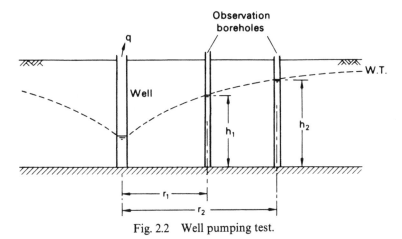

Fig. 2.2 Well pumping test.

levels in the boreholes correspond to the new water table position. The boreholes are located at distances r_1 and r_2 from the well and the respective water table levels are at heights of h_1 and h_2 above the bottom of the stratum.

The analysis is based on the assumption that the hydraulic gradient at any distance r from the well is constant with depth and is equal to the slope of the water table, i.e.:

$$i_r = \frac{dh}{dr}$$

where h is the height of the water table at radius r. This is known as the Dupuit assumption and is reasonably accurate except at points close to the well. At distance r from the well the area through which flow takes place is $2\pi rh$. Then applying Darcy's law:

$$q = 2\pi rhk \frac{dh}{dr}$$

$$\therefore \quad q \int_{r_1}^{r_2} \frac{dr}{r} = 2\pi k \int_{h_1}^{h_2} h\, dh$$

$$\therefore \quad q \ln\left(\frac{r_2}{r_1}\right) = \pi k (h_2^2 - h_1^2)$$

$$\therefore \quad k = \frac{2 \cdot 3q \log (r_2/r_1)}{\pi (h_2^2 - h_1^2)}$$

This equation is applied to each pair of boreholes and an average value of k is obtained.

Other in-situ methods include constant head and variable head *borehole tests*. In one procedure, water is allowed to flow, under constant head, into (or out of) the stratum under test through the bottom of a borehole, the sides of which are lined with a pipe casing. The arrangement is illustrated in Fig. 2.3a: the lower end of the borehole should not be less than $5d$ from either the top or bottom of the stratum, where d is the internal diameter of the casing. The water level in the borehole is maintained constant, the difference between this level and water table level being h. The rate of flow (q) required to maintain the constant water level is measured. The co-efficient of permeability can then be calculated from the following equation, developed from electrical analogy experiments:

$$k = \frac{q}{2 \cdot 75 \, dh}$$

Care must be taken to ensure that clogging of the soil face at the bottom of the borehole does not occur due to the deposition of sediment from the water. If desired, water may be pumped into the borehole under pressure.

In a variable head test the rate of flow from the stratum into the borehole is measured by observing the time (t) for the water level in the borehole, relative to water table level, to change from a value h_1 to a value h_2. Hvorslev [2.4] published formulae for the determination of permeability in a number of borehole situations: two examples are given below. A cased borehole, of internal diameter d, penetrating a short distance D (not exceeding 1·5 m) below the water table in a stratum assumed to be of infinite depth, is shown in Fig. 2.3b. The coefficient of permeability for this situation is given by:

$$k = \frac{\pi d}{11t} \ln\left(\frac{h_1}{h_2}\right)$$

A borehole with a cased length and an uncased or perforated extension of length L (where $L > 4d$) in a stratum assumed to be of infinite depth, is shown in Fig. 2.3c. The coefficient of permeability for this situation is given by:

$$k = \frac{d^2}{8Lt} \ln\left(\frac{2L}{d}\right) \ln\left(\frac{h_1}{h_2}\right)$$

The permeability of a coarse-grained soil can also be obtained from in-situ measurements of *seepage velocity*, using Equation 2.4. The method involves excavating uncased boreholes (or trial pits) at two points A and B (Fig. 2.3d), seepage taking place from A towards B. The hydraulic gradient is given by the difference in the steady state water levels in the boreholes divided by the distance AB. Dye or any other suitable tracer is inserted into borehole A and the time taken for the dye to appear in borehole B is

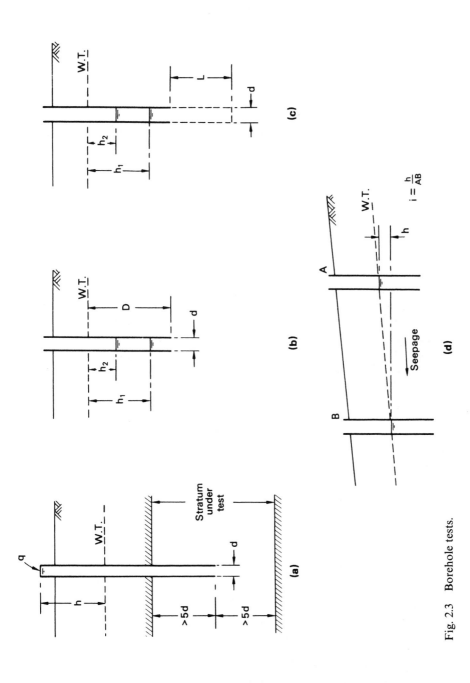

Fig. 2.3 Borehole tests.

measured. The seepage velocity is then the distance AB divided by this time. The porosity of the soil can be determined from density tests. Then:

$$k = \frac{v'n}{i}$$

2.3   Seepage Theory

The general case of seepage in two dimensions will now be considered. Initially it will be assumed that the soil is homogeneous and isotropic with respect to permeability, the coefficient of permeability being k. In the x–z plane, Darcy's law can be written in the generalised form:

$$v_x = ki_x = - k\frac{\partial h}{\partial x} \qquad (2.5a)$$

$$v_z = ki_z = - k\frac{\partial h}{\partial z} \qquad (2.5b)$$

(total head h decreasing in the directions of v_x and v_z).

An element of fully saturated soil having dimensions dx, dy and dz in the x, y and z directions respectively, with flow taking place in the x–z plane only, is shown in Fig. 2.4. The components of discharge velocity of water entering the element are v_x and v_z and the rates of change of discharge velocity in the x and z directions are $\partial v_x/\partial x$ and $\partial v_z/\partial z$ respectively. The volume of water entering the element per unit time is:

$$v_x \, dy \, dz + v_z \, dx \, dy$$

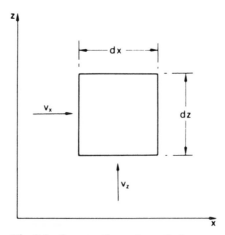

Fig. 2.4   Seepage through a soil element.

and the volume of water leaving per unit time is:

$$\left(v_x + \frac{\partial v_x}{\partial x}\, dx\right) dy\; dz + \left(v_z + \frac{\partial v_z}{\partial z}\, dz\right) dx\; dy$$

If the element is undergoing no volume change and if water is assumed to be incompressible, the difference between the volume of water entering the element per unit time and the volume leaving must be zero. Therefore:

$$\frac{\partial v_x}{\partial x} + \frac{\partial v_z}{\partial z} = 0 \tag{2.6}$$

Equation 2.6 is the *equation of continuity* in two dimensions. If, however, the volume of the element is undergoing change, the equation of continuity becomes:

$$\left(\frac{\partial v_x}{\partial x} + \frac{\partial v_z}{\partial z}\right) dx\; dy\; dz = \frac{dV}{dt} \tag{2.7}$$

where dV/dt is the volume change per unit time.

Consider, now, the function $\phi(x,z)$, called the *potential function*, such that:

$$\frac{\partial \phi}{\partial x} = v_x = -k\frac{\partial h}{\partial x} \tag{2.8a}$$

$$\frac{\partial \phi}{\partial z} = v_z = -k\frac{\partial h}{\partial z} \tag{2.8b}$$

From Equations 2.6 and 2.8 it is apparent that:

$$\frac{\partial^2 \phi}{\partial x^2} + \frac{\partial^2 \phi}{\partial z^2} = 0 \tag{2.9}$$

i.e. the function $\phi(x,z)$ satisfies the Laplace equation.

Integrating Equation 2.8:

$$\phi(x,z) = -kh(x,z) + C$$

where C is a constant. Thus, if the function $\phi(x,z)$ is given a constant value, equal to ϕ_1 (say), it will represent a curve along which the value of total head (h_1) is constant. If the function $\phi(x,z)$ is given a series of constant values, ϕ_1, ϕ_2, ϕ_3, etc., a family of curves is specified along each of which the total head is a constant value (but a different value for each curve). Such curves are called *equipotentials*.

A second function $\psi(x,z)$, called the *flow function*, is now introduced, such that:

$$-\frac{\partial \psi}{\partial x} = v_z = -k\frac{\partial h}{\partial z} \tag{2.10a}$$

$$\frac{\partial \psi}{\partial z} = v_x = -k \frac{\partial h}{\partial x}$$
(2.10b)

It can be shown that this function also satisfies the Laplace equation.

The total differential of the function $\psi(x,z)$ is:

$$d\psi = \frac{\partial \psi}{\partial x} dx + \frac{\partial \psi}{\partial z} dz$$

$$= -v_z \, dx + v_x \, dz$$

If the function $\psi(x,z)$ is given a constant value ψ_1 then $d\psi = 0$ and:

$$\frac{dz}{dx} = \frac{v_z}{v_x}$$
(2.11)

Thus the tangent at any point on the curve represented by

$$\psi(x,z) = \psi_1$$

specifies the direction of the resultant discharge velocity at that point: the curve therefore represents the flow path. If the function $\psi(x,z)$ is given a series of constant values ψ_1, ψ_2, ψ_3, etc., a second family of curves is specified, each representing a flow path. These curves are called *flow lines*.

Referring to Fig. 2.5, the flow per unit time between two flow lines for which the values of the flow function are ψ_1 and ψ_2 is given by:

$$\Delta q = \int_{\psi_1}^{\psi_2} (-v_z \, dx + v_x \, dz)$$

$$= \int_{\psi_1}^{\psi_2} \left(\frac{\partial \psi}{\partial x} dx + \frac{\partial \psi}{\partial z} dz \right) = \psi_2 - \psi_1$$

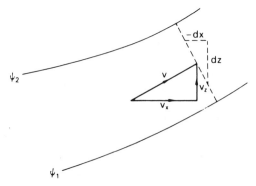

Fig. 2.5   Seepage between two flow lines.

Thus the flow through the 'channel' between the two flow lines is constant.

The total differential of the function $\phi(x,z)$ is

$$d\phi = \frac{\partial\phi}{\partial x}dx + \frac{\partial\phi}{\partial z}dz$$

$$= v_x\,dx + v_z\,dz$$

If $\phi(x,z)$ is constant then $d\phi = 0$ and:

$$\frac{dz}{dx} = -\frac{v_x}{v_z} \tag{2.12}$$

Comparing Equations 2.11 and 2.12 it is apparent that the flow lines and the equipotentials intersect each other at right angles.

Consider, now, two flow lines ψ_1 and $(\psi_1 + \Delta\psi)$ separated by the distance Δn. The flow lines are intersected orthogonally by two equipotentials ϕ_1 and $(\phi_1 + \Delta\phi)$ separated by the distance Δs, as shown in Fig. 2.6. The directions s and n are inclined at angle α to the x and z axes respectively. At point A the discharge velocity (in direction s) is v_s: the components of v_s in the x and z directions respectively are:

$$v_x = v_s \cos\alpha$$

$$v_z = v_s \sin\alpha$$

Now:

$$\frac{\partial\phi}{\partial s} = \frac{\partial\phi}{\partial x}\frac{\partial x}{\partial s} + \frac{\partial\phi}{\partial z}\frac{\partial z}{\partial s}$$

$$= v_s \cos^2\alpha + v_s \sin^2\alpha = v_s$$

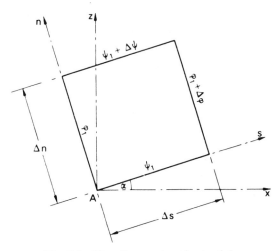

Fig. 2.6   Flow lines and equipotentials.

and:

$$\frac{\partial \psi}{\partial n} = \frac{\partial \psi}{\partial x}\frac{\partial x}{\partial n} + \frac{\partial \psi}{\partial z}\frac{\partial z}{\partial n}$$

$$= -v_s \sin \alpha \, (- \sin \alpha) + v_s \cos^2 \alpha = v_s$$

Thus:

$$\frac{\partial \psi}{\partial n} = \frac{\partial \phi}{\partial s}$$

or, approximately:

$$\frac{\Delta \psi}{\Delta n} = \frac{\Delta \phi}{\Delta s} \qquad (2.13)$$

2.4 Flow Nets

In principle, for the solution of a practical seepage problem the functions $\phi(x,z)$ and $\psi(x,z)$ must be found for the relevant boundary conditions. The solution is represented by a family of flow lines and a family of equipotentials, constituting what is referred to as a flow net. Possible methods of solution are complex variable techniques, the finite difference method, the finite element method, electrical analogy and the use of hydraulic models. However, the most widely used method of solution is by the trial-and-error sketching of the flow net, the general form of which can be deduced from a consideration of the boundary conditions.

The fundamental condition to be satisfied in a flow net is that every intersection between a flow line and an equipotential must be at right angles. In addition it is *convenient* to construct the flow net such that $\Delta \psi$ is the same value between any two adjacent flow lines and $\Delta \phi$ is the same value between any two adjacent equipotentials. It is also *convenient* to make $\Delta s = \Delta n$ in Equation 2.13, i.e. the flow lines and equipotentials form 'curvilinear squares' throughout the flow net. Then for any curvilinear square:

$$\Delta \psi = \Delta \phi$$

Now, $\Delta \psi = \Delta q$ and $\Delta \phi = k\Delta h$, therefore:

$$\Delta q = k\Delta h \qquad (2.14)$$

The hydraulic gradient is given by:

$$i = \frac{\Delta h}{\Delta s} \qquad (2.15)$$

For the entire flow net: h = difference in total head between the first and last equipotentials, N_d = number of *equipotential drops*, each representing the same total head loss Δh, and N_f = number of *flow channels*, each carrying the same flow Δq. Then,

$$\Delta h = \frac{h}{N_d} \qquad (2.16)$$

and

$$q = N_f \Delta q$$

Hence, from Equation 2.14:

$$q = kh \frac{N_f}{N_d} \qquad (2.17)$$

Equation 2.17 gives the total volume of water flowing per unit time (per unit dimension in the y direction) and is a function of the *ratio N_f/N_d*.

Example of a Flow Net

As an illustration the flow net for the problem detailed in Fig. 2.7a will be considered. The figure shows a line of sheet piling driven 6·00 m into a stratum of soil 8·60 m thick, underlain by an impermeable stratum. On one side of the piling the depth of water is 4·50 m; on the other side the depth of water (reduced by pumping) is 0·50 m.

The first step is to consider the boundary conditions of the flow region. At every point on the boundary AB the total head is constant, therefore AB is an equipotential; similarly CD is an equipotential. The datum to which total head is referred may be any level but in seepage problems it is convenient to select the downstream water level as datum. Then the total head on equipotential CD is zero (pressure head 0·50 m; elevation head − 0·50 m) and the total head on equipotential AB is 4·00 m (pressure head 4·50 m; elevation head − 0·50 m). From point B, water must flow down the upstream face BE of the piling, round the tip E and up the downstream face EC. Water from point F must flow along the impermeable surface FG. Thus BEC and FG are flow lines. The shapes of other flow lines must be between the extremes of BEC and FG.

The first trial sketching of the flow net (Fig. 2.7b) can now be attempted using a procedure suggested by Taylor [2.7]. The estimated line of flow (HJ) from a point on AB near the piling is lightly sketched. This line must start at right angles to equipotential AB and follow a smooth curve round the bottom of the piling. Trial equipotential lines are then drawn between the flow lines BEC and HJ, intersecting both flow lines at right angles and forming curvilinear squares. If necessary the position of HJ should be altered slightly so that a whole number of squares is obtained between BH and CJ. The procedure is continued by sketching the estimated line of flow (KL) from a second point on AB and extending the equipotentials already

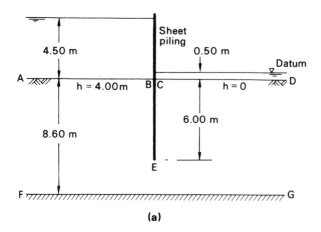

(a)

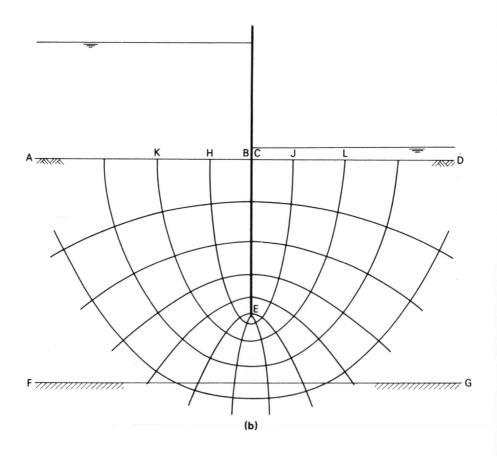

(b)

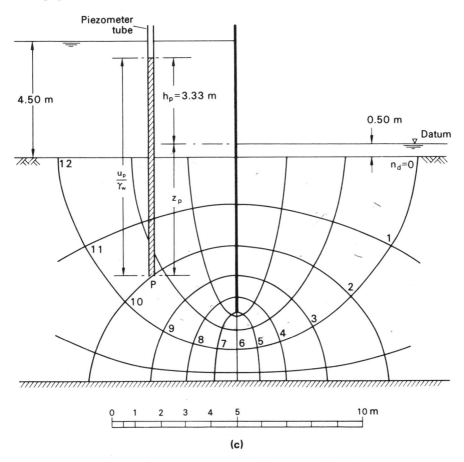

Fig. 2.7   Flow net construction: (a) section, (b) first trial (c) final flow net.

drawn. The flow line KL and the equipotential extensions are adjusted so that all intersections are at right angles and all areas are square. The procedure is repeated until the boundary FG is reached. At the first attempt it is almost certain that the last flow line drawn will be inconsistent with the boundary FG as, for example, in Fig. 2.7b. By studying the nature of this inconsistency the position of the first flow line (HJ) can be adjusted in a way that will tend to correct the inconsistency. The entire flow net is then adjusted and the inconsistency should now be small. After a third trial the last flow line should be consistent with the boundary FG, as shown in Fig. 2.7c. In general, the areas between the last flow line and the lower boundary will not be square but the length/breadth ratio of each area should be constant within this flow channel. In constructing a flow net it is a mistake to draw too many flow lines: typically 4 to 5 flow channels are sufficient.

In the flow net in Fig. 2.7c the number of flow channels is 4·3 and the number of equipotential drops is 12: thus the ratio N_f/N_d is 0·36. The equipotentials are numbered from zero at the downstream boundary: this number is denoted by n_d. The loss in total head between any two adjacent equipotentials is:

$$\Delta h = \frac{h}{N_d} = \frac{4·00}{12} = 0·33\,\mathrm{m}$$

The total head at every point on an equipotential numbered n_d is $n_d \Delta h$. The total volume of water flowing under the piling per unit time per unit length of piling is given by:

$$q = kh\frac{N_f}{N_d} = k \times 4·00 \times 0·36$$

$$= 1·44\,k\,\mathrm{m}^3/\mathrm{s}$$

A piezometer tube is shown at a point P on the equipotential denoted by $n_d = 10$. The total head at P is:

$$h_P = \frac{n_d}{N_d}h = \frac{10}{12} \times 4·00 = 3·33\,\mathrm{m}$$

i.e. the water level in the tube is 3·33 m above the datum. The point P is distance z_P *below* the datum, i.e. the elevation head is $-z_P$. The pore water pressure at P can then be calculated from Bernoulli's theorem:

$$u_P = \gamma_w\{h_P - (-z_P)\}$$

$$= \gamma_w(h_P + z_P)$$

The hydraulic gradient across any square in the flow net involves measuring the average dimension of the square (Equation 2.15). The highest hydraulic gradient (and hence the highest seepage velocity) occurs across the smallest square, and vice versa.

Example 2.1

The section through a sheet pile wall along a tidal estuary is given in Fig. 2.8. At low tide the depth of water in front of the wall is 4·00 m: the water table behind the wall lags 2·50 m behind tidal level. Plot the net distribution of water pressure on the piling.

The flow net is shown in the figure. The water level in front of the piling is selected as datum. The total head at water table level (the upstream equipotential) is 2·50 m (pressure head zero; elevation head +2·50 m). The total head on the soil surface in front of the piling (the downstream equipotential) is zero (pressure head 4·00 m; elevation head −4·00 m). There are 12 equipotential drops in the flow net.

The water pressures are calculated on both sides of the piling at selected

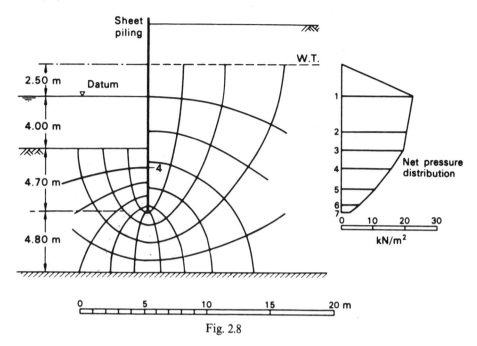

Fig. 2.8

Table 2.2

Level	z (m)	h_b (m)	u_b/γ_w (m)	h_f (m)	u_f/γ_w (m)	$u_b - u_f$ (kN/m²)
1	0	2·30	2·30	0	0	22·6
2	−2·70	2·10	4·80	0	2·70	20·6
3	−4·00	2·00	6·00	0	4·00	19·6
4	−5·50	1·83	7·33	0·21	5·71	15·9
5	−7·10	1·68	8·78	0·50	7·60	11·6
6	−8·30	1·51	9·81	0·84	9·14	6·6
7	−8·70	1·25	9·95	1·04	9·74	2·1

levels numbered 1 to 7. For example at level 4 the total head on the back of the piling is:

$$h_b = \frac{8\cdot8}{12} \times 2\cdot50 = 1\cdot83\,\text{m}$$

and the total head on the front is:

$$h_f = \frac{1}{12} \times 2\cdot50 = 0\cdot21\,\text{m}$$

The elevation head at level 4 is $-5\cdot5\,\text{m}$.

$\frac{ud}{Md} \ h = \frac{1 \cdot 4}{15} = 0.266$

Table 2.3

Point	h_p (m)	z (m)	$h - z$ (m)	$u = \gamma_w(h - z)$ (kN/m^2)
1	0·27	−1·80	2·07	20·3
2	0·53	−1·80	2·33	22·9
3	0·80	−1·80	2·60	25·5
4	1·07	−2·10	3·17	31·1
5	1·33	−2·40	3·73	36·6
6	1·60	−2·40	4·00	39·2
7	1·87	−2·40	4·27	41·9
7½	2·00	−2·40	4·40	43·1

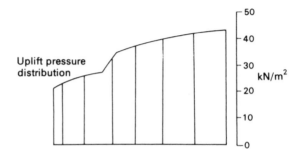

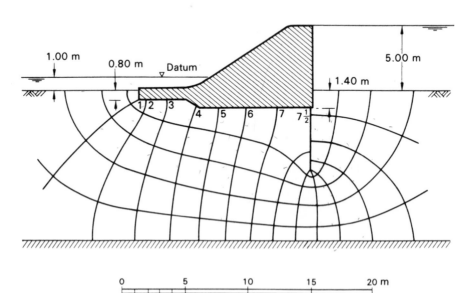

Fig. 2.9

Therefore the net pressure on the back of the piling is:

$$u_b - u_f = 9 \cdot 8\,(1 \cdot 83 + 5 \cdot 5) - 9 \cdot 8\,(0 \cdot 21 + 5 \cdot 5)$$
$$= 9 \cdot 8\,(7 \cdot 33 - 5 \cdot 71)$$
$$= 15 \cdot 9\,\text{kN/m}^2$$

The calculations for the selected points are shown in Table 2.2 and the net pressure diagram is plotted in Fig. 2.8.

Example 2.2

The section through a dam in shown in Fig. 2.9. Determine the quantity of seepage under the dam and plot the distribution of uplift pressure on the base of the dam. The coefficient of permeability of the foundation soil is $2 \cdot 5 \times 10^{-5}\,\text{m/s}$.

The flow net is shown in the figure. The downstream water level is selected as datum. Between the upstream and downstream equipotentials the total head loss is $4 \cdot 00\,\text{m}$. In the flow net there are $4 \cdot 7$ flow channels and 15 equipotential drops. The seepage is given by:

$$q = kh\,\frac{N_f}{N_d} = 2 \cdot 5 \times 10^{-5} \times 4 \cdot 00 \times \frac{4 \cdot 7}{15}$$
$$= 3 \cdot 1 \times 10^{-5}\,\text{m}^3/\text{s (per m)}$$

The pore water pressure is calculated at the points of intersection of the equipotentials with the base of the dam. The total head at each point is obtained from the flow net and the elevation head from the section. The calculations are shown in Table 2.3 and the pressure diagram is plotted in Fig. 2.9.

Example 2.3

A river bed consists of a layer of sand $8 \cdot 25$ m thick overlying impermeable rock: the depth of water is $2 \cdot 50$ m. A long cofferdam $5 \cdot 50$ m wide is formed by driving two lines of sheet piling to a depth of $6 \cdot 00$ m below the level of the river bed and excavation to a depth of $2 \cdot 00$ m below bed level is carried out within the cofferdam. The water level within the cofferdam is kept at excavation level by pumping. If the flow of water into the cofferdam is $0 \cdot 25\,\text{m}^3/\text{h}$ per unit length, what is the coefficient of permeability of the sand? What is the hydraulic gradient immediately below the excavated surface?

The section and flow net appear in Fig. 2.10. In the flow net there are $6 \cdot 0$ flow channels and 10 equipotential drops. The total head loss is $4 \cdot 50$ m. The coefficient of permeability is given by:

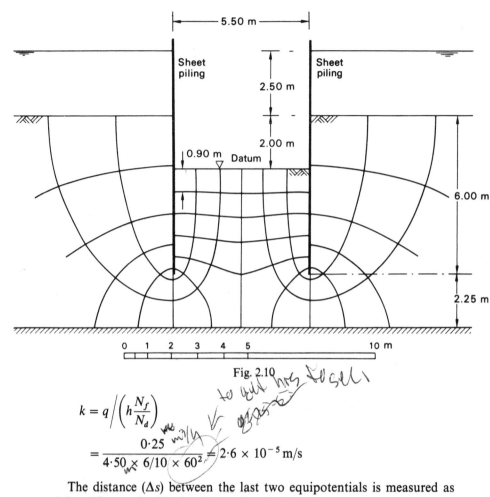

Fig. 2.10

$$k = q \Big/ \left(h\frac{N_f}{N_d} \right)$$

$$= \frac{0.25}{4.50 \times 6/10 \times 60^2} = 2.6 \times 10^{-5}\,\text{m/s}$$

The distance (Δs) between the last two equipotentials is measured as $0.9\,\text{m}$. The required hydraulic gradient is given by:

$$i = \frac{\Delta h}{\Delta s}$$

$$= \frac{4.50}{10 \times 0.9} = 0.50$$

2.5 Anisotropic Soil Conditions

It will now be assumed that the soil, although homogeneous, is anisotropic with respect to permeability. Most natural soil deposits are anisotropic, with the coefficient of permeability having a maximum value in the direction of stratification and a minimum value in the direction normal to that of stratification: these directions are denoted by x and z respectively, i.e.:

$$k_x = k_{max} \quad \text{and} \quad k_z = k_{min}$$

In this case the generalized form of Darcy's law is:

$$v_x = k_x i_x = -k_x \frac{\partial h}{\partial x} \tag{2.18a}$$

$$v_z = k_z i_z = -k_z \frac{\partial h}{\partial z} \tag{2.18b}$$

Also, in any direction s, inclined at angle α to the x direction, the coefficient of permeability is defined by the equation:

$$v_s = -k_s \frac{\partial h}{\partial s}$$

Now:

$$\frac{\partial h}{\partial s} = \frac{\partial h}{\partial x}\frac{\partial x}{\partial s} + \frac{\partial h}{\partial z}\frac{\partial z}{\partial s}$$

i.e.

$$\frac{v_s}{k_s} = \frac{v_x}{k_x}\cos\alpha + \frac{v_z}{k_z}\sin\alpha$$

The components of discharge velocity are also related as follows:

$$v_x = v_s \cos\alpha$$

$$v_z = v_s \sin\alpha$$

Hence:

$$\frac{1}{k_s} = \frac{\cos^2\alpha}{k_x} + \frac{\sin^2\alpha}{k_z}$$

or

$$\frac{s^2}{k_s} = \frac{x^2}{k_x} + \frac{z^2}{k_z} \tag{2.19}$$

The directional variation of permeability is thus described by Equation 2.19 which represents the ellipse shown in Fig. 2.11.

Given the generalized form of Darcy's law (Equation 2.18) the equation of continuity (2.6) can be written:

$$k_x \frac{\partial^2 h}{\partial x^2} + k_z \frac{\partial^2 h}{\partial z^2} = 0 \tag{2.20}$$

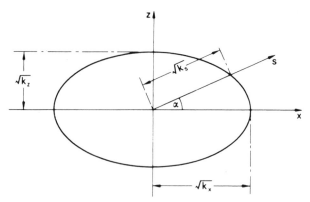

Fig. 2.11   Permeability ellipse.

or

$$\frac{\partial^2 h}{\left(\dfrac{k_z}{k_x}\right)\partial x^2} + \frac{\partial^2 h}{\partial z^2} = 0$$

Substituting:

$$x_t = x\sqrt{\frac{k_z}{k_x}} \tag{2.21}$$

the equation of continuity becomes:

$$\frac{\partial^2 h}{\partial x_t^2} + \frac{\partial^2 h}{\partial z^2} = 0 \tag{2.22}$$

which is the equation of continuity for an *isotropic* soil in an x_t, z plane.

Thus Equation 2.21 defines a scale factor which can be applied in the x direction to transform a given anisotropic flow region into a fictitious isotropic flow region in which the Laplace equation is valid. Once the flow net (representing the solution of the Laplace equation) has been drawn for the transformed section the flow net for the natural section can be obtained by applying the inverse of the scaling factor. Essential data, however, can normally be obtained from the transformed section. The necessary transformation could also be made in the z direction.

The value of coefficient of permeability applying to the transformed section, referred to as the equivalent isotropic coefficient, is

$$k' = \sqrt{(k_x k_z)} \tag{2.23}$$

A formal proof of Equation 2.23 has been given by Vreedenburgh [2.8]. The validity of Equation 2.23 can be demonstrated by considering an elemental

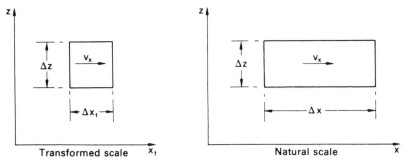

Fig. 2.12   Elemental flow net field.

flow net field through which flow is in the x direction. The flow net field is drawn to the transformed and natural scales in Fig. 2.12, the transformation being in the x direction. The discharge velocity v_x can be expressed in terms of either k' (transformed section) or k_x (natural section), i.e.:

$$v_x = -k' \frac{\partial h}{\partial x_t} = -k_x \frac{\partial h}{\partial x}$$

where

$$\frac{\partial h}{\partial x_t} = \frac{\partial h}{\sqrt{\left(\dfrac{k_z}{k_x}\right)} \partial x}$$

Thus:

$$k' = k_x \sqrt{\frac{k_z}{k_x}} = \sqrt{(k_x k_z)}.$$

2.6   Non-homogeneous Soil Conditions

Two *isotropic* soil layers of thicknesses H_1 and H_2 are shown in Fig. 2.13, the respective coefficients of permeability being k_1 and k_2: the boundary between the layers is horizontal. (If the layers are anisotropic, k_1 and k_2 represent the equivalent isotropic coefficients for the layers.) The two layers can be considered as a single homogeneous anisotropic layer of thickness $(H_1 + H_2)$ in which the coefficients in the directions parallel and normal to that of stratification are $\bar{k}_x$ and $\bar{k}_z$ respectively.

For one-dimensional seepage in the horizontal direction, the equipotentials in each layer are vertical. If h_1 and h_2 represent total head at any point

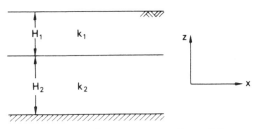

Fig. 2.13   Non-homogeneous soil conditions.

in the respective layers, then for a common point on the boundary $h_1 = h_2$. Therefore any vertical line through the two layers represents a common equipotential. Thus the hydraulic gradients in the two layers, and in the equivalent single layer, are equal: the equal hydraulic gradients are denoted by i_x.

The total horizontal flow per unit time is given by:

$$\bar{q}_x = (H_1 + H_2)\bar{k}_x i_x = (H_1 k_1 + H_2 k_2)i_x$$

$$\therefore \quad \bar{k}_x = \frac{H_1 k_1 + H_2 k_2}{H_1 + H_2} \tag{2.24}$$

For one-dimensional seepage in the vertical direction the discharge velocities in each layer, and in the equivalent single layer, must be equal if the requirement of continuity is to be satisfied. Thus:

$$v_z = \bar{k}_z \bar{i}_z = k_1 i_1 = k_2 i_2$$

where $\bar{i}_z$ is the average hydraulic gradient over the depth $(H_1 + H_2)$. Therefore:

$$i_1 = \frac{\bar{k}_z}{k_1} \bar{i}_z \quad \text{and} \quad i_2 = \frac{\bar{k}_z}{k_2} \bar{i}_z$$

Now the loss in total head over the depth $(H_1 + H_2)$ is equal to the sum of the losses in total head in the individual layers, i.e.:

$$\bar{i}_z(H_1 + H_2) = i_1 H_1 + i_2 H_2$$

$$= \bar{k}_z \bar{i}_z \left(\frac{H_1}{k_1} + \frac{H_2}{k_2} \right)$$

$$\therefore \quad \bar{k}_z = (H_1 + H_2) \Big/ \left(\frac{H_1}{k_1} + \frac{H_2}{k_2} \right) \tag{2.25}$$

Similar expressions for $\bar{k}_x$ and $\bar{k}_z$ apply in the case of any number of soil layers. It can be shown that $\bar{k}_x$ must always be greater than $\bar{k}_z$, i.e. seepage can occur more readily in the direction parallel to stratification than in the direction perpendicular to stratification.

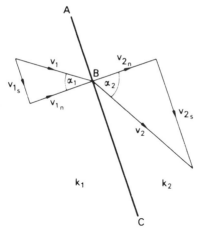

Fig. 2.14   Transfer condition.

2.7   Transfer Condition

Consideration is now given to the condition which must be satisfied when seepage takes place diagonally across the boundary between two isotropic soils 1 and 2 having coefficients of permeability k_1 and k_2 respectively. The direction of seepage approaching a point B on the boundary ABC is at angle α_1 to the normal at B, as shown in Fig. 2.14: the discharge velocity approaching B is v_1. The components of v_1 along the boundary and normal to the boundary are v_{1s} and v_{1n} respectively. The direction of seepage leaving point B is at angle α_2 to the normal, as shown: the discharge velocity leaving B is v_2. The components of v_2 are v_{2s} and v_{2n}.

For soils 1 and 2 respectively:

$$\phi_1 = -k_1 h_1 \quad \text{and} \quad \phi_2 = -k_2 h_2$$

At the common point B, $h_1 = h_2$, therefore:

$$\frac{\phi_1}{k_1} = \frac{\phi_2}{k_2}$$

Differentiating with respect to s, the direction along the boundary:

$$\frac{1}{k_1}\frac{\partial \phi_1}{\partial s} = \frac{1}{k_2}\frac{\partial \phi_2}{\partial s}$$

i.e.

$$\frac{v_{1s}}{k_1} = \frac{v_{2s}}{k_2}$$

For continuity of flow across the boundary the normal components of discharge velocity must be equal, i.e.

$$v_{1n} = v_{2n}$$

Therefore:

$$\frac{1}{k_1}\frac{v_{1s}}{v_{1n}} = \frac{1}{k_2}\frac{v_{2s}}{v_{2n}}$$

Hence it follows that:

$$\frac{\tan \alpha_1}{\tan \alpha_2} = \frac{k_1}{k_2} \tag{2.26}$$

Equation 2.26 specifies the change in direction of the flow line passing through point B. This equation must be satisfied on the boundary by every flow line crossing the boundary.

Equation 2.13 can be written:

$$\Delta\psi = \frac{\Delta n}{\Delta s}\Delta\phi$$

i.e.

$$\Delta q = \frac{\Delta n}{\Delta s}k\Delta h$$

If Δq and Δh are each to have the same values on both sides of the boundary then

$$\left(\frac{\Delta n}{\Delta s}\right)_1 k_1 = \left(\frac{\Delta n}{\Delta s}\right)_2 k_2$$

and it is clear that curvilinear squares are possible only in one soil. If

$$\left(\frac{\Delta n}{\Delta s}\right)_1 = 1$$

then

$$\left(\frac{\Delta n}{\Delta s}\right)_2 = \frac{k_1}{k_2} \tag{2.27}$$

If the permeability ratio is less than $1/10$ it is unlikely that the part of the flow net in the soil of higher permeability need be considered.

2.8 Seepage Through Earth Dams

This problem is an example of unconfined seepage, one boundary of the flow region being a phreatic surface on which the pressure is atmospheric.

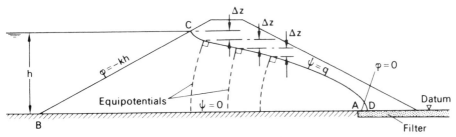

Fig. 2.15 Homogeneous earth dam section.

In section the phreatic surface constitutes the top flow line and its position must be estimated before the flow net can be drawn.

Consider the case of a homogeneous isotropic earth dam on an impermeable foundation, as shown in Fig. 2.15. The impermeable boundary AB is a flow line and CD is the required top flow line. At every point on the upstream slope BC the total head is constant, therefore BC is an equipotential. If the downstream water level is taken as datum then the total head on equipotential BC is equal to h, the difference between the upstream and downstream water levels. The discharge surface AD, for the case shown in Fig. 2.15 only, is the equipotential for zero total head. At every point on the top flow line the pressure is zero (atmospheric), therefore total head is equal to elevation head and there must be equal vertical intervals Δz between the points of intersection between successive equipotentials and the top flow line.

A suitable filter must always be constructed at the discharge surface in an earth dam. The function of the filter is to keep the seepage entirely within the dam: water seeping out onto the downstream slope would result in the gradual erosion of the slope. A horizontal underfilter is shown in Fig. 2.15. Other possible forms of filter are illustrated in Figs. 2.19a and 2.19b: in these two cases the discharge surface AD is neither a flow line nor an equipotential since there are components of discharge velocity both normal and tangential to AD.

The boundary conditions of the flow region ABCD in Fig. 2.15 can be written as follows:

Equipotential BC: $\phi = -kh$

Equipotential AD: $\phi = 0$

Flow line CD: $\psi = q$ (also, $\phi = -kz$)

Flow line AB: $\psi = 0$

The Conformal Transformation $r = w^2$

Complex variable theory can be used to obtain a solution to the earth dam problem. Let the complex number $w = \phi + i\psi$ be an analytic function of

$r = x + iz$. Consider the function:

$$r = w^2$$

Thus:

$$(x + iz) = (\phi + i\psi)^2$$
$$= (\phi^2 + 2i\phi\psi - \psi^2)$$

Equating real and imaginary parts:

$$x = \phi^2 - \psi^2 \tag{2.28}$$
$$z = 2\phi\psi \tag{2.29}$$

Equations 2.28 and 2.29 govern the transformation of points between the r and w planes.

Consider the transformation of the straight lines $\psi = n$, where $n = 0, 1, 2, 3$, (Fig. 2.16a). From Equation 2.29:

$$\phi = \frac{z}{2n}$$

and Equation 2.28 becomes:

$$x = \frac{z^2}{4n^2} - n^2 \tag{2.30}$$

Equation 2.30 represents a family of confocal parabolas. For positive values of z the parabolas for the specified values of n are plotted in Fig. 2.16b.

Consider also the transformation of the straight lines $\phi = m$, where $m = 0, 1, 2 \ldots 6$ (Fig. 2.16a). From Equation 2.29:

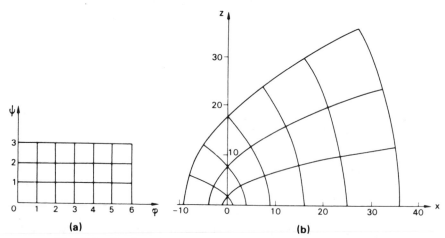

Fig. 2.16   Conformal transformation $r = w^2$: (a) w plane, (b) r plane.

$$\psi = \frac{z}{2m}$$

and Equation 2.28 becomes:

$$x = m^2 - \frac{z^2}{4m^2} \tag{2.31}$$

Equation 2.31 represents a family of confocal parabolas conjugate with the parabolas represented by Equation 2.30. For positive values of z the parabolas for the specified values of m are plotted in Fig. 2.16b. The two families of parabolas satisfy the requirements of a flow net.

Application to Earth Dam Section

The flow region in the w plane satisfying the boundary conditions for the earth dam section (Fig. 2.15) is shown in Fig. 2.17a. In this case the transformation function:

$$r = Cw^2$$

will be used, where C is a constant. Equations 2.28 and 2.29 then become:

$$x = C(\phi^2 - \psi^2)$$
$$z = 2C\phi\psi$$

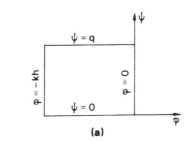

(a)

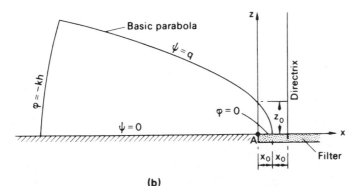

(b)

Fig. 2.17   Transformation for earth dam section: (a) w plane, (b) r plane.

The equation of the top flow line can be derived by substituting the conditions:

$$\psi = q$$
$$\phi = -kz$$

Thus:

$$z = -2Ckzq$$

$$\therefore \quad C = -\frac{1}{2kq}$$

Hence:

$$x = -\frac{1}{2kq}(k^2 z^2 - q^2)$$

$$x = \frac{1}{2}\left(\frac{q}{k} - \frac{k}{q}z^2\right) \tag{2.32}$$

The curve represented by Equation 2.32 is referred to as Kozeny's basic parabola and is shown in Fig. 2.17b, the origin and focus both being at A. When $z = 0$ the value of x is given by:

$$x_0 = \frac{q}{2k}$$

$$\therefore \quad q = 2kx_0 \tag{2.33}$$

where $2x_0$ is the directrix distance of the basic parabola. When $x = 0$ the value of z is given by:

$$z_0 = \frac{q}{k} = 2x_0$$

Substituting Equation 2.33 in Equation 2.32 yields

$$x = x_0 - \frac{z^2}{4x_0} \tag{2.34}$$

The basic parabola can be drawn using Equation 2.34 provided the coordinates of one point on the parabola are known initially.

An inconsistency arises due to the fact that the conformal transformation of the straight line $\phi = -kh$ (representing the upstream equipotential) is a parabola, whereas the upstream equipotential in the earth dam section is the upstream slope. Based on an extensive study of the earth dam problem, Casagrande [2.1] recommended that the initial point of the basic parabola should be taken at G (Fig. 2.18) where GC = 0·3 HC. The coordinates of point G, substituted in Equation 2.34, enable the value of x_0 to be determined: the basic parabola can then be plotted. The top flow line must

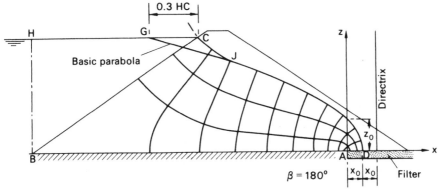

Fig. 2.18   Flow net for earth dam section.

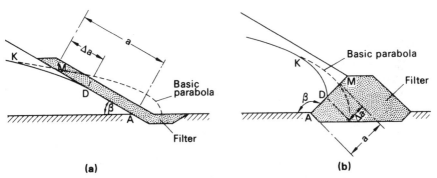

(a)   **(b)**

Fig. 2.19   Downstream correction to basic parabola.

intersect the upstream slope at right angles: a correction CJ must therefore be made (using personal judgement) to the basic parabola. The flow net can then be completed as shown in Fig. 2.18.

If the discharge surface AD is not horizontal, as in the cases shown in Fig. 2.19, a further correction KD to the basic parabola is required.

Table 2.4   Downstream Correction to Basic Parabola

Reproduced from A. Casagranda (1940) 'Seepage through Dams', in *Contributions to Soil Mechanics 1925–1940*, by permission of the Boston Society of Civil Engineers.

β	30°	60°	90°	120°	150°	180°
$\Delta a/a$	(0·36)	0·32	0·26	0·18	0·10	0°

The angle β is used to describe the direction of the discharge surface relative to AB. The correction can be made with the aid of values of the ratio MD/MA = $\Delta a/a$, given by Casagrande for the range of values of β (Table 2.4).

Seepage Control in Earth Dams

The design of an earth dam section and, where possible, the choice of soils is aimed at reducing or eliminating the detrimental effects of seeping water. Where high hydraulic gradients exist there is a possibility that the seeping water may erode channels within the dam, especially if the soil is poorly compacted: the stability of the dam may then be impaired. The process of internal erosion is referred to as *piping*. A section with a central core of low permeability, aimed at reducing the volume of seepage, is shown in Fig. 2.20a. Practically all the total head is lost in the core and if the core is narrow, high hydraulic gradients will result. There is particular danger of erosion at the boundary between the core and the adjacent soil (of higher permeability) under a high exit gradient from the core. Protection against this danger can be given by means of a 'chimney' drain (Fig. 2.20a) at the downstream boundary of the core. The drain, designed as a filter to provide a barrier to soil particles from the core, also serves as an interceptor, keeping the downstream slope in an unsaturated state.

Most earth dam sections are non-homogeneous due to zoning, making the construction of the flow net more difficult. The basic parabola construction for the top flow line applies only to homogeneous sections but the condition that there must be equal vertical distances between the points of intersection of equipotentials with the top flow line applies equally to a non-homogeneous section. The transfer condition (Equation 2.26) must be

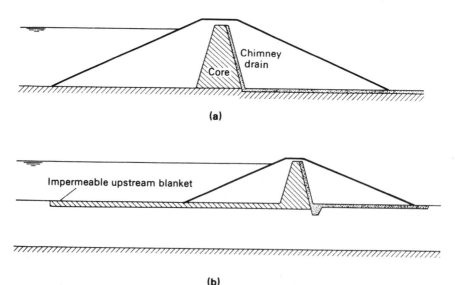

(a)

(b)

Fig. 2.20   (a) Central core and chimney drain, (b) impermeable upstream blanket.

satisfied at all zone boundaries. In the case of a section with a central core of low permeability, the application of Equation 2.26 means that the lower the permeability ratio the lower the position of the top flow line in the downstream zone (in the absence of a chimney drain).

If the foundation soil is more permeable than the dam, the control of underseepage is essential. Underseepage can be virtually eliminated by means of an 'impermeable' cut-off such as a grout curtain. Any measure designed to lengthen the seepage path, such as an impermeable upstream blanket (Fig. 2.20b), will result in a partial reduction in underseepage.

An excellent treatment of seepage control is given by Cedergren [2.2].

Filter Requirements

Filters or drains used to control seepage must satisfy two conflicting requirements:

1. The size of the pores must be small enough to prevent particles being carried in from the adjacent soil.
2. The permeability must be high enough to allow the rapid drainage of water entering the filter.

The following criteria have been found to be satisfactory for filters:

$$\frac{(D_{15})_f}{(D_{85})_s} < 4\text{--}5 \tag{2.35}$$

$$\frac{(D_{15})_f}{(D_{15})_s} > 4\text{--}5 \tag{2.36}$$

$$\frac{(D_{50})_f}{(D_{50})_s} < 25 \tag{2.37}$$

where f denotes 'filter' and s denotes 'adjacent soil'. Equation 2.35 is the requirement to prevent piping; Equations 2.36 and 2.37 are requirements to ensure that the permeability of the filter is high enough for drainage purposes. The thickness of a filter can be determined from Darcy's law.

Filters comprising two or more layers with different gradings can also be used, the finest layer being on the upstream side of the filter: such an arrangement is called a graded filter. In certain situations geotextiles can be used as an alternative to granular filters.

Example 2.4

A homogeneous anisotropic earth dam section is detailed in Fig. 2.21a, the coefficients of permeability in the x and z directions being $4\cdot5 \times 10^{-8}$ m/s and $1\cdot6 \times 10^{-8}$ m/s respectively. Construct the flow net and determine the quantity of seepage through the dam. What is the pore water pressure at point P?

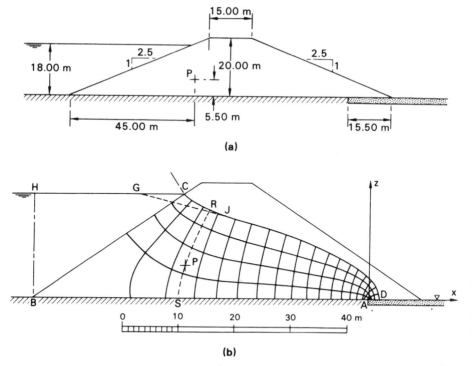

Fig. 2.21

The scale factor for transformation in the x direction is:

$$\sqrt{\frac{k_z}{k_x}} = \sqrt{\frac{1\cdot6}{4\cdot5}} = 0\cdot60$$

The equivalent isotropic permeability is:

$$k' = \sqrt{(k_x k_z)}$$
$$= \sqrt{(4\cdot5 \times 1\cdot6)} \times 10^{-8} = 2\cdot7 \times 10^{-8}\,\text{m/s}$$

The section is drawn to the transformed scale as in Fig. 2.21b. The focus of the basic parabola is at point A. The basic parabola passes through point G such that:

$$\text{GC} = 0\cdot3\,\text{HC} = 0\cdot3 \times 27\cdot00 = 8\cdot10\,\text{m}$$

i.e. the coordinates of G are:

$$x = -40\cdot80; \quad z = +18\cdot00$$

Substituting these coordinates in Equation 2.34:

$$-40\cdot80 = x_0 - \frac{18\cdot00^2}{4x_0}$$

Hence:

$$x_0 = 1.90\,\text{m}$$

Using Equation 2.34 the coordinates of a number of points on the basic parabola are now calculated:

x	1·90	0	− 5·00	− 10·00	− 20·00	− 30·00
z	0	3·80	7·24	9·51	12·90	15·57

The basic parabola is plotted in Fig. 2.21b. The upstream correction is made and the flow net completed, ensuring that there are equal vertical intervals between the points of intersection of successive equipotentials with the top flow line. In the flow net there are 3·8 flow channels and 18 equipotential drops. Hence the quantity of seepage (per unit length) is:

$$q = k'h\frac{N_f}{N_d}$$

$$= 2\cdot7 \times 10^{-8} \times 18 \times \frac{3\cdot8}{18} = 1\cdot0 \times 10^{-7}\,\text{m}^3/\text{s}$$

The quantity of seepage can also be determined from Equation 2.33 (without the necessity of drawing the flow net):

$$q = 2k'x_0$$

$$= 2 \times 2\cdot7 \times 10^{-8} \times 1\cdot90 = 1\cdot0 \times 10^{-7}\,\text{m}^3/\text{s}$$

Level AD is selected as datum. An equipotential RS is drawn through point P (transformed position). By inspection the total head at P is 15·60 m. At P the elevation head is 5·50 m, therefore the pressure head is 10·10 m and the pore water pressure is:

$$u_P = 9\cdot8 \times 10\cdot10 = 99\,\text{kN/m}^2$$

Alternatively the pressure head at P is given directly by the vertical distance of P below the point of intersection (R) of equipotential RS with the top flow line.

Example 2.5

Draw the flow net for the non-homogeneous earth dam section detailed in Fig. 2.22 and determine the quantity of seepage through the dam. Zones 1 and 2 are isotropic having coefficients of permeability $1\cdot0 \times 10^{-7}\,\text{m/s}$ and $4\cdot0 \times 10^{-7}\,\text{m/s}$ respectively.

The ratio $k_2/k_1 = 4$. The basic parabola is not applicable in this case. Three fundamental conditions must be satisfied in the flow net:

1. There must be equal vertical intervals between points of intersection of equipotentials with the top flow line.

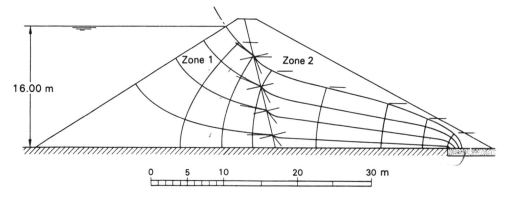

16.00 m

0   5   10   20   30 m

Fig. 2.22   (Reproduced from Cedergren, H.R. (1967) *Seepage, Drainage and Flow Nets*, © John Wiley and Sons, Inc., New York, by permission.)

   2. If the part of the flow net in zone 1 consists of curvilinear squares then the part in zone 2 must consist of curvilinear rectangles having a length/breadth ratio of 4.

   3. For each flow line the transfer condition (Equation 2.26) must be satisfied at the inter-zone boundary.

The flow net is shown in Fig. 2.22. In the flow net there are 3·6 flow channels and 8 equipotential drops. The quantity of seepage per unit length is given by:

$$q = k_1 h \frac{N_f}{N_d}$$

$$= 1{\cdot}0 \times 10^{-7} \times 16 \times \frac{3{\cdot}6}{8} = 7{\cdot}2 \times 10^{-7}\,\mathrm{m^3/s}$$

(If curvilinear squares are used in zone 2, curvilinear rectangles having a length/breadth ratio of 0·25 must be used in zone 1 and k_2 must be used in the seepage equation.)

2.9   Grouting

The permeability of coarse-grained soils can be considerably reduced by means of grouting. The process consists of injecting suitable fluids, known as grouts, into the pore space of the soil: the grout subsequently solidifies, preventing or reducing the seepage of water. Grouting also results in an increase in the strength of the soil. Fluids used for grouting include mixes of cement and water, clay suspensions, chemical solutions, such as sodium silicate or synthetic resins, and bitumen emulsion. Injection is usually effected through a pipe which is either driven into the soil or placed in a borehole and held with a packer.

The particle size distribution of the soil governs the type of grout which can be used. Particles in suspension in a grout, such as cement or clay, will only penetrate soil pores whose size is greater than a certain value; pores smaller than this size will be blocked and grout acceptability will be impaired. Cement and clay grouts are suitable only for gravels and coarse sands. For medium and fine sands, grouts of the solution or emulsion types must be used.

The extent of penetration for a given soil depends on the viscosity of the grout and the pressure under which it is injected. These factors in turn govern the required spacing of the injection points. The injection pressure must be kept below the pressure of the soil overburden otherwise heaving of the ground surface may occur and fissures may open within the soil. In soils having a wide variation of pore sizes it is expedient to use a primary injection of grout of relatively high viscosity to treat the larger pores, followed by a secondary injection of grout of relatively low viscosity for the smaller pores.

2.10 Frost Heave

Frost heave is the rise of the ground surface due to frost action. The freezing of water is accompanied by a volume increase of approximately 9%: therefore in a saturated soil the void volume above the level of freezing will increase by the same amount, representing an overall increase in the volume of the soil of $2\frac{1}{2}\%$ to 5% depending on the void ratio. However, under certain circumstances, a much greater increase in volume can occur due to the formation of ice lenses within the soil.

In a soil having a high degree of saturation the pore water freezes immediately below the surface when the temperature falls below 0 °C. The soil temperature increases with depth but during a prolonged period of sub-zero temperatures the zone of freezing gradually extends downwards. The limit of frost penetration in Great Britain is normally assumed to be 0·5 m although under exceptional conditions this depth may approach 1·0 m. The temperature at which water freezes in the pores of a soil depends on the pore size, the smaller the pores the lower the freezing temperature: water therefore freezes initially in the larger pores, remaining unfrozen in the smaller pores. As the temperature falls below zero, higher soil suction develops and water migrates towards the ice in the larger voids where it freezes and adds to the volume of ice. Continued migration gradually results in the formation of ice lenses and a rise in the ground surface. The process continues only if the bottom of the zone of freezing is within the zone of capillary rise, so that water can migrate upwards from below the water table. The magnitude of frost heave decreases as the degree of

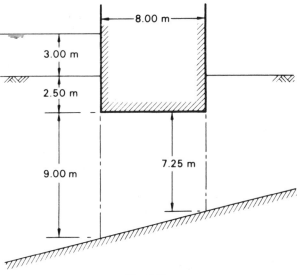

Fig. 2.23

saturation of the soil decreases. When thawing eventually takes place the soil previously frozen will contain an excess of water with the result that it will become soft and its strength will be reduced.

In the case of coarse-grained soils with little or no fines, virtually all the pores are large enough for freezing to take place throughout the soil and the only volume increase is due to the 9% increase in the volume of water on freezing. In the case of soils of very low permeability, water migration is restricted by the slow rate of flow: consequently the development of ice lenses is restricted. However, the presence of fissures can result in an increase in the rate of migration. The worst conditions for water migration occur in soils having a high percentage of silt-size particles: such soils usually have a network of small pores, yet, at the same time, the permeability is not too low. A well graded soil is reckoned to be frost-susceptible if more than 3% of the particles are smaller than 0·02 mm. A poorly graded soil is susceptible if more than 10% of the particles are smaller than 0·02 mm.

Problems

2.1   In a falling head permeability test the initial head of 1·00 m dropped to 0·35 m in 3 h, the diameter of the standpipe being 5 mm. The soil specimen was 200 mm long by 100 mm in diameter. Calculate the coefficient of permeability of the soil.

2.2   A deposit of soil is 16 m deep and overlies an impermeable stratum: the coefficient of permeability is 10^{-6} m/s. A sheet pile wall is driven to a

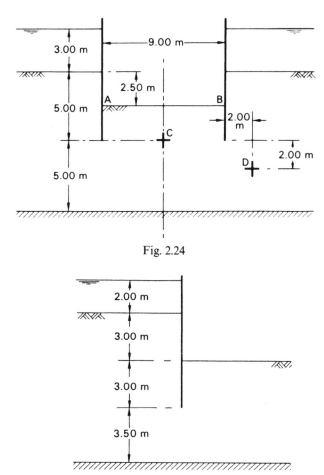

Fig. 2.24

Fig. 2.25

depth of 12·00 m in the deposit. The difference in water level between the two sides of the piling is 4·00 m. Draw the flow net and determine the quantity of seepage under the piling.

2.3 Draw the flow net for seepage under the structure detailed in Fig. 2.23 and determine the quantity of seepage. The coefficient of permeability of the soil is $5·0 \times 10^{-5}$ m/s. What is the uplift force on the base of the structure?

2.4 The section through a long cofferdam is shown in Fig. 2.24, the coefficient of permeability of the soil being $4·0 \times 10^{-7}$ m/s. Draw the flow net and determine the quantity of seepage entering the cofferdam.

2.5 The section through part of a cofferdam is shown in Fig. 2.25, the coefficient of permeability of the soil being $2·0 \times 10^{-6}$ m/s. Draw the flow net and determine the quantity of seepage.

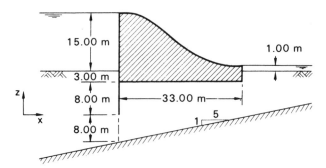

Fig. 2.26

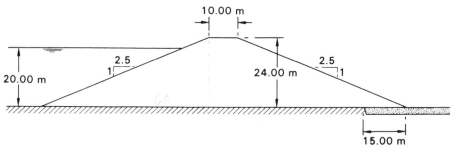

Fig. 2.27

2.6 The dam shown in section in Fig. 2.26 is located on anisotropic soil. The coefficients of permeability in the x and z directions are $5\cdot0 \times 10^{-7}$ m/s and $1\cdot8 \times 10^{-7}$ m/s respectively. Determine the quantity of seepage under the dam.

2.7 An earth dam is shown in section in Fig. 2.27 the coefficients of permeability in the horizontal and vertical directions being $7\cdot5 \times 10^{-6}$ m/s and $2\cdot7 \times 10^{-6}$ m/s respectively. Construct the top flow line and determine the quantity of seepage through the dam.

2.8 Determine the quantity of seepage under the dam shown in section in Fig. 2.28. Both layers of soil are isotropic the coefficients of permeability of the upper and lower layers being $2\cdot0 \times 10^{-6}$ m/s and $1\cdot6 \times 10^{-5}$ m/s respectively.

References

2.1 Casagrande, A. (1940): 'Seepage Through Dams', in *Contributions to Soil Mechanics 1925–1940*, Boston Society of Civil Engineers.

2.2 Cedergren, H. R. (1967): *Seepage, Drainage and Flow Nets*, John Wiley and Sons, New York.

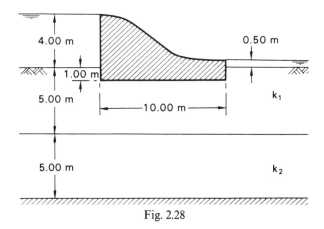

4.00 m

0.50 m

1.00 m

5.00 m

k₁

—10.00 m—

5.00 m

k₂

Fig. 2.28

2.3   Harr, M. E. (1962): *Groundwater and Seepage*, McGraw-Hill, New York.

2.4   Hvorslev, M. J. (1951): *Time Lag and Soil Permeability in Ground-Water Observations*, Bulletin No. 36, Waterways Experimental Station, U.S. Corps of Engineers, Vicksburg, Mississippi.

2.5   Ischy, E. and Glossop, R. (1962): 'An Introduction to Alluvial Grouting', *Proceedings ICE*, Vol. 21.

2.6   Sherard, J. L., Woodward, R. J., Gizienski, S. F. and Clevenger, W. A. (1963): *Earth and Earth-Rock Dams*, John Wiley and Sons, New York.

2.7   Taylor, D. W. (1948): *Fundamentals of Soil Mechanics*, John Wiley and Sons, New York.

2.8   Vreedenburgh, C. G. F. (1936): 'On the Steady Flow of Water Percolating through Soils with Homogeneous-Anisotropic Permeability', *Proceedings 1st International Conference SMFE, Cambridge, Massachusetts*, Vol. 1.

CHAPTER 3

Effective Stress

3.1 Introduction

A soil can be visualized as a skeleton of solid particles enclosing continuous voids which contain water and/or air. For the range of stresses usually encountered in practice the individual solid particles and water can be considered incompressible: air, on the other hand, is highly compressible. The volume of the soil skeleton as a whole can change due to rearrangement of the soil particles into new positions, mainly by rolling and sliding, with a corresponding change in the forces acting between particles. The actual compressibility of the soil skeleton will depend on the structural arrangement of the solid particles. In a fully saturated soil, since water is considered to be incompressible, a reduction in volume is possible only if some of the water can escape from the voids. In a dry or a partially saturated soil a reduction in volume is always possible due to compression of the air in the voids, provided there is scope for particle rearrangement.

Shear stress can be resisted only by the skeleton of solid particles, by means of forces developed at the interparticle contacts. Normal stress may be resisted by the soil skeleton through an increase in the interparticle forces. If the soil is fully saturated, the water filling the voids can also withstand normal stress by an increase in pressure.

3.2 The Principle of Effective Stress

The importance of the forces transmitted through the soil skeleton from particle to particle was recognized in 1923 when Terzaghi presented the principle of effective stress, an intuitive relationship based on experimental data. The principle applies only to *fully saturated* soils and relates the following three stresses:

1. the *total normal stress* (σ) on a plane within the soil mass, being the force per unit area transmitted in a normal direction across the plane, imagining the soil to be a solid (single-phase) material;
2. the *pore water pressure* (u), being the pressure of the water filling the void space between the solid particles;

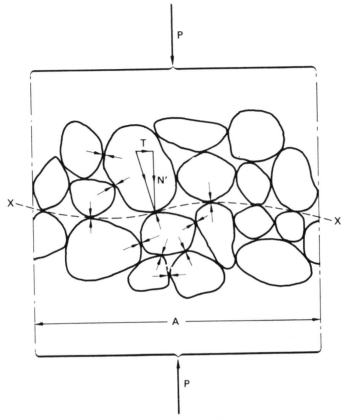

Fig. 3.1 Interpretation of effective stress.

3. the *effective normal stress* (σ') on the plane, representing the stress transmitted through the soil skeleton only.

The relationship is:

$$\sigma = \sigma' + u \tag{3.1}$$

The principle can be represented by the following physical model. Consider a 'plane' XX in a fully saturated soil, passing through points of interparticle contact only, as shown in Fig. 3.1. The wavy plane XX is really indistinguishable from a true plane on the mass scale due to the relatively small size of individual soil particles. A normal force P applied over an area A may be resisted partly by interparticle forces and partly by the pressure in the pore water. The interparticle forces are very random in both magnitude and direction throughout the soil mass but at every point of contact on the wavy plane may be split into components normal and tangential to the direction of the true plane to which XX approximates: the normal and tangential components are N' and T respectively. Then the effective normal stress is interpreted as the sum of all the components N' within the area A, divided by the area A, i.e.

$$\sigma' = \frac{\Sigma N'}{A} \qquad (3.2)$$

The total normal stress is given by

$$\sigma = \frac{P}{A} \qquad (3.3)$$

If point contact is assumed between the particles, the pore water pressure will act on the plane over the entire area A. Then for equilibrium in the direction normal to XX:

$$P = \Sigma N' + uA$$

or

$$\frac{P}{A} = \frac{\Sigma N'}{A} + u$$

i.e.

$$\sigma = \sigma' + u$$

The pore water pressure which acts equally in every direction will act on the entire surface of any particle but is assumed not to change the volume of the particle: also, the pore water pressure does not cause particles to be pressed together. The error involved in assuming point contact between particles is negligible in soils, the total contact area normally being between 1 and 3% of the cross-sectional area A. It should be understood that σ' does not represent the true contact stress between two particles, which would be the random but very much higher stress N'/a, where a is the actual contact area between the particles. Clay mineral particles, when they are present in a soil, may not be in direct contact due to their surrounding adsorbed layers but it is assumed that interparticle force can be transmitted through the highly viscous adsorbed water.

Effective Vertical Stress due to Self-Weight of Soil

Consider a soil mass having a horizontal surface and with the water table at surface level. The total vertical stress (i.e. the total normal stress on a horizontal plane) at depth z is equal to the weight of all material (solids + water) per unit area above that depth, i.e.

$$\sigma_v = \gamma_{sat} z$$

The pore water pressure at any depth will be hydrostatic since the void space between the solid particles is continuous, therefore at depth z:

$$u = \gamma_w z$$

Hence from Equation 3.1 the effective vertical stress at depth z will be:

$$\sigma'_v = \sigma_v - u$$
$$= (\gamma_{sat} - \gamma_w)z = \gamma' z$$

where γ' is the buoyant unit weight of the soil.

3.3 Response of Effective Stress to a Change in Total Stress

As an illustration of how effective stress responds to a change in total stress, consider the case of a fully saturated soil subject to an *increase* in total vertical stress and in which the lateral strain is zero, volume change being due entirely to deformation of the soil in the vertical direction. This condition may be assumed in practice when there is a change in total vertical stress over an area which is large compared with the thickness of the soil layer in question.

It is assumed initially that the pore water pressure is constant at a value governed by a constant position of the water table. This initial value is called the *static pore water pressure*. When the total vertical stress is increased the solid particles immediately try to take up new positions closer together. However, if water is incompressible and the soil is laterally confined, no such particle rearrangement, and therefore no increase in the interparticle forces, is possible unless some of the pore water can escape. Since the pore water is resisting the particle rearrangment the pore water pressure is increased above the static value immediately the increase in total stress takes place. The increase in pore water pressure will be equal to the increase in total vertical stress, i.e. the increase in total vertical stress is carried entirely by the pore water. Note that if the lateral strain were not zero some degree of particle rearrangement would be possible, resulting in an immediate increase in effective vertical stress and the increase in pore water pressure would be less than the increase in total vertical stress.

The increase in pore water pressure causes a pressure gradient in the pore water, resulting in a transient flow of pore water towards a free-draining boundary of the soil layer. This flow or *drainage* will continue until the pore water pressure again becomes equal to a value governed by a steady position of the water table. This final value is called the *steady-state pore water pressure*. In most situations the static and steady-state values of pore water pressure will be equal but it is possible for the position of the water table to change. The increase in pore water pressure above the final or steady-state value is called the *excess pore water pressure*. The reduction of the excess pore water pressure to the steady-state value is described as *dissipation* and when this has been completed the soil is said to be in a *drained* condition. Before dissipation of excess pore water pressure begins the condition of the soil is referred to as *undrained*.

As drainage of pore water takes place the solid particles become free to take up new positions with a resulting increase in the interparticle forces. In other words, as the excess pore water pressure dissipates the effective vertical stress increases, accompanied by a corresponding reduction in volume. When dissipation of excess pore water pressure is complete the increment of total vertical stress will be carried entirely by the soil skeleton. Throughout the process the soil remains in a fully saturated condition.

The time taken for drainage to be completed depends on the permeability of the soil. In soils of low permeability, such as saturated clays, drainage will be slow and the whole process is referred to as *consolidation*. With deformation taking place in one direction only, consolidation is described as one-dimensional. In soils of high permeability such as saturated sands, drainage will be very rapid.

When a soil is subject to a *reduction* in total normal stress the scope for volume increase is limited because particle rearrangement due to total stress increase is largely irreversible. As a result of increase in the interparticle forces there will be small elastic strains (normally ignored) in the solid particles especially around the contact areas and if clay mineral particles are present in the soil they may experience bending. In addition, the adsorbed water surrounding clay mineral particles will experience recoverable compression due to increase in interparticle forces, especially if there is face-to-face orientation of the particles. When a decrease in total normal stress takes place in a soil there will thus be a tendency for the soil skeleton to expand to a limited extent, especially in soils containing an appreciable proportion of clay mineral particles. As a result the pore water pressure will be reduced and the excess pore water pressure will be negative. The pore water pressure will gradually increase to the steady-state value, flow taking place into the soil, accompanied by a corresponding reduction in effective normal stress and increase in volume. In soils of low permeability, this process, the reverse of consolidation, is called *swelling*.

Consolidation Analogy

The mechanics of the one-dimensional consolidation process can be represented by means of a simple analogy. Fig. 3.2a shows a spring inside a cylinder filled with water and a piston, fitted with a valve, on top of the spring. It is assumed that there can be no leakage between the piston and the cylinder and no friction. The spring represents the compressible soil skeleton, the water in the cylinder the pore water and the bore diameter of the valve the permeability of the soil. The cylinder itself simulates the condition of no lateral strain in the soil.

Suppose a load is now placed on the piston with the valve closed, as in Fig. 3.2b. Assuming water to be incompressible, the piston will not move as long as the valve is closed, with the result that no load can be transmitted to the spring: the load will be carried by the water, the increase in pressure in

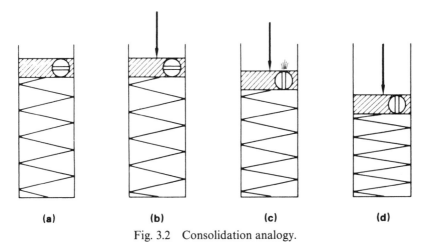

Fig. 3.2   Consolidation analogy.

the water being equal to the load divided by the piston area. This situation with the valve closed, corresponds to the undrained condition in the soil.

If the valve is now opened, water will be forced out through the valve at a rate governed by the bore diameter. This will allow the piston to move and the spring to be compressed as load is gradually transferred to it. This situation is shown in Fig. 3.2c. At any time the increase in load on the spring will correspond to the reduction in pressure in the water. Eventually, as shown in Fig. 3.2d, all the load will be carried by the spring and the piston will come to rest, this corresponding to the drained condition in the soil. At any time, the load carried by the spring represents the effective normal stress in the soil, the pressure of the water in the cylinder the pore water pressure and the load on the piston the total normal stress. The movement of the piston represents the change in volume of the soil and is governed by the compressibility of the spring (the equivalent of the compressibility of the soil skeleton). The piston and spring analogy represents only an element of soil since the stress conditions vary from point to point throughout a soil mass.

Example 3.1

A layer of saturated clay 4 m thick is overlain by sand 5 m deep, the water table being 3 m below the surface. The saturated unit weights of the clay and sand are $19 \, kN/m^3$ and $20 \, kN/m^3$ respectively: above the water table the unit weight of the sand is $17 \, kN/m^3$. Plot the values of total vertical stress and effective vertical stress against depth. If sand to a height of 1 m above the water table is saturated with capillary water, how are the above stresses affected?

The total vertical stress is the weight of all material (solids + water) per

Table 3.1

Depth (m)	$\sigma_v(kN/m^2)$		$u(kN/m^2)$	$\sigma_v' = \sigma_v - u$ (kN/m²)
3	3×17	$= 51 \cdot 0$	0	51·0
5	$(3 \times 17) + (2 \times 20)$	$= 91 \cdot 0$	$2 \times 9.8 = 19 \cdot 6$	71·4
9	$(3 \times 17) + (2 \times 20) + (4 \times 19) = 167 \cdot 0$		$6 \times 9 \cdot 8 = 58 \cdot 8$	108·2

unit area above the depth in question. Pore water pressure is the hydrostatic pressure corresponding to the depth below the water table. The effective vertical stress is the difference between the total vertical stress and the pore water pressure at the same depth. Alternatively, effective vertical stress may be calculated directly using the buoyant unit weight of the soil below the water table. The stresses need be calculated only at depths where there is a change in unit weight (Table 3.1).

The alternative calculation of σ_v' at depths of 5 and 9 m is as follows:

Buoyant unit weight of sand $= 20 - 9 \cdot 8 = 10 \cdot 2 \, kN/m^3$
Buoyant unit weight of clay $= 19 - 9 \cdot 8 = 9 \cdot 2 \, kN/m^3$
At 5 m depth: $\sigma_v' = (3 \times 17) + (2 \times 10 \cdot 2) = 71 \cdot 4 \, kN/m^2$
At 9 m depth: $\sigma_v' = (3 \times 17) + (2 \times 10 \cdot 2) + (4 \times 9 \cdot 2) = 108 \cdot 2 \, kN/m^2$

The alternative method is recommended when only the effective stress is required. In all cases the stresses would normally be rounded off to the nearest whole number. The stresses are plotted against depth in Fig. 3.3.

Effect of Capillary Rise. The water table is the level at which pore water pressure is atmospheric (i.e. $u = 0$). Above the water table, water is held under negative pressure and, even if the soil is saturated above the water table, does not contribute to hydrostatic pressure below the water table. The only effect of the 1 m capillary rise, therefore, is to increase the total unit weight of the sand between 2 m and 3 m depth from $17 \, kN/m^3$ to $20 \, kN/m^3$, an increase of $3 \, kN/m^3$. Both total and effective vertical stresses below 3 m depth are therefore increased by the constant amount $3 \times 1 = 3 \cdot 0 \, kN/m^2$, pore water pressures being unchanged.

Example 3.2

A 5 m depth of sand overlies a 6 m layer of clay, the water table being at the surface; the permeability of the clay is very low. The saturated unit weight of the sand is $19 \, kN/m^3$ and that of the clay $20 \, kN/m^3$. A 4 m depth of fill material of unit weight $20 \, kN/m^3$ is placed on the surface over an extensive area. Determine the effective vertical stress at the centre of the clay layer (a) immediately after the fill has been placed, assuming this to take place rapidly, (b) many years after the fill has been placed.

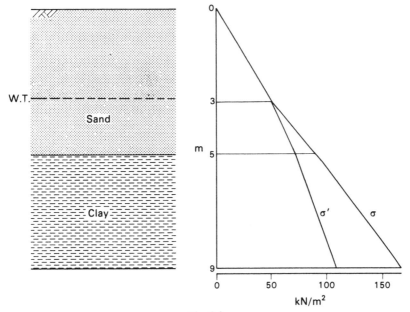

Fig. 3.3

The soil profile is shown in Fig. 3.4. Since the fill covers an extensive area it can be assumed that the condition of zero lateral strain applies. As the permeability of the clay is very low, dissipation of excess pore water pressure will be very slow: immediately after the rapid placing of the fill, no appreciable dissipation will have taken place. Therefore, the effective vertical stress at the centre of the clay layer immediately after placing will be virtually unchanged from the original value, i.e.:

$$\sigma'_v = (5 \times 9.2) + (3 \times 10.2) = 76.6 \, \text{kN/m}^2$$

(the buoyant unit weights of the sand and the clay respectively being $9.2 \, \text{kN/m}^3$ and $10.2 \, \text{kN/m}^3$).

Many years after the placing of the fill, dissipation of excess pore water pressure should be essentially complete and the effective vertical stress at the centre of the clay layer will be:

$$\sigma'_v = (4 \times 20) + (5 \times 9.2) + (3 \times 10.2) = 156.6 \, \text{kN/m}^2$$

Immediately after the fill has been placed, the total vertical stress at the centre of the clay increases by $80 \, \text{kN/m}^2$ due to the weight of the fill. Since the clay is saturated and there is no lateral strain there will be a corresponding increase in pore water pressure of $80 \, \text{kN/m}^2$. The static and steady-state pore water pressures are equal since there is no change in the level of the water table, the value being $(8 \times 9.8) = 78.4 \, \text{kN/m}^2$. Immediately after placing, the pore water pressure increases from $78.4 \, \text{kN/m}^2$

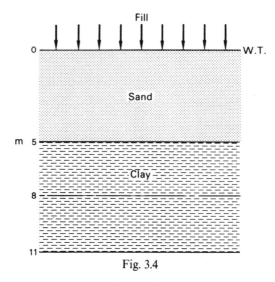

Fig. 3.4

to $158.4\,\text{kN/m}^2$ and then during subsequent consolidation gradually decreases again to $78.4\,\text{kN/m}^2$, accompanied by the gradual increase of effective vertical stress from $76.6\,\text{kN/m}^2$ to $156.6\,\text{kN/m}^2$.

3.4 Partially Saturated Soils

In the case of partially saturated soils part of the void space is occupied by water and part by air. The pore water pressure (u_w) must always be less than the pore air pressure (u_a) due to surface tension. Unless the degree of saturation is close to unity the pore air will form continuous channels through the soil and the pore water will be concentrated in the regions around the interparticle contacts. The boundaries between pore water and pore air will be in the form of menisci whose radii will depend on the size of the pore spaces within the soil. Part of any wavy plane through the soil will therefore pass through water and part through air.

In 1955 Bishop proposed the following effective stress equation for partially saturated soils:

$$\sigma = \sigma' + u_a - \chi(u_a - u_w) \tag{3.4}$$

where χ is a parameter, to be determined experimentally, related primarily to the degree of saturation of the soil. The term $(u_a - u_w)$ is a measure of the suction in the soil. For a fully saturated soil $(S_r = 1)$. $\chi = 1$ and for a completely dry soil $(S_r = 0)$, $\chi = 0$. Equation 3.4 thus degenerates to Equation 3.1 when $S_r = 1$. The value of χ is also influenced, to a lesser extent, by the soil structure and the way the particular degree of saturation

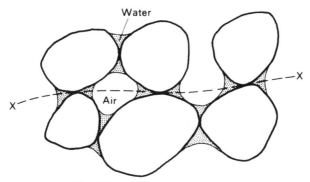

Fig. 3.5   Partially saturated soil.

was brought about. Equation 3.4 is not convenient for use in practice because of the presence of the parameter χ.

A physical model may be considered in which the parameter χ is interpreted as the average proportion of any cross-section which passes through water. Then across a given section of gross area A (Fig. 3.5) total force is given by the equation:

$$\sigma A = \sigma' A + u_w \chi A + u_a (1 - \chi) A \tag{3.5}$$

which leads to Equation 3.4.

If the degree of saturation of the soil is close to unity it is likely that the pore air will exist in the form of bubbles within the pore water and it is possible to draw a wavy plane through pore water only. The soil can then be considered as a fully saturated soil but with the pore water having some degree of compressibility due to the presence of the air bubbles: Equation 3.1 may then represent effective stress with sufficient accuracy for most practical purposes.

3.5   Influence of Seepage on Effective Stress

When water is seeping through the pores of a soil, total head is dissipated as viscous friction producing a frictional drag, acting in the direction of flow, on the solid particles. A transfer of energy thus takes place from the water to the solid particles and the force corresponding to this energy transfer is called *seepage force*. Seepage forces act on the particles of a soil in addition to gravitational forces and the combination of the forces on a soil mass due to gravity and seeping water is called the resultant body force. It is the resultant body force that governs the effective normal stress on a plane within a soil mass through which seepage is taking place.

Consider a point in a soil mass where the direction of seepage is at angle θ below the horizontal. A square element ABCD of dimension b (unit

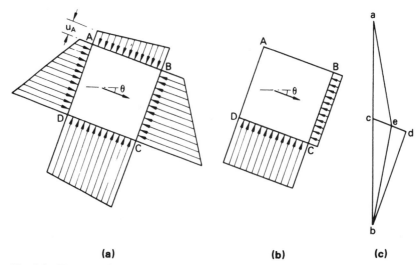

(a)   **(b)**   **(c)**

Fig. 3.6   Forces under seepage conditions. (Reproduced from D. W. Taylor (1948) *Fundamentals of Soil Mechanics*, © John Wiley & Sons Inc., by permission.)

dimension normal to the paper) is centred at the above point with sides parallel and normal to the direction of seepage, as shown in Fig. 3.6a, i.e. the square element can be considered as a flow net field. Let the drop in total head between the sides AD and BC be Δh. Consider the pore water pressures on the boundaries of the element, taking the value of pore water pressure at point A as u_A. The difference in pore water pressure between A and D is due only to the difference in elevation head between A and D, the total head being the same at A and D. However the difference in pore water pressure between A and either B or C is due to the difference in elevation head *and* the difference in total head between A and either B or C. The pore water pressures at B, C and D are as follows:

$$u_B = u_A + \gamma_w (b \sin \theta - \Delta h)$$
$$u_C = u_A + \gamma_w (b \sin \theta + b \cos \theta - \Delta h)$$
$$u_D = u_A + \gamma_w b \cos \theta$$

The following pressure differences can now be established:

$$u_B - u_A = u_C - u_D = \gamma_w (b \sin \theta - \Delta h)$$
$$u_D - u_A = u_C - u_B = \gamma_w b \cos \theta$$

These values are plotted in Fig. 3.6b, giving the distribution diagrams of net pressure across the element in directions parallel and normal to the direction of flow.

Therefore the *force* on BC due to pore water pressure acting on the boundaries of the element, called the *boundary water force*, is given by:

$$\gamma_w (b \sin \theta - \Delta h)b$$

or

$$\gamma_w b^2 \sin \theta - \Delta h \gamma_w b$$

and the boundary water force on CD by:

$$\gamma_w b^2 \cos \theta$$

If there were no seepage, i.e. if the pore water were static, the value of Δh would be zero, the forces on BC and CD would be $\gamma_w b^2 \sin \theta$ and $\gamma_w b^2 \cos \theta$ respectively and their resultant would be $\gamma_w b^2$ acting in the vertical direction. The force $\Delta h \gamma_w b$ represents the only difference between the static and seepage cases and is therefore called the seepage force (J), *acting in the direction of flow* (in this case normal to BC).

Now, the average hydraulic gradient across the element is given by

$$i = \frac{\Delta h}{b}$$

hence,

$$J = \Delta h \gamma_w b = \frac{\Delta h}{b} \gamma_w b^2 = i \gamma_w b^2$$

or

$$J = i \gamma_w V \tag{3.6}$$

where V is the volume of the soil element.

The *seepage pressure* (j) is defined as the seepage force per unit volume, i.e.

$$j = i \gamma_w \tag{3.7}$$

It should be noted that j (and hence J) depends only on the value of hydraulic gradient.

All the forces, both gravitational and due to seeping water, acting on the element ABCD may be represented in the vector diagram, Fig. 3.6c. The forces are summarized below.

Total weight of element $= \gamma_{sat} b^2 =$ vector ab

Boundary water force on CD (seepage and static cases)
$$= \gamma_w b^2 \cos \theta = \text{vector } bd$$

Boundary water force on BC (seepage case)
$$= \gamma_w b^2 \sin \theta - \Delta h \gamma_w b = \text{vector } de$$

Boundary water force on BC (static case)
$$= \gamma_w b^2 \sin \theta = \text{vector } dc$$

Resultant boundary water force (seepage case)
$$= \text{vector } be$$

Resultant boundary water force (static case)
$$= \gamma_w b^2 = \text{vector } bc$$

Seepage force $\qquad = \Delta h \gamma_w b = \text{vector } ce$

Resultant body force (seepage case)
$$= \text{vector } ae$$

Resultant body force (static case)
$$= \text{vector } ac = \gamma' b^2$$

The resultant body force can be obtained by one or other of the following force combinations:

1. Total (saturated) weight + Resultant boundary water force, i.e. vector ab + vector be
2. Effective (buoyant) weight + Seepage force, i.e. vector ac + vector ce

Only the resultant body force contributes to effective stress. A component of seepage force acting vertically upwards will therefore reduce a vertical effective stress component from the static value. A component of seepage force acting vertically downwards will increase a vertical effective stress component from the static value.

A problem may be solved using either force combination 1 or force combination 2, but it may be that one combination is more suitable than the other for a particular problem. Combination 1 involves consideration of the equilibrium of the whole soil mass (solids + water), while combination 2 involves consideration of the equilibrium of the soil skeleton only.

The Quick Condition

Consider the special case of seepage vertically upwards. The vector ce in Fig. 3.6c would then be vertically upwards and if the hydraulic gradient were high enough the resultant body force would be zero. The value of hydraulic gradient corresponding to zero resultant body force is called the *critical hydraulic gradient* (i_c). For an element of soil of volume V subject to upward seepage under the critical hydraulic gradient, the seepage force is therefore equal to the effective weight of the element, i.e.

$$i_c \gamma_w V = \gamma' V$$

Therefore

$$i_c = \frac{\gamma'}{\gamma_w} = \frac{G_s - 1}{1 + e} \tag{3.8}$$

The ratio γ'/γ_w, and hence the critical hydraulic gradient, is approximately 1·0 for most soils.

When the hydraulic gradient is i_c, the effective normal stress on any plane will be zero, gravitational forces having been cancelled out by upward seepage forces. In the case of sands the contact forces between particles will be zero and the soil will have no strength. The soil is then said to be in a *quick* condition (quick meaning 'alive') and if the critical gradient is exceeded the surface will appear to be 'boiling' as the particles are moved around in the upward flow of water. It should be realised that 'quicksand' is not a special type of soil but simply sand through which there is an upward flow of water under a hydraulic gradient equal to or exceeding i_c. It is possible for clays to have strength at zero effective normal stress, so the quick condition may not necessarily result when the hydraulic gradient reaches the critical value given by Equation 3.8.

Conditions Adjacent to Sheet Piling

High upward hydraulic gradients may be experienced in the soil adjacent to the downstream face of a sheet pile wall. Fig. 3.7 shows part of the flow net for seepage under a sheet pile wall, the embedded length on the downstream side being d. A mass of soil adjacent to the piling may become unstable and be unable to support the wall. Terzaghi has shown that failure is likely to occur within a soil mass of approximate dimensions $d \times d/2$ in section (ABCD in Fig. 3.7). Failure first shows in the form of a rise or *heave* at the surface, associated with an expansion of the soil which results in an increase in permeability. This in turn leads to increased flow, surface 'boiling' in the case of sands, and complete failure.

The variation of total head on the lower boundary CD of the soil mass can be obtained from the flow net equipotentials but for purposes of analysis it is sufficient to determine the average total head h_m by inspection. The total head on the upper boundary AB is zero.

The average hydraulic gradient is given by

$$i_m = \frac{h_m}{d}$$

Since failure due to heaving may be expected when the hydraulic gradient becomes i_c, the factor of safety (F) against heaving may be expressed as

$$F = \frac{i_c}{i_m} \tag{3.9}$$

In the case of sands, a factor of safety can also be obtained with respect to 'boiling' at the surface. The *exit* hydraulic gradient (i_e) can be determined by measuring the dimension Δs of the flow net field AEFG adjacent to the piling:

$$i_e = \frac{\Delta h}{\Delta s}$$

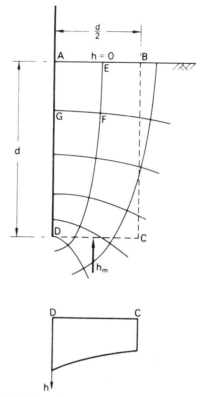

Fig. 3.7   Upward seepage adjacent to sheet piling.

where Δh is the drop in total head between equipotentials GF and AE. Then the factor of safety is:

$$F = \frac{i_c}{i_e} \tag{3.10}$$

There is unlikely to be any appreciable difference between the values of F given by Equations 3.9 and 3.10.

The sheet pile wall problem shown in Fig. 3.7 can also be used to illustrate the two methods of combining gravitational and water forces.

1. Total weight of mass ABCD $= \frac{1}{2}\gamma_{sat}d^2$

   Average total head on CD $= h_m$

   Elevation head on CD $= -d$

   Average pore water pressure on CD $= (h_m + d)\gamma_w$

   Boundary water force on CD $= \dfrac{d}{2}(h_m + d)\gamma_w$

Resultant body force of ABCD $= \frac{1}{2}\gamma_{sat}d^2 - \frac{d}{2}(h_m + d)\gamma_w$

$$= \frac{1}{2}(\gamma' + \gamma_w)d^2 - \frac{1}{2}(h_m d + d^2)\gamma_w$$
$$= \frac{1}{2}\gamma'd^2 - \frac{1}{2}h_m\gamma_w d$$

2. Effective weight of mass ABCD $= \frac{1}{2}\gamma'd^2$

Average hydraulic gradient through ABCD $= \dfrac{h_m}{d}$

Seepage force on ABCD $= \dfrac{h_m}{d}\gamma_w\dfrac{d^2}{2}$

$$= \frac{1}{2}h_m\gamma_w d$$

Resultant body force of ABCD $= \frac{1}{2}\gamma' d^2 - \frac{1}{2}h_m\gamma_w d$
as in method 1 above.

The resultant body force will be zero, leading to heaving, when:

$$\frac{1}{2}h_m\gamma_w d = \frac{1}{2}\gamma'd^2$$

The factor of safety can then be expressed as

$$F = \frac{\frac{1}{2}\gamma'd^2}{\frac{1}{2}h_m\gamma_w d} = \frac{\gamma'd}{h_m\gamma_w} = \frac{i_c}{i_m}$$

If the factor of safety against heaving is considered inadequate, the embedded length d may be increased or a surcharge load in the form of a filter may be placed on the surface AB, the filter being designed to prevent entry of soil particles, If the effective weight of the filter per unit area is w' then the factor of safety becomes

$$F = \frac{\gamma'd + w'}{h_m\gamma_w}$$

Example 3.3

The flow net for seepage under a sheet pile wall is shown in Fig. 3.8a, the saturated unit weight of the soil being $20\,\text{kN/m}^3$. Determine the values of effective vertical stress at A and B.

1. First consider the combination of total weight and resultant boundary water force. Consider the column of saturated soil of *unit area* between A and the soil surface at D. The total weight of the column is $11\gamma_{sat}$ (220 kN). Due to the change in level of the equipotentials across the column, the boundary water forces on the sides of the column will not be equal although in this case the difference will be small. There is thus a net horizontal boundary water force on the column. However, as the effective *vertical*

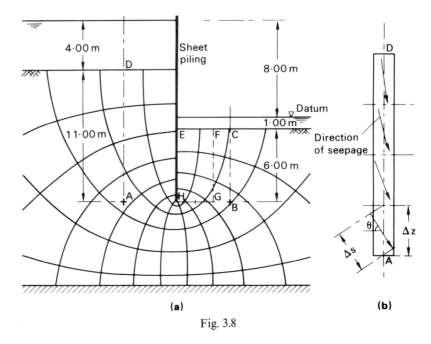

(a)   (b)

Fig. 3.8

stress is to be calculated, only the vertical component of the resultant body force is required and the net horizontal boundary water force need not be considered. The boundary water force on the top surface of the column is due only to the depth of water above D and is $4\gamma_w$ (39 kN). The boundary water force on the bottom surface of the column must be determined from the flow net, as follows:

Number of equipotential drops between the downstream soil surface and A = 8·2.

There are 12 equipotential drops between the upstream and downstream soil surfaces, representing a loss in total head of 8 m.

Total head at A, $h_A = \dfrac{8\cdot2}{12} \times 8 = 5\cdot5$ m.

Elevation head at A, $z_A = -7\cdot0$ m.

Pore water pressure at A, $u_A = \gamma_w(h_A - z_A)$
$$= 9\cdot8(5\cdot5 + 7\cdot0) = 122 \, \text{kN/m}^2.$$

i.e. Boundary water force on bottom surface = 122 kN.

Net vertical boundary water force = 122 − 39 = 83 kN.

Total weight of the column = 220 kN.

Vertical component of resultant body force $= 220 - 83 = 137\,\text{kN}$.

i.e. Effective vertical stress at A $= 137\,\text{kN/m}^2$.

It should be realized that the same result would be obtained by the direct application of the effective stress equation, the total vertical stress at A being the weight of saturated soil and water, per unit area, above A. Thus:

$$\sigma_A = 11\gamma_{\text{sat}} + 4\gamma_w = 220 + 39 = 259\,\text{kN/m}^2$$
$$u_A = 122\,\text{kN/m}^2$$
$$\sigma'_A = \sigma_A - u_A = 259 - 122 = 137\,\text{kN/m}^2$$

The only difference in concept is that the boundary water force per unit area on top of the column of saturated soil AD contributes to the total vertical stress at A. Similarly at B:

$$\sigma_B = 6\gamma_{\text{sat}} + 1\gamma_w = 120 + 9 \cdot 8 = 130\,\text{kN/m}^2$$

$$h_B = \frac{2 \cdot 4}{12} \times 8 = 1 \cdot 6\,\text{m}$$

$$z_B = -7 \cdot 0\,\text{m}$$

$$u_B = \gamma_w(h_B - z_B) = 9 \cdot 8(1 \cdot 6 + 7 \cdot 0) = 84\,\text{kN/m}^2$$
$$\sigma'_B = \sigma_B - u_B = 130 - 84 = 46\,\text{kN/m}^2$$

2. Now consider the combination of effective weight and seepage force. The direction of seepage alters over the depth of the column of soil AD as illustrated in Fig. 3.8b, the direction of seepage for any section of the column being determined from the flow net: the effective weight of the column must be combined with the vertical components of seepage force. More conveniently, the effective stress at A can be calculated using the algebraic sum of the buoyant unit weight of the soil and the average value of the vertical component of seepage pressure between A and D.

Between any two equipotentials the hydraulic gradient is $\Delta h/\Delta s$ (Equation 2.15). Hence, if θ is the angle between the direction of flow and the horizontal, the vertical component of seepage pressure ($j \sin \theta$) is

$$\frac{\Delta h}{\Delta s}\gamma_w \sin \theta = \frac{\Delta h}{\Delta z}\gamma_w$$

where $\Delta z \,(= \Delta s/\sin \theta)$ is the vertical distance between the same equipotentials. The calculation is as follows.

Number of equipotential drops between D and A $= 3 \cdot 8$.

Loss in total head between D and A $= \dfrac{3 \cdot 8}{12} \times 8 = 2 \cdot 5\,\text{m}$.

Average value of vertical component of seepage pressure between D and A, acting in the same direction as gravity

$$= \frac{2 \cdot 5}{11} \times 9 \cdot 8 = 2 \cdot 3 \, \text{kN/m}^3$$

Buoyant unit weight of soil, $\gamma' = 20 - 9 \cdot 8 = 10 \cdot 2 \, \text{kN/m}^3$.

For column AD, of unit area, resultant body force

$$= 11 \, (10 \cdot 2 + 2 \cdot 3) = 137 \, \text{kN}$$

i.e. Effective vertical stress at A $= 137 \, \text{kN/m}^2$.

The calculation is now given for point B.

Loss in total head between B and C $= \dfrac{2 \cdot 4}{12} \times 8 = 1 \cdot 6 \, \text{m}$.

Average value of vertical component of seepage pressure between B and C, acting in the opposite direction to gravity

$$= \frac{1 \cdot 6}{6} \times 9 \cdot 8 = 2 \cdot 6 \, \text{kN/m}^3$$

Hence, $\sigma'_B = 6(10 \cdot 2 - 2 \cdot 6) = 46 \, \text{kN/m}^2$.

Example 3.4

Using the flow net in Fig. 3.8a, determine the factor of safety against failure by heaving adjacent to the downstream face of the piling. The saturated unit weight of the soil is $20 \, \text{kN/m}^3$.

The stability of the soil mass EFGH in Fig. 3.8a, 6 m by 3 m in section, will be analysed.

By inspection of the flow net, the average value of total head on the base GH is given by:

$$h_m = \frac{3 \cdot 5}{12} \times 8 = 2 \cdot 3 \, \text{m}$$

The average hydraulic gradient between GH and the soil surface EF is:

$$i_m = \frac{2 \cdot 3}{6} = 0 \cdot 39$$

Critical hydraulic gradient, $i_c = \dfrac{\gamma'}{\gamma_w} = \dfrac{10 \cdot 2}{9 \cdot 8} = 1 \cdot 04$

Factor of safety, $F = \dfrac{i_c}{i_m} = \dfrac{1 \cdot 04}{0 \cdot 39} = 2 \cdot 7$

Problems

3.1   The bed of a river 5 m deep consists of sand of saturated unit weight $19 \cdot 5 \, \text{kN/m}^3$. Calculate the effective vertical stress 5 m below the surface of the sand.

3.2 A layer of clay 4 m thick lies between two layers of sand each 4 m thick, the top of the upper layer of sand being ground level. The water table is 2 m below ground level but the lower layer of sand is under artesian pressure, the piezometric surface being 4 m above ground level. The saturated unit weight of the clay is 20 kN/m^3 and that of the sand 19 kN/m^3: above the water table the unit weight of the sand is 16·5 kN/m^3. Calculate the effective vertical stresses at the top and bottom of the clay layer.

3.3 In a deposit of fine sand the water table is 3·5 m below the surface but sand to a height of 1·0 m above the water table is saturated by capillary water: above this height the sand may be assumed to be dry. The saturated and dry unit weights, respectively, are 20 kN/m^3 and 16 kN/m^3. Calculate the effective vertical stress in the sand 8 m below the surface.

3.4 A layer of sand extends from ground level to a depth of 9 m and overlies a layer of clay, of very low permeability, 6 m thick. The water table is 6 m below the surface of the sand. The saturated unit weight of the sand is 19 kN/m^3 and that of the clay 20 kN/m^3: the unit weight of the sand above the water table is 16 kN/m^3. Over a short period of time the water table rises by 3 m and is expected to remain permanently at this new level. Determine the effective vertical stress at depths of 8 m and 12 m below ground level (a) immediately after the rise of the water table, (b) several years after the rise of the water table.

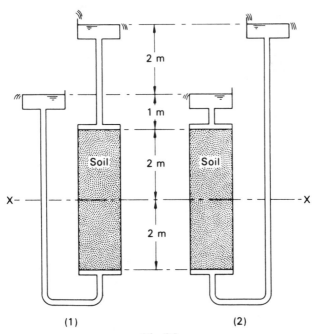

Fig. 3.9

3.5   An element of soil with sides horizontal and vertical measures 1 m in each direction. Water is seeping through the element in a direction inclined upwards at 30° above the horizontal under a hydraulic gradient of 0·35. The saturated unit weight of the soil is 21 kN/m³. Draw a force diagram to scale showing the following: total and effective weights, resultant boundary water force, seepage force. What is the magnitude and direction of the resultant body force?

3.6   For the seepage situations shown in Fig. 3.9, determine the effective normal stress on plane XX in each case (a) by considering pore water pressure, (b) by considering seepage pressure. The saturated unit weight of the soil is 20 kN/m³.

3.7   The section through a long cofferdam is shown in Fig. 2.24, the saturated unit weight of the soil being 20 kN/m³. Determine the factor of safety against 'boiling' at the surface AB and the values of effective vertical stress at C and D.

3.8   The section through part of a cofferdam is shown in Fig. 2.25, the saturated unit weight of the soil being 19·5 kN/m³. Determine the factor of safety against a heave failure in the excavation adjacent to the sheet piling. What depth of filter (unit weight 21 kN/m³) would be required to ensure a factor of safety of 3·0?

References

3.1   Skempton, A. W. (1961): 'Effective Stress in Soils, Concrete and Rocks', *Proceedings of Conference on Pore Pressure and Suction in Soils*, Butterworths, London.

3.2   Taylor, D. W. (1948): *Fundamentals of Soils Mechanics*, John Wiley and Sons, New York.

3.3   Terzaghi, K. (1943): *Theoretical Soil Mechanics*, John Wiley and Sons, New York.

CHAPTER 4

Shear Strength

4.1 The Mohr-Coulomb Failure Criterion

This chapter is concerned with the resistance of a soil to failure in shear. A knowledge of shear strength is required in the solution of problems concerning the stability of soil masses. If at a point on any plane within a soil mass the shear stress becomes equal to the shear strength of the soil, failure will occur at that point. The shear strength (τ_f) of a soil at a point on a particular plane was originally expressed by Coulomb as a linear function of the normal stress (σ_f) on the plane at the same point:

$$\tau_f = c + \sigma_f \tan \phi \tag{4.1}$$

where c and ϕ are the *shear strength parameters*, now described as the *cohesion intercept* (or the *apparent cohesion*) and the *angle of shearing resistance*, respectively. In accordance with Terzaghi's fundamental concept that shear stress in a soil can be resisted only by the skeleton of solid particles, shear strength is expressed as a function of *effective* normal stress:

$$\tau_f = c' + \sigma'_f \tan \phi' \tag{4.2}$$

where c' and ϕ' are the shear strength parameters in terms of effective stress. Failure will thus occur at any point where a critical combination of shear stress and effective normal stress develops.

The shear strength of a soil can also be expressed in terms of the effective major and minor principal stresses σ'_1 and σ'_3 *at failure* at the point in question. At failure the straight line represented by Equation 4.2 will be tangential to the Mohr circle representing the state of stress, as shown in Fig. 4.1, compressive stress being taken as positive. The coordinates of the tangent point are τ_f and σ'_f, where:

$$\tau_f = \tfrac{1}{2}(\sigma'_1 - \sigma'_3) \sin 2\theta \tag{4.3}$$

$$\sigma'_f = \tfrac{1}{2}(\sigma'_1 + \sigma'_3) + \tfrac{1}{2}(\sigma'_1 - \sigma'_3) \cos 2\theta \tag{4.4}$$

and θ is the theoretical angle between the major principal plane and the plane of failure. It is apparent that

$$\theta = 45° + \frac{\phi'}{2} \tag{4.5}$$

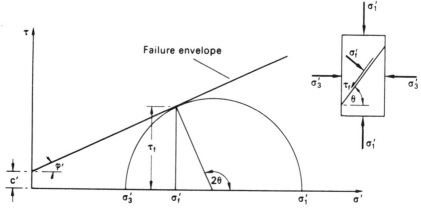

Fig. 4.1 Stress conditions at failure.

From Fig. 4.1 the relationship between the effective principal stresses at failure and the shear strength parameters can also be obtained. Now:

$$\sin \phi' = \frac{\frac{1}{2}(\sigma_1' - \sigma_3')}{c' \cot \phi' + \frac{1}{2}(\sigma_1' + \sigma_3')}$$

Therefore

$$(\sigma_1' - \sigma_3') = (\sigma_1' + \sigma_3') \sin \phi' + 2c' \cos \phi' \tag{4.6a}$$

or

$$\sigma_1' = \sigma_3' \tan^2\left(45° + \frac{\phi'}{2}\right) + 2c' \tan\left(45° + \frac{\phi'}{2}\right) \tag{4.6b}$$

Equation 4.6 is referred to as the Mohr-Coulomb failure criterion. If a number of states of stress are known, each producing shear failure in the soil, the criterion assumes that a common tangent, represented by Equation 4.2, can be drawn to the Mohr circles representing the states of stress: the common tangent is called the *failure envelope* of the soil. A state of stress plotting above the failure envelope is impossible. The criterion does not involve consideration of strains at, or prior to, failure and implies that the effective intermediate principal stress σ_2' has no influence on the shear strength of the soil. The Mohr-Coulomb failure criterion, because of its simplicity, is widely used in practice although it is by no means the only possible failure criterion for soils. The failure envelope for a particular soil may not necessarily be a straight line but a straight line approximation can be taken over the stress range of interest and the shear strength parameters determined for that range.

By plotting $\frac{1}{2}(\sigma_1' - \sigma_3')$ against $\frac{1}{2}(\sigma_1' + \sigma_3')$ any state of stress can be represented by a *stress point* rather than by a Mohr circle, as shown in

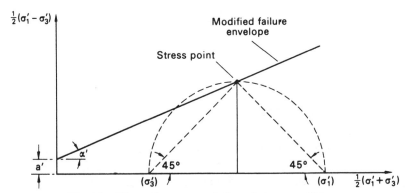

Fig. 4.2 Alternative representation of stress conditions.

Fig. 4.2, and on this plot a modified failure envelope is obtained, represented by the equation:

$$\tfrac{1}{2}(\sigma'_1 - \sigma'_3) = a' + \tfrac{1}{2}(\sigma'_1 + \sigma'_3)\tan\alpha' \qquad (4.7)$$

where a' and α' are the modified shear strength parameters. The parameters c' and ϕ' are then given by:

$$\phi' = \sin^{-1}(\tan\alpha') \qquad (4.8)$$

$$c' = \frac{a'}{\cos\phi'} \qquad (4.9)$$

Lines drawn from the stress point at angles of 45° to the horizontal, as shown in Fig. 4.2, intersect the horizontal axis at points representing the values of the principal stresses σ'_1 and σ'_3. Fig. 4.2 could also be drawn in terms of total stress, the vertical and horizontal coordinates being $\tfrac{1}{2}(\sigma_1 - \sigma_3)$ and $\tfrac{1}{2}(\sigma_1 + \sigma_3)$ respectively. It should be noted that:

$$\tfrac{1}{2}(\sigma'_1 - \sigma'_3) = \tfrac{1}{2}(\sigma_1 - \sigma_3)$$
$$\tfrac{1}{2}(\sigma'_1 + \sigma'_3) = \tfrac{1}{2}(\sigma_1 + \sigma_3) - u$$

In the case of axial symmetry a state of effective stress can also be plotted with respect to vertical and horizontal coordinates q' and p' respectively, where:

$$q' = (\sigma'_1 - \sigma'_3) \qquad (4.10)$$

$$p' = \tfrac{1}{3}(\sigma'_1 + 2\sigma'_3) \qquad (4.11)$$

The values of these stresses (being functions of the principal stresses) are independent of the orientation of the coordinate axes, such stresses being known as stress invariants. The corresponding total stresses are:

$$q = (\sigma_1 - \sigma_3)$$
$$p = \tfrac{1}{3}(\sigma_1 + 2\sigma_3)$$

In this case the relationships between effective and total stresses are:

$$q' = q$$
$$p' = p - u$$

4.2 Shear Strength Tests

The shear strength parameters for a particular soil can be determined by means of laboratory tests on specimens taken from *representative* samples of the in-situ soil. Great care and judgement are required in the sampling operation and in the storage and handling of samples prior to testing, especially in the case of undisturbed samples where the object is to preserve the in-situ structure and water content of the soil. In the case of clays, test specimens may be obtained from tube or block samples, the latter normally being subjected to the least disturbance. Swelling of a clay specimen will occur due to the release of the in-situ total stresses.

The Direct Shear Test

The specimen is confined in a metal box of square or circular cross-section split horizontally at mid-height, a small clearance being maintained between the two halves of the box. Porous plates are placed below and on top of the specimen if it is fully or partially saturated to allow free drainage: if the specimen is dry, solid metal plates may be used. The essential features of the apparatus are shown diagrammatically in Fig. 4.3. A vertical force (N) is applied to the specimen through a loading plate and shear stress is gradually applied on a horizontal plane by causing the two halves of the box to move relative to each other, the shear force (T) being measured together with the corresponding shear displacement (Δl). Normally the change in thickness (Δh) of the specimen is also measured. A number of specimens of the soil are tested, each under a different vertical force, and the value of shear stress at failure is plotted against the normal stress for each

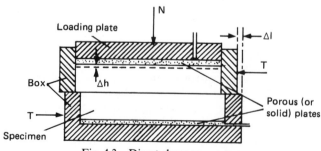

Fig. 4.3   Direct shear apparatus.

test. The shear strength parameters are then obtained from the best line fitting the plotted points.

The test suffers from several disadvantages, the main one being that drainage conditions cannot be controlled. As pore water pressure cannot be measured, only the total normal stress can be determined, although this is equal to the effective normal stress if the pore water pressure is zero. Only an approximation to the state of pure shear is produced in the specimen and shear stress on the failure plane is not uniform, failure occurring progressively from the edges towards the centre of the specimen. The area under the shear and vertical loads does not remain constant throughout the test. The advantages of the test are its simplicity and, in the case of sands, the ease of specimen preparation.

The Triaxial Test

This is the most widely used shear strength test and is suitable for all types of soil. The test has the advantages that drainage conditions can be controlled, enabling saturated soils of low permeability to be consolidated, if required, as part of the test procedure, and pore water pressure measurements can be made. A cylindrical specimen, generally having a length/diameter ratio of 2, is used in the test and is stressed under conditions of axial symmetry in the manner shown in Fig. 4.4. The main features of the apparatus are shown in Fig. 4.5. The circular base has a central pedestal on which the specimen is placed, there being access through the pedestal for drainage or for the measurement of pore water pressure. Alternatively, there may be two access holes in the pedestal, one for drainage, the other for subsequent measurement of pore water pressure. A perspex cylinder, sealed between a ring and a circular top cap, forms the body of the cell. The top cap has a central bush through which the loading ram passes. The cylinder and top cap clamp onto the base, a seal being made by means of an O-ring.

The specimen is placed on either a porous or a solid disc on the pedestal of the apparatus. A loading cap is placed on top of the specimen and the specimen is then sealed in a rubber membrane, O-rings under tension being used to seal the membrane to the pedestal and the loading cap. In the case of sands, the specimen must be prepared in a rubber membrane inside a rigid former which fits around the pedestal: a small negative pressure is applied to the pore water to maintain the stability of the specimen while the former is removed prior to the application of the all-round pressure. A drainage connection may also be made through the loading cap to the top of the specimen, a flexible plastic tube leading from the loading cap to the base of the cell. The top of the loading cap and the lower end of the loading ram both have coned seatings, the load being transmitted through a steel ball. The specimen is subjected to an all-round fluid pressure in the cell, consolidation is allowed to take place if appropriate, then the axial stress is

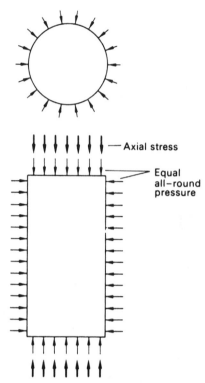

Fig. 4.4   Stress system in triaxial test.

gradually increased by the application of compressive load through the ram until failure of the specimen takes place, usually on a diagonal plane. The system for applying the all-round pressure must be capable of compensating for pressure changes due to cell leakage or specimen volume change.

In the triaxial test, consolidation takes place under equal increments of total stress normal to the end and circumferential surfaces of the specimen. Lateral strain in the specimen is *not* equal to zero during consolidation under these conditions. Dissipation of excess pore water pressure takes place due to drainage through the porous disc at the bottom (or top) of the specimen. Drainage results in a flow of water which is collected in a burette, enabling the volume of water expelled from the specimen to be measured. The datum for excess pore water pressure is therefore atmospheric pressure, assuming the water level in the burette is at the same height as the centre of the specimen. Filter paper drains, in contact with the end porous disc, are sometimes placed around the circumference of the specimen: both vertical and radial drainage then take place and the rate of dissipation of excess pore water pressure is increased.

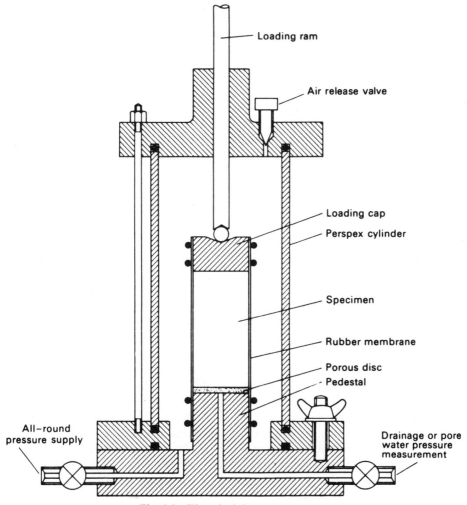

Fig. 4.5 The triaxial apparatus.

The all-round pressure is taken to be the minor principal stress and the sum of the all-round pressure and the applied axial stress as the major principal stress, on the basis that there are no shear stresses on the surfaces of the specimen. The applied axial stress is thus referred to as the *principal stress difference*. The intermediate principal stress is taken as being equal to the minor principal stress. The stress conditions can be represented by a Mohr circle or a stress point at any stage of the test and in particular for the failure condition. If several specimens are tested, each under a different value of all-round pressure, the failure envelope can be drawn and the shear strength parameters for the soil determined. In calculating the principal stress difference, account must be taken of the fact that the average cross-sectional area (A) of the specimen does not remain constant throughout the test. If the original cross-sectional area of the specimen is A_0 and the original volume is V_0 then, if the volume of the specimen decreases during the test,

$$A = A_0 \frac{1 - \varepsilon_v}{1 - \varepsilon_a} \qquad (4.12)$$

where ε_v is the volumetric strain $(\Delta V/V_0)$ and ε_a is the axial strain $(\Delta l/l_0)$. If the volume of the specimen increases during the test the sign of ΔV will change and the numerator in Equation 4.12 becomes $(1 + \varepsilon_v)$. If required, the radial strain (ε_r) could be obtained from the equation:

$$\varepsilon_v = \varepsilon_a + 2\varepsilon_r \qquad (4.13)$$

In the case of saturated soils the volume change ΔV is usually determined by measuring the volume of pore water draining from the specimen. The change in axial length Δl corresponds to the movement of the loading ram, which can be measured by a dial gauge.

The above interpretation of the stress conditions in the triaxial test is approximate only. The principal stresses in a cylindrical specimen are in fact the axial, radial and circumferential stresses, σ_z, σ_r and σ_θ respectively, as shown in Fig. 4.6, and the state of stress throughout the specimen is statically indeterminate. If it is assumed that $\sigma_r = \sigma_\theta$ the indeterminancy is overcome and σ_r then becomes constant, equal to the radial stress on the boundary of the specimen. In addition, the strain conditions in the specimen are not uniform due to frictional restraint produced by the loading cap and pedestal disc: this results in dead zones at each end of the specimen which becomes barrel-shaped as the test proceeds. Non-uniform deformation of the specimen can be largely eliminated by lubrication of the end surfaces. It has been shown, however, that non-uniform deformation

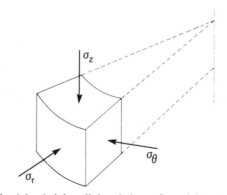

Fig. 4.6   Axial, radial and circumferential stresses.

has no significant effect on the measured strength of the soil provided the length/diameter ratio of the specimen is not less than 2.

A special case of the triaxial test is the *unconfined compression test* in which axial stress is applied to a specimen under zero (atmospheric) all-round pressure, no rubber membrane being required. The unconfined test, however, is applicable only for testing intact, fully saturated clays.

A triaxial *extension* test can also be carried out in which an upward load is applied to a ram connected to the loading cap on the specimen. The all-round pressure then becomes the major principal stress and the net vertical stress the minor principal stress.

Pore Water Pressure Measurement. The pore water pressure in a triaxial specimen can be measured, enabling the results to be expressed in terms of effective stress. Pore water pressure must be measured under conditions of *no flow* either out of or into the specimen, otherwise the correct pressure will be modified. It is, of course, possible to measure pore water pressure at one end of the specimen while drainage is taking place at the other end. The no-flow condition is maintained by the use of a null indicator, essentially a U-tube partly filled with mercury. One limb of the indicator is connected through to the porous disc under the specimen. The other limb is connected to a pressure control cylinder and a pressure gauge or manometer. The layout is illustrated in Fig. 4.7. The whole system is filled with de-aired water and it is important that the connection between the specimen and the null indicator should undergo negligible volume change under pressure.

Any change in pore water pressure in the specimen will tend to cause a movement of the mercury level in the indicator. The no-flow condition is maintained by making a corresponding change in pressure in the other half of the system by means of the control cylinder so that the mercury level remains unaltered: at the same time the balancing pressure, which will equal the pore water pressure in the specimen, is recorded by the pressure gauge or the manometer.

Pore water pressure can also be measured by means of transducers which can be constructed with very low volume change characteristics and the null indicator system is not then required. A change in pressure produces a small deflection of the transducer diaphragm and the corresponding strain is calibrated against the pressure.

If the specimen is partially saturated a fine porous ceramic disc must be sealed into the pedestal of the cell if the correct pore water pressure is to be measured. Depending on the pore size of the ceramic, only pore water can flow through the disc provided the difference between the pore air and pore water pressures is below a certain value known as the *air entry value* of the disc. Under undrained conditions the ceramic disc will remain fully saturated with water provided the air entry value is high enough, enabling the correct pore water pressure to be measured. The use of a coarse porous disc, as normally used for a fully saturated soil, would result in the measurement of the pore air pressure in a partially saturated soil.

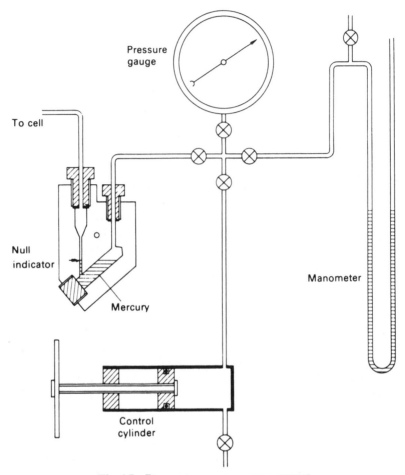

Fig. 4.7   Pore water pressure measurement.

Types of Test. Many variations of test procedure are possible with the triaxial apparatus but the three principal types of test are as follows:

1. *Unconsolidated-Undrained.*   The specimen is subjected to a specified all-round pressure then the principal stress difference is applied immediately with no drainage being permitted at any stage of the test. (The procedure for the unconsolidated-undrained triaxial test has been standardized in BS 1377 [4.5]. Details of the procedure for the unconfined compression test using a portable apparatus are also given in BS 1377.)
2. *Consolidated-Undrained.*   Drainage of the specimen is permitted under a specified all-round pressure until consolidation is complete: the principal stress difference is then applied with no drainage being

permitted. Pore water pressure measurements may be made during the undrained part of the test.

3. *Drained.* Drainage of the specimen is permitted under a specified all-round pressure until consolidation is complete: with drainage still being permitted, the principal stress difference is then applied at a rate slow enough to ensure that the excess pore water pressure is maintained at zero.

Shear strength parameters determined by means of the above test procedures are relevant only in situations where the field drainage conditions correspond to the test conditions. The shear strength of a soil under undrained conditions is different from that under drained conditions. The undrained strength can, under certain conditions, be expressed in terms of total stress, the shear strength parameters being denoted by c_u and ϕ_u. The drained strength is expressed in terms of the effective stress parameters c' and ϕ'.

The vital consideration in practice is the rate at which the changes in total stress (due to construction operations) are applied in relation to the rate of dissipation of excess pore water pressure, which in turn is related to the permeability of the soil. Undrained conditions apply if there has been no significant dissipation during the period of total stress change; this would be the case in soils of low permeability such as clays immediately after the completion of construction. Drained conditions apply in situations where the excess pore water pressure is zero; this would be the case in soils of low permeability after consolidation is complete and would represent the situation a long time, perhaps many years, after the completion of construction. The drained condition would also be relevant if the rate of dissipation were to keep pace with the rate of change of total stress; this would be the case in soils of high permeability such as sands. The drained condition is therefore relevant for sands both immediately after construction and in the long term. Only if there were extremely rapid changes in total stress (e.g. as the result of an explosion or an earthquake) would the undrained condition be relevant for a sand. In some situations, partially drained conditions may apply at the end of construction, due perhaps to a very long construction period or to the soil in question being of intermediate permeability. In such cases the excess pore water pressure would have to be estimated and the shear strength would then be calculated in terms of effective stress using the parameters c' and ϕ'.

Testing Under Back Pressure. Testing under back pressure involves raising the pore water pressure artificially by connecting a source of constant pressure through a porous disc to one end of a triaxial specimen. In a drained test this connection remains open throughout the test, drainage taking place against the back pressure: the back pressure is the datum for excess pore water pressure. In a consolidated-undrained test the connection to the back pressure source is closed at the end of the consolidation stage, before application of the principal stress difference is commenced.

The object of applying a back pressure is to ensure full saturation of the specimen or to simulate in-situ pore water pressure conditions. During sampling the degree of saturation of a clay may fall below 100% owing to swelling on the release of in-situ stresses. Compacted specimens will have a degree of saturation below 100%. In both cases a back pressure is applied which is high enough to drive the pore air into solution in the pore water.

It is essential to ensure that the back pressure does not by itself change the effective stresses in the specimen. It is necessary, therefore, to raise the all-round pressure simultaneously with the application of the back pressure and by an equal increment.

The Vane Shear Test

This test is used for the in-situ determination of the undrained strength of intact, fully saturated clays: the test is not suitable for other types of soil. In particular the test is very suitable for soft clays, the shear strength of which may be significantly altered by the sampling process and subsequent handling. Generally the test is only used in clays having undrained strengths less than $100 \, \text{kN/m}^2$. The test may not give reliable results if the clay contains sand or silt laminations.

Details of the test are given in BS 1377. The equipment consists of a stainless steel vane (Fig. 4.8) of four thin rectangular blades, carried on the end of a high tensile steel rod: the rod is enclosed by a sleeve packed with grease. The length of the vane is equal to twice its overall width, typical

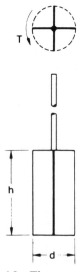

Fig. 4.8   The vane test.

dimensions being 150 mm by 75 mm and 100 mm by 50 mm. Preferably the diameter of the rod should not exceed 12.5 mm.

The vane and rod are pushed into the clay below the bottom of a borehole to a depth of at least three times the borehole diameter: if care is taken this can be done without appreciable disturbance of the clay. Steady bearings are used to keep the rod and sleeve central in the borehole casing. The test can also be carried out in soft clays, without a borehole, by direct penetration of the vane from ground level: in this case a shoe is required to protect the vane during penetration.

Torque is applied gradually to the upper end of the rod by means of suitable equipment until the clay fails in shear due to rotation of the vane. Shear failure takes place over the surface and ends of a cylinder having a diameter equal to the overall width of the vane. The rate of rotation of the vane should be within the range of 6° to 12° per minute. The shear strength is calculated from the expression

$$T = \pi c_u \left(\frac{d^2 h}{2} + \frac{d^3}{6} \right) \tag{4.14}$$

where T = torque at failure, d = overall vane width, and h = vane length. However the shear strength over the cylindrical vertical surface may be different from that over the two horizontal end surfaces, as a result of anisotropy. The shear strength is normally determined at intervals over the depth of interest. If, after the initial test, the vane is rotated rapidly through several revolutions the clay will become remoulded and the shear strength in this condition could then be determined if required.

Small, hand-operated vane testers are also available for use in exposed clay strata.

Special Tests

In practice there are very few problems in which a state of axial symmetry exists as in the triaxial test. In practical states of stress the intermediate principal stress is not usually equal to the minor principal stress and the principal stress directions can undergo rotation as the failure condition is approached. A common condition is that of plane strain in which the strain in the direction of the intermediate principal stress is zero due to restraint imposed by virtue of the length of the structure in question. In the triaxial test, consolidation proceeds under equal all-round pressure (i.e. isotropic consolidation) whereas in-situ consolidation takes place under anisotropic stress conditions.

Tests of a more complex nature, generally employing adaptions of triaxial equipment, have been devised to simulate the more complex states of stress encountered in practice but these are used principally in research. The plane strain test uses a prismatic specimen in which strain in one direction (that of the intermediate principal stress) is maintained at zero throughout the test by means of two rigid side plates tied together. The all-round pressure is the minor principal stress and the sum of the applied axial stress and the all-round pressure the major principal stress. A more sophisticated test, also using a prismatic specimen, enables the values of all three principal stresses to be controlled independently, two side pressure bags or jacks being used to apply the intermediate principal stress. Independent control of the three principal stresses can also be achieved by means of tests on soil specimens in the form of hollow cylinders in which different values of external and internal fluid pressure can be applied in addition to axial stress. Torsion applied to the hollow cylinders results in the rotation of the principal stress directions.

Because of its relative simplicity it seems likely that the triaxial test will continue to be the main test for the determination of shear strength characteristics. If considered necessary, corrections can be applied to the results of triaxial tests to obtain the characteristics under more complex states of stress.

4.3   Shear Strength of Sands

The shear strength characteristics of a sand can be determined from the results of either drained triaxial tests or direct shear tests. The characteristics of dry and saturated sands are the same provided there is zero excess pore water pressure in the case of saturated sands. Typical curves relating principal stress difference and axial strain (i.e. major principal strain) for dense and loose sand specimens in drained triaxial compression tests are shown in Fig. 4.9a. Similar curves are obtained relating shear stress and shear displacement in direct shear tests.

In a dense sand there is a considerable degree of interlocking between particles, and before shear failure can take place this interlocking must be overcome in addition to the frictional resistance at the points of contact. The degree of interlocking will be greatest in the case of very dense, well-graded sands consisting of angular particles. The characteristic stress-strain curve for a dense sand shows a peak stress at a relatively low strain and thereafter, as interlocking is progressively overcome, the stress necessary for additional strain decreases. The reduction in the degree of interlocking produces an increase in the volume of the specimen during shearing, as characterized by the relationship between volumetric strain and axial strain, shown in Fig. 4.9c (compressive strain being taken as positive): in the direct shear test a similar relationship would be obtained between change in specimen thickness and shear displacement. The change in volume is also shown in terms of void ratio (e) in Fig. 4.9d. Eventually the specimen would become loose enough to allow particles to move over and around their

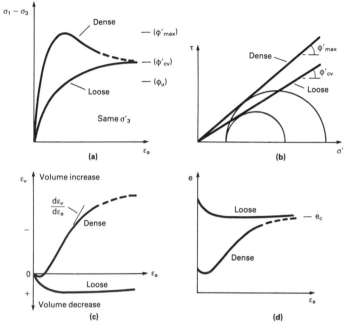

Fig. 4.9 Drained triaxial test results for sand.

neighbours without any further net volume change and the principal stress difference would reach an ultimate value.

The term *dilatancy* is used to describe the increase in volume of a dense sand during shearing and the rate of dilation can be represented by the gradient $d\varepsilon_v/d\varepsilon_a$. The concept of dilatancy can be appreciated more clearly in the context of the direct shear test. During shearing of a dense sand, the macroscopic shear plane is horizontal, but sliding between individual particles takes place on numerous microscopic planes inclined at various angles above the horizontal as the particles move up and over their neighbours. The loading plate of the apparatus is thus forced upwards and work is done against the normal stress.

For a dense sand the maximum angle of shearing resistance (ϕ'_{max}) determined from peak stresses (Fig. 4.9b) is significantly greater than the true angle of friction (ϕ_μ) between the surfaces of individual particles, the difference representing the energy required to overcome interlocking and rearrange the particles. In most practical problems, where a factor of safety in terms of shear strength is employed and strains are relatively low, peak stresses are used to define failure.

In the case of a loose sand there is no significant particle interlocking to be overcome, and the principal stress difference increases gradually to an ultimate value without a prior peak. The increase in stress is accom-

Table 4.1   Ranges of ϕ' for Sands

	Loose (ϕ'_{cv})	Dense (ϕ'_{max})
Uniform sand, rounded particles	27°	− 35°
Well graded sand, angular particles	33°	− 45°
Sandy gravel	35°	− 50°
Silty sand	(27° − 30°)	− (30° − 34°)

panied by a decrease in volume. The ultimate values of principal stress difference and void ratio for dense and loose specimens under the same all-round pressure in the triaxial test, or under the same normal stress in the direct shear test, are essentially equal as indicated in Figs. 4.9a and 4.9d. Thus at the ultimate (or critical) state shearing takes place at constant volume: the corresponding angle of shearing resistance is denoted ϕ'_{cv}, being the relevant parameter in practice for a loose sand. The difference between ϕ'_{cv} and ϕ_μ represents the energy required to rearrange the particles. It should be appreciated that the void ratio at the ultimate or critical state is a function of either effective all-round pressure or normal stress.

Only the drained strength of a sand is normally relevant in practice and typical values of the shear strength parameter ϕ' (c' being zero) are given in Table 4.1. In the case of dense sands it has been shown that the peak value of ϕ' in plane strain can be 4° or 5° higher than the corresponding value obtained by conventional triaxial tests: the increase in the case of loose sands is negligible.

In triaxial tests at very high all-round pressures (in excess of 500 kN/m²) some fracturing or crushing of particles may occur, resulting in the failure envelope becoming curved (and a reduction in the value of ϕ'). However in most practical situations the stresses are not high enough to produce this effect.

Stress–Dilatancy Theory

Rowe [4.12] developed a stress–dilatancy theory relating the ratio of the principal stresses, the geometry of ideal particle arrangements and the relative rates of change of volumetric and major principal strains. It was shown that:

$$\frac{\sigma'_1}{\sigma'_3} = \left\{ \tan^2 \left(45° + \frac{\phi'_f}{2} \right) \right\} \left(1 - \frac{d\varepsilon_v}{d\varepsilon_1} \right) \qquad (4.15)$$

where $d\varepsilon_v$ and $d\varepsilon_1$ are corresponding small changes in volumetric and major principal strains respectively (compressive strain being taken as positive) and ϕ'_f is a value of angle of shearing resistance between the limits of ϕ_μ and ϕ'_{cv} depending on the strain conditions imposed by the test. The value of ϕ'_f is a function of the instantaneous directions of local interparticle slip as rearrangement takes place: the preferred direction of local slip would be at $(45° + \phi_\mu/2)$ to the major principal plane. The actual directions depend on the degree of freedom for particle rearrangement within the test apparatus. Under conditions of axial symmetry, as in the triaxial test, there is considerable freedom for local slip to occur at angles close to the preferred direction and, in the case of a dense sand up to peak stress ratio, ϕ'_f approaches the lower limit ϕ_μ. However under conditions of plane strain the test apparatus imposes severe restrictions on the development of local slip in the preferred direction and ϕ'_f approaches the upper limit of ϕ'_{cv}. At the ultimate or critical state the value of ϕ'_f will, of course, always be equal to ϕ'_{cv}. In the case of a dense sand the peak strength normally corresponds to the maximum rate of dilation, i.e. the maximum value of $d\varepsilon_v/d\varepsilon_1$.

A parameter known as the angle of dilation (ψ) is defined either in terms of the maximum and minimum principal strain increments $d\varepsilon_1$ and $d\varepsilon_3$ or in terms of increments of volumetric strain $(d\varepsilon_v)$ and maximum shear strain $(d\gamma)$, as follows:

$$\sin \psi = \frac{d\varepsilon_1 + d\varepsilon_3}{d\varepsilon_1 - d\varepsilon_3} = -\frac{d\varepsilon_v}{d\gamma} \tag{4.16}$$

The parameter indicates the relative magnitudes of the strain increments.

4.4 Shear Strength of Saturated Clays

Isotropic Consolidation

If a saturated clay specimen is allowed to consolidate in the triaxial apparatus under a sequence of equal all-round pressures, sufficient time being allowed between successive increments to ensure that consolidation is complete, the relationship between void ratio (e), or specific volume $(v = 1 + e)$, and effective stress (σ'_3) can be obtained. Consolidation in the triaxial apparatus under equal all-round pressure is referred to as isotropic consolidation. It should be noted that p' (Equation 4.11) is equal to σ'_3 under equal all-round pressure. Void ratio or specific volume can be calculated from Equation 1.17.

The relationship between void ratio and effective stress depends on the stress history of the clay. If the present effective stress is the maximum to which the clay has ever been subjected, the clay is said to be *normally consolidated*. If, on the other hand, the effective stress at some time in the past has been greater than the present value, the clay is said to be *overconsolidated*. The maximum value of effective stress in the past divided by the present value is defined as the *overconsolidation ratio* (OCR). A normally consolidated clay thus has an overconsolidation ratio of unity; an overconsolidated clay has an overconsolidation ratio greater than unity. Overconsolidation is usually the result of geological factors, for example the erosion of overburden, the melting of ice sheets after glaciation and the permanent rise of the water table. Overconsolidation can also be due to higher stresses previously applied to a specimen in the triaxial apparatus.

The characteristic relationship between e (or v) and σ'_3 (or p') is shown in Fig. 4.10a. AB is the curve for a clay in the normally consolidated condition. If after consolidation to point B the effective stress is reduced, the clay will swell or expand and the relationship will be represented by curve BC. During consolidation from A to B, changes in soil structure continuously take place but the clay does not revert to its original structure during swelling. A clay existing at a state represented by point C is now in the overconsolidated condition, the overconsolidation ratio being the effective stress at point B divided by that at point C. If the effective stress is again increased the consolidation curve is CD, known as the recompression curve, eventually becoming the continuation of the normal consolidation curve AB. It should be realized that a state represented by a point to the right of the normal consolidation curve is impossible.

If effective stress is represented on a logarithmic scale the relationship between void ratio and effective stress for a normally consolidated clay is linear, as shown in Fig. 4.10b. It is also possible to approximate the swelling and recompression relationships to a single straight line (BC).

The equation of the normal consolidation line (AB), in terms of v and p', is

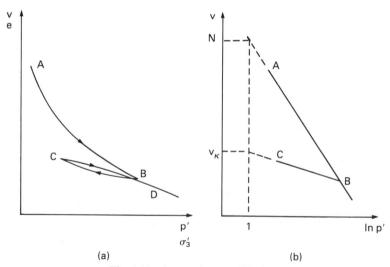

(a)   (b)

Fig. 4.10   Isotropic consolidation.

$$v = N - \lambda \ln p' \tag{4.17}$$

where λ is the (negative) slope of AB and N is the value of v at $p' = 1\,\mathrm{kN/m^2}$.
  The equation of the swelling and recompression line (BC) is

$$v = v_\kappa - \kappa \ln p' \tag{4.18}$$

where κ is the slope of BC and v_κ is the value of v at $p' = 1\,\mathrm{kN/m^2}$.

Undrained Strength

In principle the *unconsolidated-undrained* triaxial test enables the undrained strength of the clay in its in-situ condition to be determined, the
void ratio of the specimen at the start of the test being unchanged from the
in-situ value at the depth of sampling. In practice, however, the effects of
sampling and preparation result in a small increase in void ratio. There is
also strong evidence that the undrained strength in-situ is significantly
anisotropic, the strength depending on the direction of the major principal
stress relative to the in-situ orientation of the specimen.
  The effective stresses in the specimen remain unchanged after the
application of the all-round pressure, regardless of its value, because for a
fully saturated soil under undrained conditions any increase in all-round
pressure results in an equal increase in pore water pressure (see Section 4.7).
Assuming all specimens to be identical, a number of unconsolidated-
undrained tests, each at a different value of all-round pressure, should
result, therefore, in equal values of principal stress difference at failure. The
results are expressed in terms of total stress as shown in Fig. 4.11, the failure
envelope being horizontal, i.e. $\phi_u = 0$ and the shear strength is given by
$\tau_f = c_u$. It should be noted that if the values of pore water pressure at failure were measured in a series of tests only one effective stress circle
(shown dotted in Fig. 4.11) would be obtained.
  In the case of fissured clays the failure envelope at low values of all-round
pressure is curved, as shown in Fig. 4.11. This is due to the fact that the
fissures open to some extent on sampling, resulting in a lower strength: only
when the all-round pressure becomes high enough to close the fissures
again does the strength become constant. The unconfined compression test

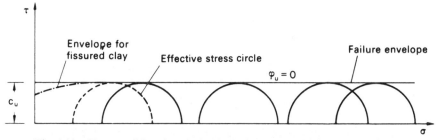

Fig. 4.11   Unconsolidated-undrained triaxial test results for saturated clay.

is not appropriate, therefore, in the case of fissured clays. The size of a fissured clay specimen should be large enough to represent the mass structure otherwise the measured strength will be greater than the in-situ strength. Large specimens are also required for clays exhibiting other features of macro-fabric.

The results of unconsolidated-undrained tests are usually presented as a plot of c_u against the corresponding depth from which the specimen originated. Considerable scatter can be expected on such a plot as the result of sampling disturbance and macro-fabric features if present. For normally consolidated clays the undrained strength is generally taken to increase linearly with increase in effective vertical stress σ'_v (i.e. with depth if the water table is at the surface): this is comparable to the variation of c_u with σ'_3 (Fig. 4.12) in consolidated-undrained triaxial tests. If the water table is below the surface of the clay the undrained strength between the surface and the water table will be significantly higher than that immediately below the water table due to drying of the clay.

The following correlation between the ratio c_u/σ'_v and plasticity index (I_P) for normally consolidated clays was proposed by Skempton:

$$\frac{c_u}{\sigma'_v} = 0.11 + 0.0037 I_p \tag{4.19}$$

The *consolidated-undrained* triaxial test enables the undrained strength of the clay to be determined after the void ratio has been changed from the initial value by consolidation. The undrained strength is thus a function of this void ratio or of the corresponding all-round pressure (σ'_3) under which consolidation took place. The all-round pressure during the undrained part of the test (i.e. when the principal stress difference is applied) has no influence on the strength of the clay, although it is normally the same pressure as that under which consolidation took place. The results of a series of tests can be represented by plotting the value of c_u (ϕ_u being zero) against the corresponding consolidation pressure σ'_3, as shown in Fig. 4.12.

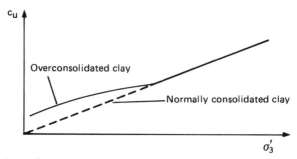

Fig. 4.12   Consolidated-undrained triaxial test: variation of undrained strength with consolidation pressure.

For clays in the normally consolidated state the relationship between c_u and σ'_3 is linear, passing through the origin. For clays in the overconsolidated state the relationship is non-linear, as shown in Fig. 4.12.

The unconsolidated-undrained test and the undrained part of the consolidated-undrained test can be carried out rapidly (provided no pore water pressure measurements are to be made), failure normally being produced within a period of 10–15 min. However, a slight decrease in strength can be expected if the time to failure is significantly increased and there is evidence that this decrease is more pronounced the greater the plasticity index of the clay. Each test should be continued until the maximum value of principal stress difference has been passed or until an axial strain of 20% has been attained.

It should be realized that clays in-situ have been consolidated under conditions of zero lateral strain, the effective vertical and horizontal stresses being unequal, i.e. the clay has been consolidated anisotropically: a stress release then occurs on sampling. In the consolidated-undrained triaxial test the specimen is consolidated again under equal all-round pressure, normally equal to the value of the effective vertical stress in-situ, i.e. the specimen is consolidated isotropically. Isotropic consolidation in the triaxial test under a pressure equal to the in-situ effective vertical stress results in a void ratio lower than the in-situ value and therefore an undrained strength higher than the in-situ value. In the case of soft clays, which are very sensitive to sample disturbance, it has been suggested that specimens should be reconsolidated anisotropically under the same stresses as those acting in-situ prior to sampling.

The undrained strength of intact soft and firm clays can be measured in-situ by means of the *vane test*. However Bjerrum [4.6] has presented evidence that undrained strength as measured by the vane test is generally greater than the average strength mobilized along a failure surface in a field situation (undrained strength values for field situations being obtained by back calculation). The discrepancy between vane and field strengths was found to be greater the higher the plasticity index of the clay. The discrepancy is attributed primarily to the rate effect mentioned earlier in this section, i.e. the more rapidly the stresses are applied the greater is the shear strength. In the vane test shear failure occurs within a few minutes whereas in a field situation the stresses are usually applied over a period of a few weeks or months. A secondary factor may be anisotropy, the undrained strength varying with the direction along which shearing takes place. Bjerrum presented a correction factor (μ), correlated empirically with plasticity index, as shown in Fig. 4.13, the vane strength being multiplied by the factor to give the probable field strength.

Clays may be classified on the basis of undrained shear strength as in Table 4.2.

Shear strength depends fundamentally on effective stresses but these are related to the particular void ratio at which shearing takes place. Thus it is

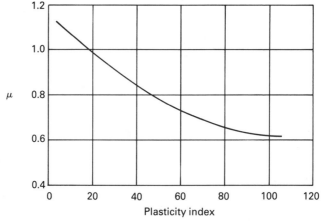

Fig. 4.13 Correction factor for undrained strength measured by the vane test.
(After Bjerrum [4.6])

Table 4.2   Undrained strength classification
(Reproduced from BS 8004:1986 by permission of the British Standards
Institution.)

Consistency	Undrained strength (kN/m^2)
Very stiff or hard	> 150
Stiff	100–150
Firm to stiff	75–100
Firm	50–75
Soft to firm	40–50
Soft	20–40
Very soft	< 20

only correct to express shear strength in terms of total stress for the case of
saturated soils under undrained conditions in which the void ratio remains
constant. A particular value of c_u therefore applies only to a saturated clay
at a particular void ratio (i.e. at a particular depth in situ).

Sensitivity of Clays. Some clays are very sensitive to remoulding, suffering
considerable loss of strength due to their natural structure being damaged
or destroyed. The *sensitivity* of a clay is defined as the ratio of the undrained
strength in the undisturbed state to the undrained strength, at the same
water content, in the remoulded state. Remoulding for test purposes is
normally brought about by the process of kneading. The sensitivity of most
clays is between 1 and 4. Clays with sensitivities between 4 and 8 are referred
to as *sensitive* and those with sensitivities between 8 and 16 as *extra-*

sensitive. Quick clays are those having sensitivities greater than 16: the sensitivities of some quick clays may be of the order of 100.

Strength in Terms of Effective Stress

The strength of a clay in terms of effective stress can be determined by means of either the consolidated-undrained triaxial test with pore water pressure measurement during the undrained part of the test, or the drained triaxial test. The undrained part of the consolidated-undrained test must be run at a rate of strain slow enough to allow equalization of pore water pressure throughout the specimen, this rate being a function of the permeability of the clay. If the pore water pressure at failure is known, the effective principal stresses σ'_1 and σ'_3 can be calculated and the corresponding Mohr circle or stress point drawn. A number of tests, each performed at a different value of all-round pressure, enables the failure envelope to be drawn and the shear strength parameters c' and ϕ' determined as shown in Fig. 4.14.

For a normally consolidated clay the value of c' is zero, whereas for an overconsolidated clay c' has a value not usually exceeding $30 \, kN/m^2$. The value of ϕ' generally lies between 20° and 35°, lower values being associated with high plasticity index and vice versa.

The parameters c' and ϕ' can also be obtained by means of drained triaxial tests (or direct shear tests). The rate of strain must be slow enough to ensure full dissipation of excess pore water pressure at any time during the application of the principal stress difference: total and effective stresses will thus be equal at any time. The rate of strain must again be related to the permeability of the soil. The volume change taking place during the application of the principal stress difference must be measured in the drained test so that the corrected cross-sectional area of the specimen can be calculated.

Typical results for specimens of normally consolidated and overconsolidated clays are shown in Fig. 4.15. In consolidated-undrained tests,

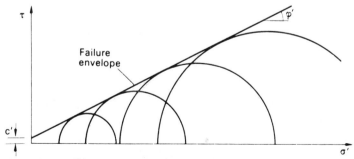

Fig. 4.14 Strength of saturated clay in terms of effective stress. Series of consolidated-undrained or drained tests.

failure of both normally consolidated and overconsolidated specimens occurs at relatively large strains. For normally consolidated specimens the pore water pressure increases steadily during shearing. For overconsolidated specimens the pore water pressure increases initially then decreases, the higher the overconsolidation ratio the greater the decrease. The pore water pressure may become negative in the case of heavily overconsolidated clays, as shown by the dotted line in Fig. 4.15b. In drained tests, failure of normally consolidated specimens occurs at relatively large strains; a decrease in volume takes place during shearing and the clay hardens. Failure of overconsolidated specimens in drained tests occurs at a relatively low strain, a definite peak in the stress-strain graph usually being apparent; subsequently the axial stress decreases with increasing strain.

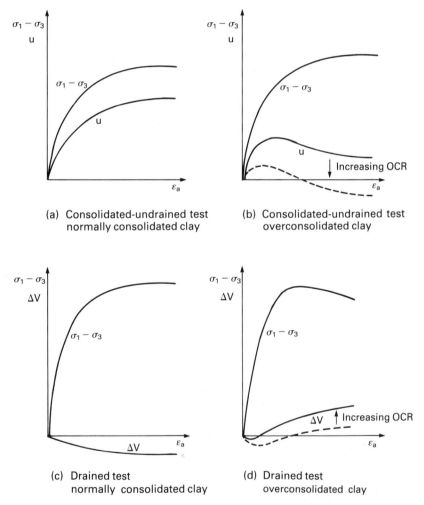

Fig. 4.15   Typical results from consolidated-undrained and drained triaxial tests.

After an initial decrease, the volume of overconsolidated specimens increases prior to and after failure and the clay softens.

The shear strength of a partially saturated soil is a function of the relevant effective stresses but it is extremely difficult to apply the principle of effective stress because of the parameter χ in Equation 3.4. Effective stress parameters can be obtained from a series of consolidated-undrained tests with pore water pressure measurement or drained tests on specimens brought to a condition of full saturation by the application of back pressure.

Stress Paths

The successive states of stress in a test specimen or an in-situ element of soil can be represented by a series of Mohr circles or, in a less confusing way, by a series of stress points. The curve or straight line connecting the relevant stress points is called the *stress path*, giving a clear representation of the successive states of stress. Stress paths may be drawn in terms of either effective or total stresses. The horizontal distance between the effective and total stress paths is the value of pore water pressure at the stresses in question. In general the horizontal distance between the two stress paths is the sum of the pore water pressure due to the change in total stress and the static pore water pressure. In the normal triaxial test procedure the static pore water pressure (u_s) is zero. However, if a triaxial test is performed under back pressure, the static pore water pressure is equal to the back pressure. The static pore water pressure of an in-situ element is the pressure resulting from the water table level.

The effective and total stress paths for the triaxial tests represented in Fig. 4.15 are shown in Fig. 4.16, the coordinates being $\frac{1}{2}(\sigma'_1 - \sigma'_3)$ and $\frac{1}{2}(\sigma'_1 + \sigma'_3)$ or the total stress equivalents. All the total stress paths and the effective stress paths for the drained tests are straight lines at a slope of 45°. The detailed shape of the effective stress paths for the consolidated-undrained tests depends on the clay in question. The effective and total stress paths for the drained tests coincide provided no back pressure has been applied. The dotted line in Fig. 4.16c is the effective stress path for a heavily overconsolidated clay in which the pore water pressure at failure is negative. The stress paths could also be drawn with reference to coordinates q' and p' (Equation 4.10 and 4.11) or their total stress equivalents. In this case all the total stress paths and the effective stress paths for the drained tests would be straight lines at a slope of 3 vertical to 1 horizontal: this is because there is no change in σ_3 and changes in q' and p' (or q and p) are then in the ratio of 3 to 1.

Example 4.1

The following results were obtained from direct shear tests on specimens of a sand compacted to the in-situ density. Determine the values of the shear strength parameters.

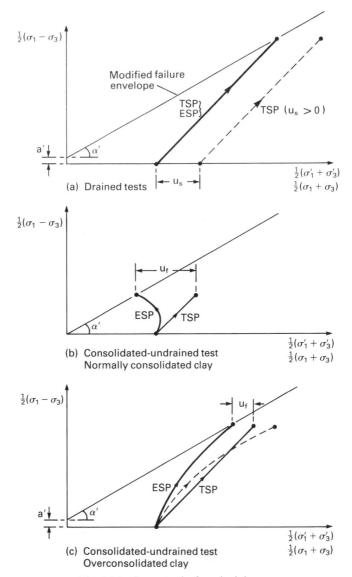

Fig. 4.16   Stress paths for triaxial tests.

Normal stress (kN/m²)	50	100	200	300
Shear stress at failure (kN/m²)	36	80	154	235

Would failure occur on a plane within a mass of this sand at a point where the shear stress is $122 \, \text{kN/m}^2$ and the effective normal stress $246 \, \text{kN/m}^2$?

The values of shear stress at failure are plotted against the corresponding values of normal stress as shown in Fig. 4.17. The failure envelope is the line

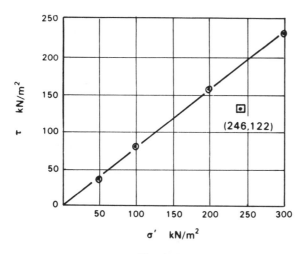

Fig. 4.17

having the best fit to the plotted points. In this case the envelope is a straight line through the origin, therefore the value of c' is zero. If the stress scales are the same, the value of ϕ' can be measured directly and is 38°.

The stress state $\tau = 122\,kN/m^2$, $\sigma' = 246\,kN/m^2$ plots below the failure envelope and therefore would not produce failure.

Example 4.2

The results shown in Table 4.3 were obtained at failure in a series of triaxial tests on specimens of a saturated clay initially 38 mm in diameter by 76 mm long. Determine the values of the shear strength parameters with respect to (a) total stress, (b) effective stress.

The principal stress difference at failure in each test is obtained by

Table 4.3

	Type of test	All-round pressure (kN/m²)	Axial load (N)	Axial deformation (mm)	Volume change (ml)
(a)	Undrained	200	222	9·83	—
		400	215	10·06	—
		600	226	10·28	—
(b)	Drained	200	467	10·81	6·6
		400	848	12·26	8·2
		600	1265	14·17	9·5

Table 4.4

	σ_3 (kN/m²)	$\Delta l/l_0$	$\Delta V/V_0$	Area (mm²)	$\sigma_1 - \sigma_3$ (kN/m²)	σ_1 (kN/m²)
(a)	200	0·129	—	1304	170	370
	400	0·132	—	1309	164	564
	600	0·135	—	1312	172	772
(b)	200	0·142	0·077	1222	382	582
	400	0·161	0·095	1225	691	1091
	600	0·186	0·110	1240	1020	1620

dividing the axial load by the cross-sectional area of the specimen at failure (Table 4.4). The corrected cross-sectional area is calculated from Equation 4.12. There is, of course, no volume change during an undrained test on a saturated clay. The intial values of length, area and volume for each specimen are:

$$l_0 = 76\,\text{mm} \qquad A_0 = 1135\,\text{mm}^2 \qquad V_0 = 86 \times 10^3\,\text{mm}^3$$

The Mohr circles at failure and the corresponding failure envelopes for both series of tests are shown in Fig. 4.18. In both cases the failure envelope is the line nearest to a common tangent to the Mohr circles. The total stress parameters, representing the undrained strength of the clay, are:

$$c_u = 85\,\text{kN/m}^2 \qquad \phi_u = 0$$

The effective stress parameters, representing the drained strength of the clay, are:

$$c' = 20\,\text{kN/m}^2 \qquad \phi' = 26°$$

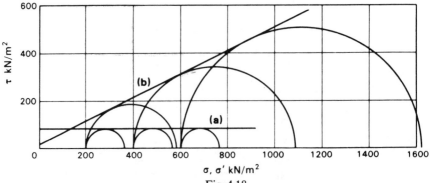

Fig. 4.18

Example 4.3

The results shown in Table 4.5 were obtained at failure in a series of consolidated-undrained tests, with pore water pressure measurement, on specimens of a saturated clay. Determine the values of the effective stress parameters c' and ϕ'.

Table 4.5

All-round pressure (kN/m²)	Principal stress difference (kN/m²)	Pore water pressure (kN/m²)
150	192	80
300	341	154
450	504	222

The values of the effective principal stresses σ'_3 and σ'_1 at failure are calculated by subtracting the pore water pressure at failure from the total principal stresses (see Table 4.6: all stresses kN/m²).

Table 4.6

σ_3	σ_1	σ'_3	σ'_1	$\frac{1}{2}(\sigma_1 - \sigma_3)$	$\frac{1}{2}(\sigma'_1 + \sigma'_3)$
150	342	70	262	96	166
300	641	146	487	170	316
450	954	228	732	252	480

The Mohr circles in terms of effective stress and the failure envelope are drawn in Fig. 4.19a. The shear strength parameters are measured as

$$c' = 16\,\text{kN/m}^2 \qquad \phi' = 29°$$

An alternative procedure is to represent the states of stress at failure by stress points, plotting $\frac{1}{2}(\sigma_1 - \sigma_3)$ against $\frac{1}{2}(\sigma'_1 + \sigma'_3)$ as shown in Fig. 4.19b, and drawing the line best fitting the points. The values of the modified parameters are:

$$a' = 13\,\text{kN/m}^2; \quad \alpha' = 26°$$

then

$$\phi' = \sin^{-1}(\tan 26°) = 29°$$

$$c' = \frac{13}{\cos 29°} = 15\,\text{kN/m}^2$$

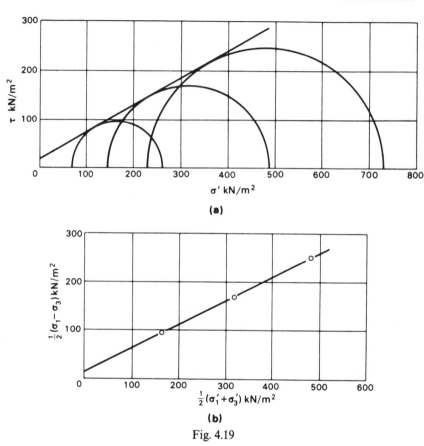

(a)

(b)

Fig. 4.19

4.5 The Critical State Concept

The critical state concept, due to Roscoe, Schofield and Wroth [4.11] relates the effective stresses and the corresponding specific volume ($v = 1 + e$) of a clay during shearing under drained or undrained conditions: it thus unifies the characteristics of shear strength and deformation. The concept represents an idealization of the observed patterns of behaviour of saturated remoulded clays in triaxial compression tests but is assumed to apply also to undisturbed clays. It was demonstrated that a characteristic surface exists which limits all possible states of the clay and that all effective stress paths reach or approach a line, on that surface, which defines the state at which the clay will continue to yield at constant volume under constant effective stresses. The concept can be extended to cover general three-dimensional states of stress. However, it should be realized that different characteristic surfaces will exist for different stress systems.

Effective stress paths for a consolidated-undrained and a drained triaxial test (C′A′ and C′B′ respectively) on specimens of a *normally consolidated* clay are shown in Fig. 4.20a, the coordinate axes being q' and p' (Equations 4.10 and 4.11). Each specimen was allowed to consolidate under the same all-round pressure p'_c and failure occurs at A′ and B′ respectively, these points lying on or close to a straight line OS′ through the origin, i.e. failure occurs if the stress path reaches this line. If a series of consolidated-undrained tests were carried out on specimens each consolidated to a different value of p'_c, the stress paths would all have similar shapes to that shown in Fig. 4.20a. The stress paths for a series of drained tests would be straight lines rising from the points representing p'_c at a slope of 3 vertical to 1 horizontal. In all these tests the state of stress at failure would lie on or close to the straight line OS′.

The isotropic consolidation curve (NN) for the normally consolidated

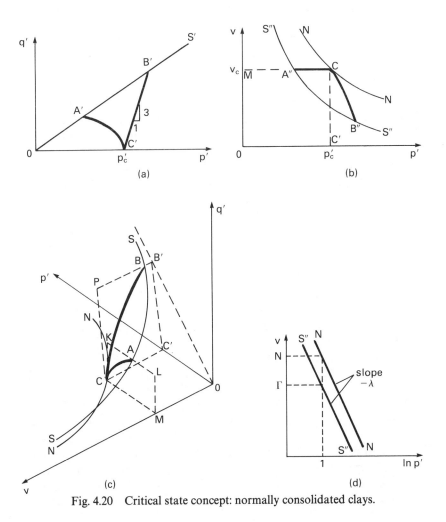

Fig. 4.20   Critical state concept: normally consolidated clays.

clay would have the form shown in Fig. 4.20b, the coordinate axes being v and p'. The volume of the specimen during the application of the principal stress difference in a consolidated-undrained test on a saturated clay remains constant, therefore the relationship between v and p' will be represented by a horizontal line starting from the point (C) on the consolidation curve corresponding to p_c' and finishing at the point (A″) representing the value of p' at failure. During a drained test the volume of the specimen decreases and the relationship between v and p' will be represented by a curve CB″. If a series of consolidated-undrained and drained tests were carried out on specimens each consolidated to a different value of p_c', the points representing the values of v and p' at failure would lie on or close to a curve S″S″ of similar shape to the consolidation curve.

The data represented in Figs. 4.20a and 4.20b can be combined in a three-dimensional plot with coordinates q', p' and v, as shown in Fig. 4.20c. On this plot the line OS′ and the curve S″S″ combine as the single curve SS. The curve SS is known as the *critical state line*, points on this line representing combinations of q', p' and v at which shear failure and subsequent yielding at constant effective stresses occur. In Figs. 4.20a and 4.20b, OS′ and S″S″ are the projections of the critical state line on the q'–p' and v–p' planes respectively. The stress paths for a consolidated-undrained test (CA) and a drained test (CB), both consolidated to the same pressure p_c', are also shown in Fig. 4.20c. The stress path for the consolidated-undrained test lies on a plane CKLM parallel to the q'–p' plane, the value of v being constant throughout the undrained part of the test. The stress path for the drained test lies on a plane normal to the q'–p' plane and inclined at a slope of 3:1 to the direction of the q' axis. Both stress paths start at point C on the normal consolidation curve NN which lies on the v–p' plane.

The stress paths for a series of consolidated-undrained and drained tests on specimens each consolidated to different values of p_c' would all lie on a curved surface, spanning between the normal consolidation curve NN and the critical state line SS, called the *state boundary surface*. It is impossible for a specimen to reach a state represented by a point beyond this surface.

The stress paths for a consolidated-undrained and a drained triaxial test (D′E′ and D′F′ respectively) on specimens of a *heavily overconsolidated* clay are shown in Fig. 4.21a. The stress paths start from a point (D′) on the expansion (or recompression) curve for the clay. The consolidated-undrained specimen reaches failure at point E′ on the line U′H′, above the projection (OS′) of the critical state line. If the test were continued after failure, the stress path would be expected to continue along U′H′ and to approach point H′ on the critical state line. However, the higher the overconsolidation ratio the higher the strain required to reach the critical state. The deformation of the consolidated-undrained triaxial specimen would become non-uniform at high strains and it is unlikely that the specimen as a whole would reach the critical state. The drained specimen reaches failure at point F′ also on the line U′H′. After failure, the stresses decrease along the same stress path, approaching the critical state line at point X'. However, heavily overconsolidated specimens increase in volume (and hence soften) prior to and after failure in a drained test. Narrow zones adjacent to the failure planes become weaker than the remainder of the clay and the specimen as a whole does not reach the critical state. The corresponding relationships between v and p' are represented by the lines DE″ and DF″ respectively in Fig. 4.21b: these lines approach but do not reach the critical state line (S″S″) at points H″ and X″ respectively. The volume of the undrained specimen remains constant during shearing but that of the drained specimen, after decreasing initially, increases up to and beyond failure.

The line U′H′ is the projection of the state boundary surface, known as the *Hvorslev surface*, for heavily overconsolidated clays. However, it is assumed that the soil cannot withstand tensile effective stresses, i.e. the effective minor principal stress (σ'_3) cannot be less than zero. A line (OU′) through the origin at a slope of 3:1 ($q'/p' = 3$ for $\sigma'_3 = 0$ in Equations 4.10 and 4.11) is therefore a limit to the state boundary. On the $q' - p' - v$ plot, shown in Fig. 4.21c, this line becomes a plane lying between the line TT (referred to as the 'no tension' cut-off) and the v axis. Thus the state boundary surface for heavily overconsolidated clays lies between TT and the critical state line SS. In Fig. 4.21c the undrained stress path (DE) lies on a plane RHUV parallel to the $q' - p'$ plane. The drained stress path (DF) lies on a plane WXYD′ normal to the $q' - p'$ plane and inclined at a slope of 3:1 to the direction of the q' axis.

Also shown in Figs. 4.21a and 4.21c are the stress paths for consolidated-undrained and drained tests (G′H′ and G′J′ respectively) on *lightly overconsolidated* specimens of the same clay, starting at the same value of specific volume as the heavily overconsolidated specimens. The initial point on the stress paths (G′) is on the expansion (or recompression) curve to the right of the projection S″S″ of the critical state line in Fig. 4.21b. In both tests, failure is reached at points on or close to the critical state line. During a drained test, a lightly overconsolidated specimen decreases in volume and hardens; no decrease in stresses therefore occurs after failure. As a result the deformation of the specimen is relatively uniform and the critical state is likely to be reached.

The section of the complete state boundary surface, for normally consolidated and overconsolidated clays, on a plane of constant specific volume is RHU in Fig. 4.21c. The shape of the section will be similar on all planes of constant specific volume. A single section (TSN) can therefore be drawn with respect to coordinate axes q'/p'_e and p'/p'_e, as shown in Fig. 4.22, where p'_e is the value of p' at the intersection of a given plane of constant specific volume with the normal consolidation curve. In Fig. 4.22, point N is on the normal consolidation line, S is on the critical state line and T is on the 'no tension' cut-off. A specimen whose state is represented by a point lying between N and the vertical through S is said to be *wet of critical* (i.e. its

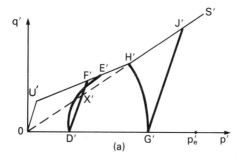

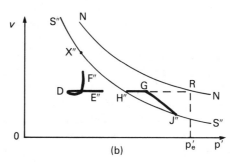

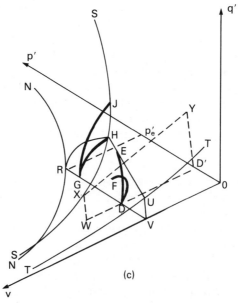

Fig. 4.21 Critical state concept: overconsolidated clays.

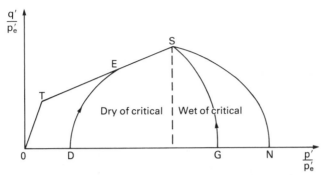

Fig. 4.22 Section of the complete state boundary surface.

water content is higher than that of clay at the critical state, at the same value of p'). A specimen whose state is represented by a point lying between the origin and the vertical through S is said to be *dry of critical*.

To summarize, the state boundary surface joins the lines NN, SS and TT in Fig. 4.21c and marks the limit to all possible combinations of stresses q' and p' and specific volume v. The plane between TT and the v axis is the boundary for no tension failure. The critical state line SS defines all possible states of ultimate failure, i.e. of continuing strain at constant volume under constant stresses. In the case of normally consolidated clays the stress paths for both drained and undrained tests lie entirely on the state boundary surface, failure being reached at a point on the critical state line; the state of the clay remains wet of critical. In the case of overconsolidated clays, the stress paths prior to failure for both drained and undrained tests lie inside the state boundary surface. A distinction must be made between heavily overconsolidated and lightly overconsolidated clays. Heavily overconsolidated clays reach failure at a point on the state boundary surface on the dry side of the critical state line; subsequently the stress path moves along the state boundary surface but is unlikely to reach the critical state line. Lightly overconsolidated clays remain wet of critical and reach failure on the critical state line.

The characteristics of both loose and dense sands during shearing under drained conditions are broadly similar to those for overconsolidated clays, failure occurring on the state boundary surface on the dry side of the critical state line.

The equation of the projection of the critical state line (OS' in Fig. 4.20a) on the $q' - p'$ plane is:

$$q' = Mp' \tag{4.20}$$

where M is the slope of OS'. If the projection of the critical state line on the $v - p'$ plane is replotted on a $v - \ln p'$ plane it will approximate to a straight line parallel to the corresponding normal consolidation line (of slope $-\lambda$)

as shown in Fig. 4.20d. The equation of the critical state line, with respect to v and p', can therefore be written as

$$v = \Gamma - \lambda \ln p' \qquad\qquad (4.21)$$

where Γ is the value of v on the critical state line at $p' = 1\,\text{kN/m}^2$.

Example 4.4

The following parameters are known for a saturated normally consolidated clay: $N = 2.48$, $\lambda = 0.12$, $\Gamma = 2.41$ and $M = 1.35$. Estimate the values of principal stress difference and void ratio at failure in undrained and drained triaxial tests on specimens of the clay consolidated under an all-round pressure of $300\,\text{kN/m}^2$.

After normal consolidation to $300\,\text{kN/m}^2$ (p'_c), the specific volume (v_c) is given by:

$$v_c = N - \lambda \ln p'_c = 2.48 - 0.12 \ln 300 = 1.80$$

In an undrained test on a saturated clay the volume change is zero, therefore the specific volume at failure (v_f) will also be 1.80, i.e. the void ratio at failure (e_f) will be 0.80.

Assuming failure to take place on the critical state line,

$$q'_f = M p'_f$$

and the value of p'_f can be obtained from Equation 4.21. Therefore

$$q'_f = M \exp\left(\frac{\Gamma - v_f}{\lambda}\right)$$

$$= 1.35 \exp\left(\frac{2.41 - 1.80}{0.12}\right)$$

$$= 218\,\text{kN/m}^2 = (\sigma_1 - \sigma_3)_f$$

For a drained test the slope of the stress path on a $q' - p'$ plot is 3, i.e.

$$q'_f = 3(p'_f - p'_c) = 3\left(\frac{q'_f}{M} - p'_c\right)$$

Therefore

$$q'_f = \frac{3M p'_c}{3 - M} = \frac{3 \times 1.35 \times 300}{3 - 1.35} = 736\,\text{kN/m}^2$$

$$= (\sigma_1 - \sigma_3)_f$$

Then,

$$p'_f = \frac{q'_f}{M} = \frac{736}{1.35} = 545\,\text{kN/m}^2$$

$$v_f = \Gamma - \lambda \ln p'_f = 2.41 - 0.12 \ln 545 = 1.65$$

Therefore

$$e_f = 0.65$$

4.6 Residual Strength

In the drained triaxial test, most clays would eventually show a decrease in shear strength with increasing strain after the peak strength has been reached. However in the triaxial test there is a limit to the strain which can be applied to the specimen. The most satisfactory method of investigating the shear strength of clays at large strains is by means of the specialized ring shear apparatus [4.3], an annular direct shear apparatus. The annular specimen (Fig. 4.23) is sheared, under a given normal stress, on a horizontal plane by the rotation of one half of the apparatus relative to the other; there is no restriction to the magnitude of shear displacement between the two halves of the specimen. The rate of rotation must be slow enough to ensure that the specimen remains in a drained condition. Shear stress, which is calculated from the applied torque, is plotted against shear displacement as shown in Fig. 4.24.

The shear strength falls below the peak value and the clay in a narrow zone adjacent to the failure plane will soften and reach the critical state. However, because of non-uniform strain in the specimen, the exact point on the curve corresponding to the critical state is uncertain. With continuing shear displacement the shear strength continues to decrease, below the critical state value, and eventually reaches a residual value at a relatively large displacement. If the clay contains a relatively high proportion of plate-like particles a reorientation of these particles parallel to the failure plane will occur (in the narrow zone adjacent to the failure plane) as the strength decreases towards the residual value. However reorientation may not occur if the plate-like particles exhibit high interparticle friction. In this case, and in the case of soils containing a relatively high proportion of bulky particles, rolling and translation of particles takes place as the

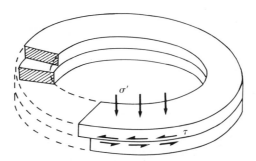

Fig. 4.23   Ring shear test.

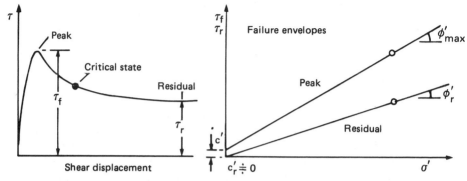

Fig. 4.24   Residual strength.

residual strength is approached. It should be appreciated that the critical state concept envisages continuous deformation of the specimen as a whole whereas in the residual condition there is preferred orientation or translation of particles in a narrow shear zone.

The results from a series of tests, under a range of values of normal stress, enable the failure envelopes for both peak and residual strength to be obtained. Peak strength is defined by:

$$\tau_f = c' + \sigma'_f \tan \phi'_{max} \qquad (4.22a)$$

and residual strength by:

$$\tau_r = c'_r + \sigma'_f \tan \phi'_r \qquad (4.22b)$$

where c'_r and ϕ'_r are the residual shear strength parameters in terms of effective stress. Residual strength data for a large range of soils has been published [4.8]. For many soils the value of c'_r is relatively low and can be taken to be zero. In general, the value of ϕ'_r decreases with increasing clay content. The relative values of peak and residual strength can be expressed by the *brittleness index* (I_B), defined as follows:

$$I_B = \frac{\tau_f - \tau_r}{\tau_f} \qquad (4.23)$$

The brittleness index for a particular soil is dependent on the level of effective normal stress.

4.7   Pore Pressure Coefficients

Pore pressure coefficients are used to express the response of pore pressure to *changes* in total stress *under undrained conditions*. Values of the coefficients may be determined in the laboratory and can be used to predict pore pressures in the field under similar stress conditions.

(1) Increment of Isotropic Stress

Consider an element of soil, of volume V and porosity n, in equilibrium under total principal stresses σ_1, σ_2 and σ_3, as shown in Fig. 4.25, the pore pressure being u_0. The element is subjected to equal increases in total stress $\Delta\sigma_3$ in each direction, resulting in an immediate increase Δu_3 in pore pressure.

The increase in effective stress in each direction $= \Delta\sigma_3 - \Delta u_3$

Reduction in volume of the soil skeleton $= C_s V(\Delta\sigma_3 - \Delta u_3)$

where C_s = compressibility of soil skeleton under an isotropic effective stress increment.

Reduction in volume of the pore space $= C_v nV\Delta u_3$

where C_v = compressibility of pore fluid under an isotropic pressure increment.

If the soil *particles* are assumed to be incompressible and if no drainage of pore fluid takes place then the reduction in volume of the soil skeleton must equal the reduction in volume of the pore space, i.e.

$$C_s V(\Delta\sigma_3 - \Delta u_3) = C_v nV\Delta u_3$$

Therefore

$$\Delta u_3 = \Delta\sigma_3 \left(\frac{1}{1 + n\dfrac{C_v}{C_s}} \right)$$

Writing $1/[1 + n(C_v/C_s)] = B$, defined as a pore pressure coefficient,

$$\Delta u_3 = B\Delta\sigma_3 \tag{4.24}$$

In fully saturated soils the compressibility of the pore fluid (water only) is considered negligible compared with that of the soil skeleton, therefore $C_v/C_s \to 0$ and $B \to 1$. Equation 4.24 with $B = 1$ has already been assumed in the discussion on undrained strength earlier in the present chapter. In partially saturated soils the compressibility of the pore fluid is high due to

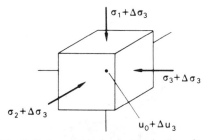

Fig. 4.25   Soil element under isotropic stress increment.

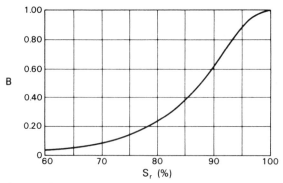

Fig. 4.26   Typical relationship between B and degree of saturation.

the presence of pore air, therefore $C_v/C_s > 0$ and $B < 1$. The variation of B with degree of saturation for a particular soil is shown in Fig. 4.26.

The value of B can be measured in the triaxial apparatus. A specimen is set up under any value of all-round pressure and the pore water pressure measured (after consolidation if desired). Under undrained conditions the all-round pressure is then increased (or reduced) by an amount $\Delta\sigma_3$ and the *change* in pore water pressure (Δu) from the initial value is measured, enabling the value of B to be calculated from Equation 4.24.

(2)   Major Principal Stress Increment

Consider now an increase $\Delta\sigma_1$ in the total major principal stress only, as shown in Fig. 4.27, resulting in an immediate increase Δu_1 in pore pressure.

The increases in effective stress are:

$$\Delta\sigma_1' = \Delta\sigma_1 - \Delta u_1$$
$$\Delta\sigma_3' = \Delta\sigma_2' = -\Delta u_1$$

If the soil behaved as an elastic material then the reduction in volume of the soil skeleton would be

$$\tfrac{1}{3}C_s V(\Delta\sigma_1 - 3\Delta u_1)$$

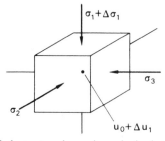

Fig. 4.27   Soil element under major principal stress increment.

The reduction in volume of the pore space is

$$C_v n V \Delta u_1$$

Again, these two volume changes will be equal for undrained conditions, i.e.

$$\tfrac{1}{3} C_s V (\Delta \sigma_1 - 3\Delta u_1) = C_v n V \Delta u_1$$

Therefore

$$\Delta u_1 = \tfrac{1}{3} \left(\frac{1}{1 + n \dfrac{C_v}{C_s}} \right) \Delta \sigma_1$$

$$= \tfrac{1}{3} B \Delta \sigma_1$$

Soils, however, are not elastic and the above equation is rewritten in the general form:

$$\Delta u_1 = AB \Delta \sigma_1 \qquad (4.25)$$

where A is a pore pressure coefficient to be determined experimentally. AB may also be written as $\bar{A}$. In the case of fully saturated soils ($B = 1$):

$$\Delta u_1 = A \Delta \sigma_1 \qquad (4.26)$$

The value of A for a fully saturated soil can be determined from measurements of pore water pressure during the application of principal stress difference under undrained conditions in a triaxial test. The *change* in total major principal stress is equal to the value of the principal stress difference applied and if the corresponding *change* in pore water pressure is measured the value of A can be calculated from Equation 4.26. The value of the coefficient at any stage of the test can be obtained but the value at failure is of most interest.

For highly compressible soils such as normally consolidated clays the value of A is found to lie within the range 0·5 to 1·0. In the case of clays of high sensitivity the increase in major principal stress may cause collapse of the soil structure, resulting in very high pore water pressures and values of A greater than 1. For soils of lower compressibility such as lightly overconsolidated clays the value of A lies within the range 0 to 0·5. If the clay is heavily overconsolidated there is a tendency for the soil to dilate as the major principal stress is increased but under undrained conditions no water can be drawn into the element and a negative pore water pressure may result. The value of A for heavily overconsolidated soils may lie between $-0{\cdot}5$ and 0. A typical relationship between the value of A at failure (A_f) and overconsolidation ratio (OCR) for a fully saturated clay is shown in Fig. 4.28.

For the condition of zero lateral strain in the soil element, reduction in volume is possible in the direction of the major principal stress only. If C_{so}

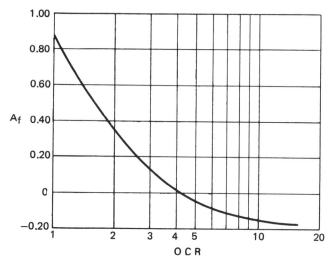

Fig. 4.28   Typical relationship between A at failure and overconsolidation ratio.

is the uni-axial compressibility of the soil skeleton then under undrained conditions:

$$C_{s0}V(\Delta\sigma_1 - \Delta u_1) = C_v nV\Delta u_1$$

Therefore

$$\Delta u_1 = \Delta\sigma_1 \left(\frac{1}{1 + n\dfrac{C_v}{C_{s0}}}\right)$$

$$= A\Delta\sigma_1$$

where $A = 1/[1 + n(C_v/C_{s0})]$. For a fully saturated soil, $C_v/C_{s0} \to 0$ and $A \to 1$, for the condition of zero lateral strain only. This was assumed in the discussion on consolidation in Chapter 3.

(3)   Combination of Increments

Cases 1 and 2 above may be combined to give the equation for the pore pressure response Δu to an isotropic stress increase $\Delta\sigma_3$ together with an axial stress increase $(\Delta\sigma_1 - \Delta\sigma_3)$ as occurs in the triaxial test. Combining Equations 4.24 and 4.25:

$$\Delta u = \Delta u_3 + \Delta u_1$$

$$= B[\Delta\sigma_3 + A(\Delta\sigma_1 - \Delta\sigma_3)] \tag{4.27}$$

An overall coefficient $\bar{B}$ can be obtained by dividing Equation 4.27 by $\Delta\sigma_1$.

$$\frac{\Delta u}{\Delta \sigma_1} = B\left[\frac{\Delta \sigma_3}{\Delta \sigma_1} + A\left(1 - \frac{\Delta \sigma_3}{\Delta \sigma_1}\right)\right]$$

Therefore

$$\frac{\Delta u}{\Delta \sigma_1} = B\left[1 - (1 - A)\left(1 - \frac{\Delta \sigma_3}{\Delta \sigma_1}\right)\right]$$

or

$$\frac{\Delta u}{\Delta \sigma_1} = \bar{B} \tag{4.28}$$

Since soils are not elastic materials the pore pressure coefficients are not constants, their values depending on the stress levels over which they are determined.

Example 4.5

The following results refer to a consolidated-undrained triaxial test on a saturated clay specimen under an all-round pressure of 300kN/m^2:

$\Delta l/l_0$	0	0·01	0·02	0·04	0·08	0·12
$\sigma_1 - \sigma_3 (\text{kN/m}^2)$	0	138	240	312	368	410
$u (\text{kN/m}^2)$		0	108	158	178	182	172

Draw the total and effective stress paths and plot the variation of the pore pressure coefficient A during the test.

From the data, the values in Table 4.7 are calculated. For example, when the strain is 0·01, $A = 108/138 = 0·78$. The stress paths and the variation of A are plotted in Fig. 4.29 in terms of (a) $\frac{1}{2}(\sigma_1 - \sigma_3)$ and $\frac{1}{2}(\sigma_1 + \sigma_3)$, (b) q and p, or the effective stress equivalents. From the shape of the effective stress path and the value of A at failure it can be concluded that the clay is overconsolidated.

Table 4.7

$\Delta l/l_0$	0	0·01	0·02	0·04	0·08	0·12
$\frac{1}{2}(\sigma_1 - \sigma_3)$	0	69	120	156	184	205
$\frac{1}{2}(\sigma_1 + \sigma_3)$	300	369	420	456	484	505
$\frac{1}{2}(\sigma_1' + \sigma_3')$	300	261	262	278	302	333
q	0	138	240	312	368	410
p	300	346	380	404	423	437
p'	300	238	222	226	241	265
A	—	0·78	0·66	0·57	0·50	0·42

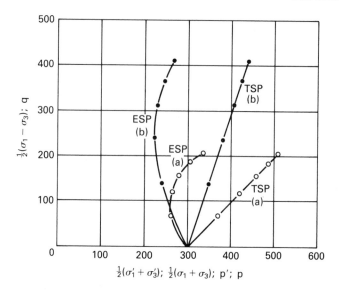

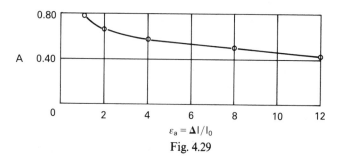

Fig. 4.29

Example 4.6

In a triaxial test a soil specimen was consolidated under an all-round pressure of $800 \, \text{kN/m}^2$ and a back pressure of $400 \, \text{kN/m}^2$. Thereafter, under undrained conditions, the all-round pressure was raised to $900 \, \text{kN/m}^2$, resulting in a pore water pressure reading of $495 \, \text{kN/m}^2$; then (with the all-round pressure remaining at $900 \, \text{kN/m}^2$) axial load was applied to give a principal stress difference of $585 \, \text{kN/m}^2$ and a pore water pressure reading of $660 \, \text{kN/m}^2$. Calculate the values of the pore pressure coefficients B, $\bar{A}$ and $\bar{B}$.

Corresponding to an increase in all-round pressure from 800 to $900 \, \text{kN/m}^2$ the pore pressure increases from the value of the back pressure, $400 \, \text{kN/m}^2$, to $495 \, \text{kN/m}^2$. Therefore

$$B = \frac{\Delta u_3}{\Delta \sigma_3} = \frac{495 - 400}{900 - 800} = \frac{95}{100} = 0 \cdot 95$$

The total major principal stress increases from $900 \, \text{kN/m}^2$ to $(900 + 585) \, \text{kN/m}^2$: the corresponding increase in pore pressure is from 495 to $660 \, \text{kN/m}^2$. Therefore

$$\bar{A} = \frac{\Delta u_1}{\Delta \sigma_1} = \frac{660 - 495}{585} = \frac{165}{585} = 0.28$$

The overall increase in pore pressure is from 400 to $660 \, \text{kN/m}^2$, corresponding to an increase in total major principal stress from $800 \, \text{kN/m}^2$ to $(800 + 100 + 585) \, \text{kN/m}^2$. Therefore

$$\bar{B} = \frac{\Delta u}{\Delta \sigma_1} = \frac{660 - 400}{100 + 585} = \frac{260}{685} = 0.38$$

4.8   In-Situ Piezometer Measurements

Whenever there is a change in pore water pressure in situ there will be corresponding changes, in the opposite sense, in the values of effective stress, resulting in a change in the shear strength of the soil. In-situ pore water pressures can be measured by means of piezometers.

If the soil has a relatively high permeability and is fully saturated, pore water pressure can be determined by measuring the water level in an open standpipe placed in a borehole (Fig. 4.30) since the water level will react almost immediately to any change in pressure. The lower end of the standpipe is either perforated or fitted with a porous element. Sand or fine gravel is packed around the lower end of the standpipe and the standpipe must be efficiently sealed in the borehole with clay or mortar immediately above the level of the stratum in which the pore water pressure is required. The remainder of the borehole is backfilled with sand except for a second seal near the surface to prevent the inflow of surface water. The top of the standpipe must remain accessible at all times and should be fitted with a cap.

If the soil has a relatively low permeability the pressure at the point of measurement will be altered if even a small flow of water is required to operate the measuring device and it will then take some time for the original pressure to be re-established. Thus in the case of soils of low permeability the piezometer is required to respond rapidly to a change in pore water pressure but without any significant flow of water: to achieve these requirements a closed hydraulic system is essential. The closed system employs a piezometer tip consisting of a plastic body into which a porous stone or ceramic is sealed: three types of tip are shown in Fig. 4.31. Two tubes lead from the tip to a Bourdon pressure gauge located as near as possible to the tip: the tubes are of nylon coated with polythene, nylon being impermeable to air and polythene to water. A change in pore water

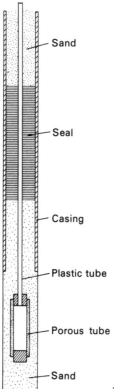

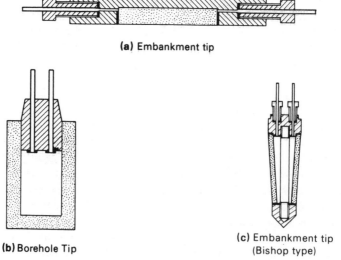

Fig. 4.30   Casagrande piezometer.

(a) Embankment tip

(b) Borehole Tip

(c) Embankment tip
(Bishop type)

Fig. 4.31   Piezometer tips.

pressure in the soil adjacent to the tip will result in a small flow of water through the porous stone, continuing until an equal change has taken place in the closed measuring system. The time for this change to take place is called the response time of the piezometer and should be as short as possible. The hydraulic system should thus be as stiff as possible and no air should be tolerated in the system. The two tubes enable the system to be kept air free by the periodic circulation of de-aired water. The pressure recorded by the gauge must be corrected for the difference in elevation between the tip and the gauge, which should be sited below the level of the tip whenever possible.

Problems

4.1 What is the shear strength in terms of effective stress on a plane within a saturated soil mass at a point where the total normal stress is $295 \, kN/m^2$ and the pore water pressure $120 \, kN/m^2$? The effective stress parameters for the soil are $c' = 12 \, kN/m^2$ and $\phi' = 30°$.

4.2 A series of drained triaxial tests was carried out on specimens of a sand prepared at the same porosity and the following results were obtained at failure. Determine the value of the angle of shearing resistance ϕ'.

All-round pressure (kN/m^2)	100	200	400	800
Principal stress difference (kN/m^2)	452	908	1810	3624

4.3 In a series of unconsolidated-undrained triaxial tests on specimens of a fully saturated clay the following results were obtained at failure. Determine the values of the shear strength parameters c_u and ϕ_u.

All-round pressure (kN/m^2)	200	400	600
Principal stress difference (kN/m^2)	222	218	220

4.4 The effective stress parameters for a fully saturated clay are known to be $c' = 15 \, kN/m^2$ and $\phi' = 29°$. In an unconsolidated-undrained triaxial test on a specimen of the same clay the all-round pressure was $250 \, kN/m^2$ and the principal stress difference at failure $134 \, kN/m^2$. What was the value of pore water pressure in this specimen at failure?

4.5 The results in Table 4.8 were obtained at failure in a series of consolidated-undrained triaxial tests, with pore water pressure measurement, on specimens of a fully saturated clay. Determine the values of the shear strength parameters c' and ϕ'. If a specimen of the same soil were consolidated under an all-round pressure of $250 \, kN/m^2$ and the principal stress difference applied with the all-

Table 4.8

σ_3 (kN/m²)	$\sigma_1 - \sigma_3$ (kN/m²)	u (kN/m²)
150	103	82
300	202	169
450	305	252
600	410	331

round pressure changed to 350 kN/m², what would be the expected value of principal stress difference at failure?

4.6 The following results were obtained at failure in a series of drained triaxial tests on fully saturated clay specimens originally 38 mm diameter by 76 mm long. Determine the values of the shear strength parameters c' and ϕ'.

All-round pressure (kN/m²)	200	400	600
Axial compression (mm)	7·22	8·36	9·41
Axial load (N)	480	895	1300
Volume change (ml)	5·25	7·40	9·30

4.7 Derive Equation 4.14.

In an in-situ vane test on a saturated clay a torque of 35 Nm is required to shear the soil. The vane is 50 mm wide by 100 mm long. What is the undrained strength of the clay?

4.8 A consolidated-undrained triaxial test on a specimen of saturated clay was carried out under an all-round pressure of 600 kN/m². Consolidation took place against a back pressure of 200 kN/m². The following results were recorded during the test.

$\sigma_1 - \sigma_3$ (kN/m²)	0	80	158	214	279	319
u (kN/m²)	200	229	277	318	388	433

Draw the stress paths and give the value of the pore pressure coefficient A at failure.

4.9 In a triaxial test a soil specimen is allowed to consolidate fully under an all-round pressure of 200 kN/m². Under undrained conditions the all-round pressure is increased to 350 kN/m², the pore water pressure then being measured as 144 kN/m². Axial load is then applied under undrained conditions until failure takes place, the following results being obtained.

Axial strain (%)	0	2	4	6	8	10
Principal stress difference (kN/m²)	0	201	252	275	282	283
Pore water pressure (kN/m²)	144	244	240	222	212	209

Determine the value of the pore pressure coefficient B and plot the variation of coefficient $\bar{A}$ with axial strain, stating the value at failure.

References

4.1 Bishop, A. W. (1966): 'The Strength of Soils as Engineering Materials', *Geotechnique*, Vol. 16, No. 2.

4.2 Bishop, A. W., Alpan, I., Blight, G. E. and Donald, I. B. (1960): 'Factors Controlling the Strength of Partly Saturated Cohesive Soils', *Proc. A.S.C.E. Conference on Shear Strength of Cohesive Soils, Boulder, Colorado, U.S.A.*

4.3 Bishop, A. W., Green, G. E., Garga, V. K., Andresen, A. and Brown, J. D. (1971): 'A New Ring Shear Apparatus and its Application to the Measurement of Residual Strength', *Geotechnique*, Vol. 21, No. 4.

4.4 Bishop, A. W. and Henkel, D. J. (1962): *The Measurement of Soil Properties in the Triaxial Test* (2nd Edition), Edward Arnold. London.

4.5 British Standard 1377 (1975): *Methods of Test for Soils for Civil Engineering Purposes*, British Standards Institution, London.

4.6 Bjerrum, L. (1973): 'Problems of Soil Mechanics and Construction on Soft Clays'. *Proc. 8th International Conference S.M.F.E.*, Moscow, Vol. 3.

4.7 Cornforth, D. H. (1964): 'Some Experiments on the Influence of Strain Conditions on the Strength of Sand', *Geotechnique*, Vol. 14, No. 2.

4.8 Lupini, J. F., Skinner, A. E. and Vaughan, P. R. (1981): 'The Drained Residual Strength of Cohesive Soils', *Geotechnique*, Vol. 31, No. 2.

4.9 Penman, A. D. M. (1956): 'A Field Piezometer Apparatus', *Geotechnique*, Vol. 6, No. 2.

4.10 Penman, A. D. M. (1961): 'A Study of the Response Time of Various Types of Piezometer', *Proceedings of Conference on Pore Pressure and Suction in Soils*, Butterworth, London.

4.11 Roscoe, K. H., Scchofield, A. N. and Wroth, C. P. (1958): 'On the Yielding of Soils', *Geotechnique*, Vol. 8, No. 1.

4.12 Rowe, P. W. (1962): 'The Stress–Dilatancy Relation for Static Equilibrium of an Assembly of Particles in Contact', *Proc. Roy. Soc. A*, Vol. 269.

4.13 Skempton, A. W. (1954): 'The Pore Pressure Coefficients A and B', *Geotechnique*, Vol. 4, No. 4.

4.14 Skempton, A. W. and Sowa, V. A. (1963): 'The Behaviour of Saturated Clays During Sampling and Testing', *Geotechnique*, Vol. 13, No. 4.

4.15 Skempton, A. W. (1985): 'Residual Strength of Clays in Landslides, Folded Strata and the Laboratory', *Geotechnique*, Vol. 35, No. 1

CHAPTER 5

Stresses and Displacements

5.1 Elasticity and Plasticity

The stresses and displacements in a soil mass due to applied loading are considered in this chapter. Many problems can be treated by analysis in two dimensions, i.e. only the stresses and displacements in a single plane need to be considered. The total normal stresses and shear stresses in the x and z directions on an element of soil are shown in Fig. 5.1, the stresses being positive as shown: the stresses vary across the element. The rates of change of the normal stresses in the respective directions are $\partial\sigma_x/\partial x$ and $\partial\sigma_z/\partial z$; the rates of change of the shear stresses are $\partial\tau_{xz}/\partial x$ and $\partial\tau_{zx}/\partial z$. Every such element in a soil mass must be in static equilibrium. By equating moments about the centre point of the element, and neglecting higher-order differentials, it is apparent that $\tau_{xz} = \tau_{zx}$. By equating forces in the x and z directions the following equations are obtained:

$$\frac{\partial\sigma_x}{\partial x} + \frac{\partial\tau_{zx}}{\partial z} - X = 0 \tag{5.1a}$$

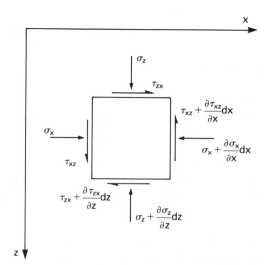

Fig. 5.1   Two-dimensional state of stress in an element.

$$\frac{\partial \tau_{xz}}{\partial x} + \frac{\partial \sigma_z}{\partial z} - Z = 0 \tag{5.1b}$$

where X and Z are the respective body forces per unit volume. These are the equations of equilibrium in two dimensions: they can also be written in terms of effective stress. In terms of total stress the body forces are $X = 0$ and $Z = \gamma$ (or γ_{sat}). In terms of effective stress the body forces are $X' = 0$ and $Z' = \gamma'$; however, if seepage is taking place these become $X' = i_x \gamma_w$ and $Z' = \gamma' + i_z \gamma_w$ where i_x and i_z are the hydraulic gradients in the x and z directions respectively.

Due to the applied loading, points within the soil mass will be displaced relative to the axes and to one another. If the components of displacement in the x and z directions are denoted by u and w respectively then the normal strains are given by:

$$\varepsilon_x = \frac{\partial u}{\partial x} \qquad \varepsilon_z = \frac{\partial w}{\partial z}$$

and the shear strain by:

$$\gamma_{xz} = \frac{\partial u}{\partial z} + \frac{\partial w}{\partial x}$$

However these strains are not independent; they must be compatible with each other if the soil mass as a whole is to remain continuous. This requirement leads to the following relationship, known as the equation of compatibility in two dimensions:

$$\frac{\partial^2 \varepsilon_x}{\partial z^2} + \frac{\partial^2 \varepsilon_z}{\partial x^2} - \frac{\partial \gamma_{xz}}{\partial x \partial z} = 0 \tag{5.2}$$

Equations 5.1 and 5.2, being independent of material properties, can be applied to both elastic and plastic behaviour.

The rigorous solution of a particular problem requires that the equations of equilibrium and compatibility are satisfied for the given boundary conditions. However an appropriate stress-strain relationship is also required. In the theory of elasticity [5.17] a linear stress-strain relationship is combined with the above equations.

A typical stress-strain relationship for a soil during shearing is shown in Fig. 5.2. The relationship is non-linear, therefore strains will have both elastic (recoverable) and plastic (non-recoverable) components. Initially (between O and Y) the relationship is approximately linear for most soils then (beyond Y) significant plastic strain becomes apparent, i.e. Y is the point at which yielding takes place. The peak shear strength is reached at point F. Between the points of yielding and failure (Y and F) further plastic strain will occur only if the stress is increased, but a progressively smaller stress increment is required to produce a given plastic strain increment: this characteristic is known as strain (or work) hardening. Hardening occurs

150

Soil Mechanics

only if plastic work is done. In certain cases the resistance to shear shows a decrease after the peak strength has been reached, this characteristic being referred to as strain (or work) softening. In analysis a simplification can be made by assuming that the soil behaves as an elastic-perfectly plastic material, the stress-strain relationship being represented by OY′P in Fig. 5.2: elastic behaviour between O and Y′ (the assumed yield point) is followed by unrestricted plastic strain at constant stress (Y′P).

The state of stress (denoted by σ) at which yielding occurs is defined by means of a *yield criterion*. The general form of the function which defines yielding is:

$$f(\sigma) = 0 \qquad (5.3)$$

in which σ may be defined in terms of stress components or principal stresses. Plastic strain is not related uniquely to the state of stress, unlike elastic strain. The plastic strains occurring subsequent to yielding are defined by means of a *strain hardening law* (if appropriate) and a *flow rule*. The hardening law, which describes how the yield criterion changes due to strain hardening, is expressed in the form:

$$f(\sigma,h) = 0 \qquad (5.4)$$

where h is a function of the plastic strain components. The flow rule, which determines the direction of plastic strain, takes the form:

$$d\varepsilon^p = d\lambda \frac{\partial f(\sigma)}{\partial \sigma} \qquad (5.5)$$

where ε^p represents the components of plastic strain and $d\lambda$ is a proportionality coefficient. In the context of the flow rule, $f(\sigma)$ is known as the plastic potential. When the yield function also serves as the plastic potential (as indicated in Equation 5.5) the flow rule is said to be 'associated'. Otherwise, if the plastic potential function differs from the yield function, the flow rule is said to be 'non-associated'. The flow rule relates *increments* of plastic strain to the corresponding *state* of stress (not the stress increments). A given state of stress can be represented by a point on the line or surface represented by the yield function. It can be shown that the vector

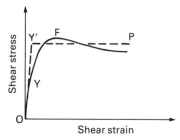

Fig. 5.2   Stress-strain relationship.

of plastic strain increment is in a direction normal to the line or surface at that point: this is known as the normality condition. The rigorous solution of a plasticity problem is only possible in relatively few cases.

When the stresses are low relative to the failure value, the stress-strain relationship of most soils can be assumed to be linear, the main exceptions being loose sands and soft clays. A factor of safety of 3 with respect to shear failure is applied in many problems and, in such cases, solutions from the theory of elasticity are normally used for the estimation of stresses and displacements. A comprehensive collection of solutions has been published by Poulos and Davis [5.14]. If justified, the finite element method [5.2], in which it is possible to specify either linear or non-linear stress-strain relationships, can be used. One method of representing non-linear behaviour in a finite element analysis is to approximate the stress-strain curve beyond the yield point by a second straight line. Another method is to use the finding, due to Kondner [5.8], that for many soils the stress-strain curve could be represented, to an acceptable degree of accuracy, by a hyperbola.

With reference to the critical state concept it is assumed that, in the case of an overconsolidated clay, if an effective stress path lies entirely within the state boundary surface the resulting strains are elastic. This assumption implies that the stress path rises vertically towards the state boundary surface. Plastic strains are assumed to occur only in respect of that part of the stress path which lies on the state boundary surface (i.e. the state boundary surface is also a yield surface). Thus in the case of a normally consolidated clay both elastic and plastic components of strain will result for any stress path. Displacements in normally consolidated clays, therefore, are relatively large in comparison with those in overconsolidated clays for the same stress changes.

Displacement solutions from elastic theory require a knowledge of the values of Young's modulus (E) and Poisson's ratio (v) for the soil, either for undrained conditions or in terms of effective stress. Poisson's ratio is required for certain stress solutions. It should be noted that the shear modulus (G), where

$$G = \frac{E}{2(1 + v)} \tag{5.6}$$

is independent of the drainage conditions, assuming that the soil is isotropic.

The volumetric strain of an element of linearly elastic material under three principal stresses is given by:

$$\frac{\Delta V}{V} = \frac{1 - 2v}{E} (\sigma_1 + \sigma_2 + \sigma_3)$$

If this expression is applied to soils over the initial part of the stress-strain curve then for undrained conditions $\Delta V/V = 0$, hence $v = 0.5$. The undrained value of Young's modulus is then related to the shear modulus by the expression $E_u = 3G$. If consolidation takes place then $\Delta V/V > 0$ and $\nu < 0.5$ for drained or partially drained conditions.

The value of E can be estimated from the curve relating principal stress difference and axial strain in an appropriate triaxial test. The value is usually determined as the secant modulus between the origin and one-third of the peak stress, or over the actual stress range in the particular problem. However, because of the effects of sampling disturbance, it is preferable to determine E (or G) from the results of in-situ tests. One such method is to apply load increments to a test plate, either in a shallow pit or at the bottom of a large-diameter borehole, and to measure the resulting vertical displacements. The value of E is then calculated using the relevant displacement solution, an appropriate value of ν being assumed.

The Pressuremeter

The shear modulus (G) can be determined in-situ by means of the pressure-meter. The original pressuremeter was developed in the 1950s by Menard in an attempt to overcome the problem of sampling disturbance and to ensure that the macro-fabric of the soil is adequately represented. Menard's original design, illustrated in Fig. 5.3a, consists of three cylindrical rubber cells of equal diameter arranged coaxially. The device is lowered into a (slightly oversize) borehole to the required depth and the central measuring cell is expanded against the borehole wall by means of water pressure, measurements of the applied pressure and the corresponding increase in volume of the cell being recorded. Pressure is applied to the water by compressed gas (usually nitrogen) in a control cylinder at the surface. The increase in volume of the measuring cell is determined from the movement of the gas/water interface in the control cylinder, readings normally being taken at times of 15 s, 30 s, 60 s and 120 s after a pressure increment has been applied. The pressure is corrected for (a) the head difference between the water level in the cylinder and the test level in the borehole, (b) the pressure required to stretch the rubber cell and (c) the expansion of the control cylinder and tubing under pressure. The two outer guard cells are expanded under the same pressure as in the measuring cell but using compressed gas: the increase in volume of the guard cells is not measured. The function of the guard cells is to eliminate end effects, ensuring a state of plane strain adjacent to the measuring cell.

The results of a test using the Menard pressuremeter are represented by a plot of corrected pressure (p) against volume (V) as shown in Fig. 5.4a. On this plot a linear section occurs between pressures p_i and p_f. The value p_i is the pressure necessary to achieve initial contact between the cell and the borehole wall and to recompress soil disturbed or softened as a result of boring. The value p_f is the pressure corresponding to the onset of plastic strain in the soil. Eventually a limit pressure (p_l) is approached at which continuous expansion of the borehole cavity would occur. A 'creep' curve,

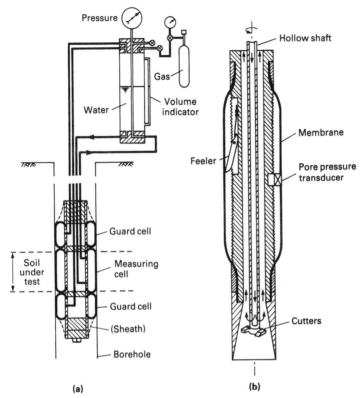

Fig. 5.3 Basic features of (a) Menard pressuremeter, (b) self-boring pressure-meter.

obtained by plotting the volume change between the 30 s and 120 s readings against the corresponding pressure may be a useful aid in fixing the values p_i and p_f, significant breaks occurring at these pressures.

The datum or reference pressure for the interpretation of pressuremeter results is a value (p_0) equal to the in-situ total horizontal stress in the soil before boring. Originally this value was assumed to be equal to p_i but the use of a pre-formed borehole means that the soil is being stressed from an unloaded condition, not from the initial undisturbed state: consequently the value of p_0 should be greater than p_i. (It should be appreciated that it is normally very difficult to obtain an independent value of in-situ total horizontal stress.) The reference volume V_0 (corresponding to the pressure p_0) is taken to be the initial volume of the borehole cavity over the test length. At any stage during a test the volume V, corresponding to the pressure p, is referred to as the current volume.

Alternatively the results of a pressuremeter test can be represented by plotting corrected pressure against the circumferential strain (ε_c) at the borehole wall. The circumferential strain is given by the ratio of the increase

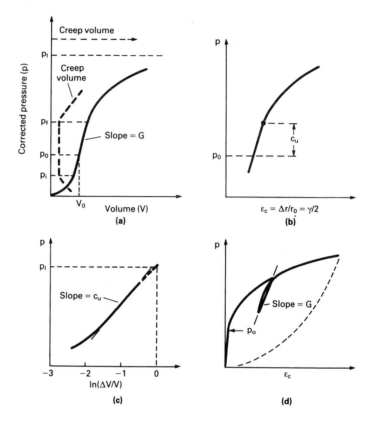

Fig. 5.4   Pressuremeter test results.

in radius of the borehole cavity (Δr) to the radius at the reference state (r_0). The relationship between current volumetric strain and circumferential strain is:

$$\Delta V/V = 1 - (1 + \varepsilon_c)^{-2}$$

(Shear strain (γ) is equal to twice circumferential strain).

Marsland and Randolph [5.11] proposed a procedure, using the $p-\varepsilon_c$ plot, for the determination of p_0, applicable to soils such as stiff clays which exhibit essentially linear stress-strain behaviour up to peak strength. The linear section of the $p-\varepsilon_c$ plot should terminate when the shear stress at the borehole wall is equal to the (peak) undrained strength of the clay, i.e. when the pressure becomes equal to ($p_0 + c_u$). The value of c_u is determined using Equation 5.10, for which a value of the reference volume V_0 is required. The method involves an iterative process in which estimates of p_0, and hence V_0, are made and the corresponding value of c_u determined until the point representing ($p_0 + c_u$) corresponds with the point on the plot at which significant curvature begins, as shown in Fig. 5.4b.

The value of the limit pressure (p_l) can be determined by plotting pressure against the logarithm of current volumetric strain and extrapolating to a strain of unity, representing continuous expansion, as shown in Fig. 5.4c.

An analysis of the expansion of the borehole cavity during a pressuremeter test was presented by Gibson and Anderson [5.5], the soil being considered as an elastic – perfectly plastic material. Within the linear section of the $p–V$ plot the shear modulus is given by:

$$G = V \frac{dp}{dV} \tag{5.7}$$

where dp/dV is the slope of the linear section and V is the current volume of the borehole cavity. However it is recommended that the modulus is determined from an unloading/reloading cycle to minimize the effect of soil disturbance.

In the case of saturated clays it is possible to obtain the value of the undrained shear strength (c_u), by iteration, from the following expression:

$$p_l - p_0 = \left[\ln\left(\frac{G}{c_u}\right) + 1 \right] c_u \tag{5.8}$$

In modern developments of the pressuremeter the measuring cell is expanded directly by gas pressure. This pressure and the radial expansion of the rubber membrane are recorded by means of electrical transducers within the cell. In addition a pore water pressure transducer is fitted into the cell wall such that it is in contact with the soil during the test. A considerable increase in accuracy is obtained with these pressuremeters compared with the original Menard device. It is also possible to adjust the cell pressure continuously, using electronic control equipment, to achieve a constant rate of increase in circumferential strain (i.e. a strain-controlled test), rather than to apply the pressure in increments (a stress-controlled test).

Some soil disturbance adjacent to a borehole is inevitable and the results of pressuremeter tests in pre-formed holes can be sensitive to the method of boring. The self-boring pressuremeter was developed to overcome this problem and is suitable for use in most types of soil: however special insertion techniques are required in the case of sands. This device, illustrated in Fig. 5.3b, is jacked slowly into the ground and the soil is broken up by a rotating cutter fitted inside a cutting head at the lower end, the optimum position of the cutter being a function of the shear strength of the soil. Water or drilling mud is pumped down the hollow shaft to which the cutter is attached and the resulting slurry is carried to the surface through the annular space adjacent to the shaft: the device is thus inserted with minimal disturbance of the soil. The only correction required is for the pressure required to stretch the membrane.

A 'push-in' penetrometer has also been developed for insertion below the bottom of a borehole, for use particularly in off-shore work. This pressuremeter is fitted with a cutting shoe, a soil core passing upwards inside the device.

The membrane of a pressuremeter may be protected against possible damage (particularly in coarse soils) by a thin stainless steel sheath with longitudinal cuts, designed to cause only negligible resistance to the expansion of the cell.

Results from a strain-controlled test in clay using the self-boring pressuremeter are of the form shown in Fig. 5.4d, the pressure (p) being plotted against the circumferential strain (ε_c). Use of the self-boring pressuremeter overcomes the difficulty in determining the initial in-situ total horizontal stress: because soil disturbance is minimal the pressure at which the membrane starts to expand (referred to as the 'lift-off' pressure) should be equal to p_0, as shown in Fig. 5.4d.

The value of the shear modulus is given by the following equation, derived in a later analysis by Palmer [5.13] in which no assumption is made regarding the stress-strain characteristics of the soil. For expansion of a borehole cavity at small strains it was shown that:

$$G = \tfrac{1}{2}\frac{dp}{d\varepsilon_c}$$ (5.9)

The modulus should be obtained from the slope of an unloading/reloading cycle as shown in Fig. 5.4d, ensuring that the soil remains in the 'elastic' state during unloading. Wroth [5.19] has shown that, in the case of a clay, this requirement will be satisfied if the reduction in pressure during the unloading stage is less than $2c_u$.

For a saturated clay the undrained shear strength (c_u) can also be obtained from the following equation derived from the analysis of Gibson and Anderson:

$$p = p_l + c_u \ln\left(\frac{\Delta V}{V}\right)$$ (5.10)

where $\Delta V/V$ is the current volumetric strain.

It should be noted that equation 5.10 is relevant only after the plastic state has been reached in the soil (i.e. when $p_f < p < p_l$). The plot of p against $\ln(\Delta V/V)$ should become essentially linear for the final stage of the test as shown in Fig. 5.4c, and the value of c_u is given by the slope of the line.

In Palmer's analysis it was shown that at small strains the shear stress in the soil at the wall of the expanding cavity is given by:

$$\tau = \varepsilon_c \frac{dp}{d\varepsilon_c}$$ (5.11)

and at larger strains by:

$$\tau = \frac{dp}{d[\ln(\Delta V/V)]}$$ (5.12)

both the circumferential and volumetric strains being defined with respect to the reference state. Equations 5.11 and 5.12 can be used to derive the entire stress-strain curve for the soil.

An analysis for the interpretation of pressuremeter tests in sands has been given by Hughes, Wroth and Windle [5.6]. The analysis enables values for the angle of shearing resistance (ϕ') and the angle of dilation (ψ) to be determined. A comprehensive review of the use of pressuremeters, including examples of test results and their application in design, has been given by Mair and Wood [5.10].

5.2 Stresses from Elastic Theory

The stresses within a semi-infinite, homogeneous, isotropic mass, with a linear stress-strain relationship, due to a point load on the surface, were determined by Boussinesq in 1885. The vertical, radial, circumferential and shear stresses at a depth z and a horizontal distance r from the point of application of the load were given. The stresses due to surface loads distributed over a particular area can be obtained by integration from the point load solutions. The stresses at a point due to more than one surface load are obtained by superposition. In practice, loads are not usually applied directly on the surface but the results for surface loading can be applied conservatively in problems concerning loads at a shallow depth.

A range of solutions, suitable for determining the stresses below foundations, is given in the following sections. Negative values of loading can be used if the stresses due to excavation are required or in problems in which the principle of superposition is used. The stresses due to surface loading act in addition to the in-situ stresses due to the self-weight of the soil.

Point Load

Referring to Fig. 5.5a, the stresses at point X due to a point load Q on the surface are as follows:

$$\sigma_z = \frac{3Q}{2\pi z^2} \left\{ \frac{1}{1 + (r/z)^2} \right\}^{5/2} \tag{5.13}$$

$$\sigma_r = \frac{Q}{2\pi} \left\{ \frac{3r^2 z}{(r^2 + z^2)^{5/2}} - \frac{1 - 2v}{r^2 + z^2 + z(r^2 + z^2)^{1/2}} \right\} \tag{5.14}$$

$$\sigma_\theta = -\frac{Q}{2\pi} (1 - 2v) \left\{ \frac{z}{(r^2 + z^2)^{3/2}} - \frac{1}{r^2 + z^2 + z(r^2 + z^2)^{1/2}} \right\} \tag{5.15}$$

$$\tau_{rz} = \frac{3Q}{2\pi} \left\{ \frac{rz^2}{(r^2 + z^2)^{5/2}} \right\} \tag{5.16}$$

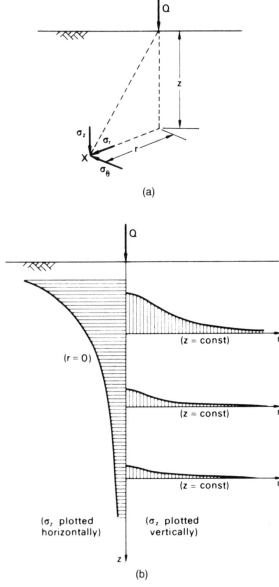

Fig. 5.5   (a) Stresses due to point load. (b) Variation of vertical stress due to point load.

It should be noted that when $v = 0.5$ the second term in Equation 5.14 vanishes and Equation 5.15 gives $\sigma_\theta = 0$.

Equation 5.13 is used most frequently in practice and can be written in terms of an influence factor I_p, where

Table 5.1   Influence Factors for Vertical Stress due to Point Load

r/z	I_p	r/z	I_p	r/z	I_p
0·00	0·478	0·80	0·139	1·60	0·020
0·10	0·466	0·90	0·108	1·70	0·016
0·20	0·433	1·00	0·084	1·80	0·013
0·30	0·385	1·10	0·066	1·90	0·011
0·40	0·329	1·20	0·051	2·00	0·009
0·50	0·273	1·30	0·040	2·20	0·006
0·60	0·221	1·40	0·032	2·40	0·004
0·70	0·176	1·50	0·025	2·60	0·003

$$I_p = \frac{3}{2\pi} \left\{ \frac{1}{1 + (r/z)^2} \right\}^{5/2}$$

Then

$$\sigma_z = \frac{Q}{z^2} I_p$$

Values of I_p in terms of r/z are given in Table 5.1. The form of the variation of σ_z with z and r is illustrated in Fig. 5.5b. The left-hand side of the figure shows the variation of σ_z with z on the vertical through the point of application of the load Q (i.e. for $r = 0$): the right-hand side of the figure shows the variation of σ_z with r for three different values of z.

In another solution to the problem, due to Westergaard, the elastic mass is assumed to be reinforced laterally by horizontal inelastic sheets of negligible thickness, spaced at close intervals in the vertical direction, preventing lateral strain in the mass as a whole. The solution simulates an extreme condition of anisotropy and gives stresses less than the Boussinesq values. The conditions in most soil masses probably lie between the two extremes represented by the Boussinesq and Westergaard solutions. The Boussinesq solution is used more extensively in practice.

Line Load

Referring to Fig. 5.6a, the stresses at point X due to a line load of Q per unit length on the surface are as follows:

$$\sigma_z = \frac{2Q}{\pi} \frac{z^3}{(x^2 + z^2)^2} \tag{5.17}$$

$$\sigma_x = \frac{2Q}{\pi} \frac{x^2 z}{(x^2 + z^2)^2} \tag{5.18}$$

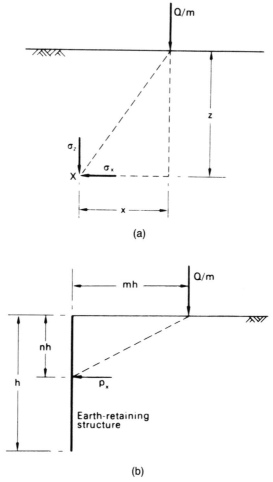

Fig. 5.6   (a) Stresses due to line load. (b) Lateral pressure due to line load.

$$\tau_{xz} = \frac{2Q}{\pi} \frac{xz^2}{(x^2 + z^2)^2} \tag{5.19}$$

Equation 5.18 can be used to estimate the lateral pressure on an earth-retaining structure due to a line load on the surface of the backfill. In terms of the dimensions given in Fig. 5.6b, Equation 5.18 becomes

$$\sigma_x = \frac{2Q}{\pi h} \frac{m^2 n}{(m^2 + n^2)^2}$$

However the structure will tend to interfere with the lateral strain due to the load Q and to obtain the lateral pressure on a relatively rigid structure a second load Q must be imagined at an equal distance on the other side of the structure. Then the lateral pressure is given by:

$$p_x = \frac{4Q}{\pi h} \frac{m^2 n}{(m^2 + n^2)^2} \tag{5.20}$$

The total thrust on the structure is given by:

$$P_x = \int_0^1 p_x h \, dn = \frac{2Q}{\pi} \frac{1}{m^2 + 1} \tag{5.21}$$

Strip Area Carrying Uniform Pressure

The stresses at point X due to a uniform pressure q on a strip area of width B and infinite length are given in terms of the angles α and β defined in Fig. 5.7a.

$$\sigma_z = \frac{q}{\pi} \{\alpha + \sin \alpha \cos (\alpha + 2\beta)\} \tag{5.22}$$

$$\sigma_x = \frac{q}{\pi} \{\alpha - \sin \alpha \cos (\alpha + 2\beta)\} \tag{5.23}$$

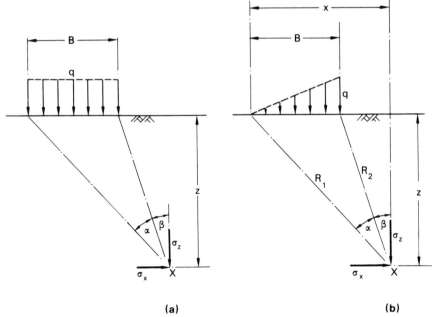

(a)   (b)

Fig. 5.7   Stresses due to (a) uniform pressure, (b) linearly increasing pressure, on strip area.

$$\tau_{xz} = \frac{q}{\pi} \{\sin \alpha \sin (\alpha + 2\beta)\} \qquad (5.24)$$

Contours of equal vertical stress in the vicinity of a strip area carrying a uniform pressure are plotted in Fig. 5.8a. The zone lying inside the vertical stress contour of value $0 \cdot 2q$ is described as the *bulb of pressure*.

Strip Area Carrying Linearly Increasing Pressure

The stresses at point X due to pressure increasing linearly from zero to q on a strip area of width B are given in terms of the angles α and β and the lengths R_1 and R_2, as defined in Fig. 5.7b.

$$\sigma_z = \frac{q}{\pi} \left(\frac{x}{B} \alpha - \tfrac{1}{2} \sin 2\beta \right) \qquad (5.25)$$

$$\sigma_x = \frac{q}{\pi} \left(\frac{x}{B} \alpha - \frac{z}{B} \ln \frac{R_1^2}{R_2^2} + \tfrac{1}{2} \sin 2\beta \right) \qquad (5.26)$$

$$\tau_{xz} = \frac{q}{2\pi} \left(1 + \cos 2\beta - 2\frac{z}{B} \alpha \right) \qquad (5.27)$$

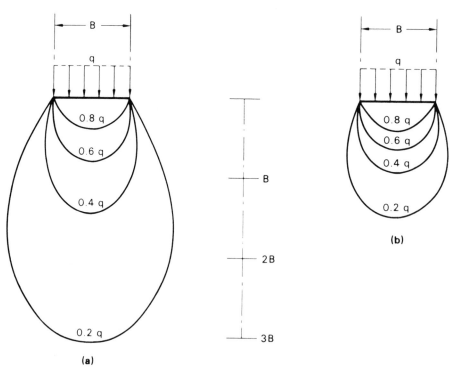

Fig. 5.8   Contours of equal vertical stress: (a) under strip area, (b) under square area.

Circular Area Carrying Uniform Pressure

The vertical stress at depth z under the *centre* of a circular area of diameter $D = 2R$ carrying a uniform pressure q is given by:

$$\sigma_z = q\left[1 - \left\{\frac{1}{1 + (R/z)^2}\right\}^{3/2}\right] = qI_c \tag{5.28}$$

Values of the influence factor I_c in terms of D/z are given in Fig. 5.9.

The radial and circumferential stresses under the centre are equal and are given by:

$$\sigma_r = \sigma_\theta = \frac{q}{2}\left[(1 + 2v) - \frac{2(1 + v)}{\{1 + (R/z)^2\}^{1/2}} + \frac{1}{\{1 + (R/z)^2\}^{3/2}}\right] \tag{5.29}$$

Rectangular Area Carrying Uniform Pressure

A solution has been obtained for the vertical stress at depth z under a *corner* of a rectangular area of dimensions mz and nz (Fig. 5.10) carrying a uniform pressure q. The solution can be written in the form:

$$\sigma_z = qI_r$$

Values of the influence factor I_r in terms of m and n are given in the chart due to Fadum [5.4] shown in Fig. 5.10. The factors m and n are interchangeable. The chart can also be used for a strip area, considered as a rectangular area

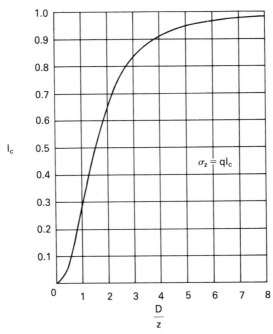

Fig. 5.9   Vertical stress under centre of circular area carrying a uniform pressure.

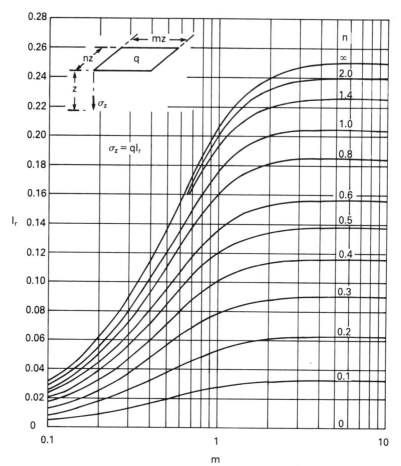

Fig. 5.10 Vertical stress under corner of rectangular area carrying a uniform pressure. (Reproduced from R. E. Fadum (1948) *Proceedings 2nd International Conference SMFE*, Rotterdam, Vol. 3, by permission of Professor Fadum.)

of infinite length. Superposition enables any area based on rectangles to be dealt with and enables the vertical stress under any point within or outside the area to be obtained.

Contours of equal vertical stress in the vicinity of a square area carrying a uniform pressure are plotted in Fig. 5.8b. Influence factors for σ_x and σ_y (which depend on v) are given in Poulos and Davis [5.14].

Influence Chart for Vertical Stress

Newmark [5.12] constructed an influence chart, based on the Boussinesq solution, enabling the vertical stress to be determined at any point below an area of any shape carrying a uniform pressure q. The chart (Fig. 5.11) consists of influence areas, the boundaries of which are two radial lines and

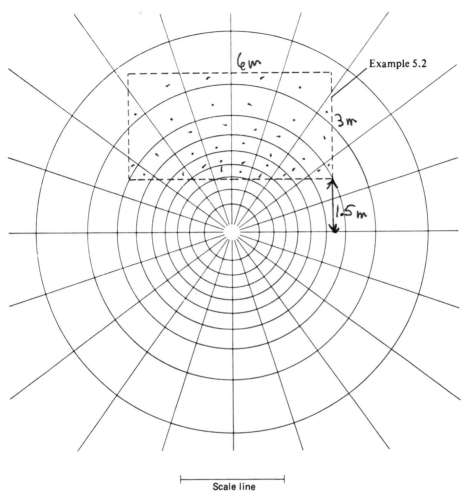

Scale line

Fig. 5.11 Newmark's influence chart for vertical stress. Influence value per unit pressure = 0.005. (Reproduced from N.M. Newmark (1942) *Influence Charts for Computation of Stresses in Elastic Foundations*, University of Illinois Bulletin No. 338, by permission of Professor Newmark.)

two circular arcs. The loaded area is drawn on tracing paper to a scale such that the length of the scale line on the chart represents the depth z at which the vertical stress is required. The position of the loaded area on the chart is such that the point at which the vertical stress is required is at the centre of the chart. For the chart shown in Fig. 5.11 the influence value is 0·005, i.e. each influence area represents a vertical stress of 0·005q. Hence, if the number of influence areas covered by the scale drawing of the loaded area is N, the required vertical stress is given by

$$\sigma_z = 0·005\,Nq$$

oad of 1500 kN is carried on a foundation 2 m square at a shallow depth
a soil mass. Determine the vertical stress at a point 5 m below the centre
of the foundation (a) assuming the load is uniformly distributed over the
foundation, (b) assuming the load acts as a point load at the centre of the
foundation.

(a) Uniform pressure,

$$q = \frac{1500}{2^2} = 375 \, \text{kN/m}^2$$

The area must be considered as four quarters to enable Fig. 5.10 to be used.
In this case:

$$mz = nz = 1 \, \text{m}$$

Then, for $z = 5$ m,

$$m = n = 0 \cdot 2$$

From Fig. 5.10,

$$I_r = 0 \cdot 018$$

Hence,

$$\sigma_z = 4qI_r = 4 \times 375 \times 0 \cdot 018 = 27 \, \text{kN/m}^2$$

(b) From Table 5.1, $I_p = 0 \cdot 478$ since $r/z = 0$ vertically below a point load.
Hence,

$$\sigma_z = \frac{Q}{z^2} I_p = \frac{1500}{5^2} \times 0 \cdot 478 = 29 \, \text{kN/m}^2$$

The point load assumption should not be used if the depth to the point X
(Fig. 5.5a) is less than three times the larger dimension of the foundation.

Example 5.2

A rectangular foundation 6 m × 3 m carries a uniform pressure of
300 kN/m² near the surface of a soil mass. Determine the vertical stress at a
depth of 3 m below a point (A) on the centre line 1·5 m outside a long edge of
the foundation, (a) using influence factors, (b) using Newmark's influence
chart.

(a) Using the principle of superposition the problem is dealt with in the
manner shown in Fig. 5.12. For the two rectangles (1) carrying a *positive*
pressure of 300 kN/m², $m = 1 \cdot 00$ and $n = 1 \cdot 50$, therefore

$$I_r = 0 \cdot 193$$

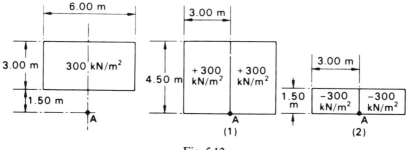

Fig. 5.12

For the two rectangles (2) carrying a *negative* pressure of $300\,\mathrm{kN/m^2}$, $m = 1 \cdot 00$ and $n = 0 \cdot 50$, therefore

$$I_r = 0 \cdot 120$$

Hence,

$$\sigma_z = (2 \times 300 \times 0 \cdot 193) - (2 \times 300 \times 0 \cdot 120)$$
$$= 44\,\mathrm{kN/m^2}$$

(b) Using Newmark's influence chart (Fig. 5.11) the scale line represents 3 m, fixing the scale to which the rectangular area must be drawn. The area is positioned such that the point A is at the centre of the chart. The number of influence areas covered by the rectangle is approximately 30 (i.e. $N = 30$), hence

$$\sigma_z = 0 \cdot 005 \times 30 \times 300$$
$$= 45\,\mathrm{kN/m^2}$$

Example 5.3

A strip footing 2 m wide carries a uniform pressure of $250\,\mathrm{kN/m^2}$ on the surface of a deposit of sand. The water table is at the surface. The saturated unit weight of the sand is $20\,\mathrm{kN/m^3}$ and $K_0 = 0 \cdot 40$. Determine the effective vertical and horizontal stresses at a point 3 m below the centre of the footing before and after the application of the pressure.

Before loading:

$$\sigma_z' = 3\gamma' = 3 \times 10 \cdot 2 = 30 \cdot 6\,\mathrm{kN/m^2}$$
$$\sigma_x' = K_0\sigma_z' = 0 \cdot 40 \times 30 \cdot 6 = 12 \cdot 2\,\mathrm{kN/m^2}$$

After loading: Referring to Fig. 5.7a, for a point 3 m below the centre of the footing,

$$\alpha = 2\tan^{-1}(\tfrac{1}{3}) = 36° \, 52' = 0 \cdot 643 \text{ radians}$$

$$\sin \alpha = 0\cdot600$$

$$\beta = -\alpha/2 \qquad \therefore \cos(\alpha + 2\beta) = 1$$

The increases in total stress due to the applied pressure are:

$$\Delta\sigma_z = \frac{q}{\pi}(\alpha + \sin \alpha) = \frac{250}{\pi}(0\cdot643 + 0\cdot600) = 99\cdot0\,\text{kN/m}^2$$

$$\Delta\sigma_x = \frac{q}{\pi}(\alpha - \sin \alpha) = \frac{250}{\pi}(0\cdot643 - 0\cdot600) = 3\cdot4\,\text{kN/m}^2$$

Hence,

$$\sigma_z' = 30\cdot6 + 99\cdot0 = 129\cdot6\,\text{kN/m}^2$$

$$\sigma_x' = 12\cdot2 + 3\cdot4 = 15\cdot6\,\text{kN/m}^2$$

5.3   Displacements from Elastic Theory

The vertical displacement (s_i) under an area carrying a uniform pressure q on the surface of a semi-infinite, homogeneous, isotropic mass, with a linear stress-strain relationship, can be expressed as

$$s_i = \frac{qB}{E}(1 - v^2)I_s \qquad\qquad (5.30)$$

where I_s is an influence factor depending on the shape of the loaded area. In the case of a rectangular area, B is the lesser dimension (the greater dimension being L) and in the case of a circular area, B is the diameter. The loaded area is assumed to be flexible. Values of influence factors are given in Table 5.2 for displacements under the centre and a corner (the edge in the case of a circle) of the area and for the average displacement under the area as a whole. According to Equation 5.30, vertical displacement increases in direct proportion to both the pressure and the width of the loaded area. The distribution of vertical displacement is of the form shown in Fig. 5.13a,

Table 5.2   Influence Factors for Vertical Displacement under Flexible Area Carrying Uniform Pressure

	I_s		
Shape of area	Centre	Corner	Average
Square	1·12	0·56	0·95
Rectangle $L/B = 2$	1·52	0·76	1·30
Rectangle $L/B = 5$	2·10	1·05	1·83
Circle	1·00	0·64	0·85

Fig. 5.13 Distributions of vertical displacement: (a) clay, (b) sand.

extending beyond the edges of the area. The contact pressure between the loaded area and the supporting mass is uniform.

In the case of an extensive, homogeneous deposit of saturated clay, it is a reasonable approximation to assume that E is constant throughout the deposit and the distribution of Fig. 5.13a applies. In the case of sands, however, the value of E varies with confining pressure and, therefore, will increase with depth and vary across the width of the loaded area, being greater under the centre of the area than at the edges. As a result, the distribution of vertical displacement will be of the form shown in Fig. 5.13b: the contact pressure will again be uniform if the area is flexible. Due to the variation of E, and to heterogeneity, elastic theory is little used in practice in the case of sands.

If the loaded area is *rigid* the vertical displacement will be uniform across the width of the area and its magnitude will be only slightly less than the *average* displacement under a corresponding flexible area. For example the value of I_s for a rigid circular area is $\pi/4$, this value being used in the calculation of E from the results of in-situ plate loading tests. The contact pressure under a rigid area is not uniform: for a circular area the forms of the distributions of contact pressure on clay and sand respectively are shown in Fig. 5.14a and Fig. 5.14b.

In most cases in practice, the soil deposit will be of limited thickness and will be underlain by a hard stratum. For cases in which Poisson's ratio is equal to $0\cdot5$, a solution for the *average* vertical displacement under a flexible area carrying a uniform pressure q was presented by Janbu, Bjerrum and Kjaernsli [5.7]. The solution covers circular, rectangular and strip areas at any depth below the surface. The vertical displacement is given by

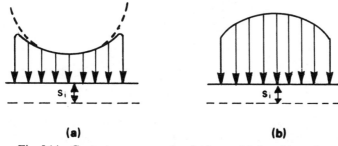

Fig. 5.14 Contact pressure under rigid area: (a) clay, (b) sand.

$$s_i = \mu_0 \mu_1 \frac{qB}{E} \tag{5.31}$$

values of the coefficients μ_0 and μ_1 being obtained from Fig. 5.15. The principle of superposition can be used in cases of a number of soil layers each having a different value of E (see Example 5.4).

The above solutions for vertical displacement are used mainly to estimate the immediate settlement of foundations on saturated clays: such settlement occurs under undrained conditions, the appropriate value of Poisson's ratio being 0·5. The value of the undrained modulus E_u is therefore required and the main difficulty in predicting immediate settlement is in the determination of this parameter. The value of E_u can be

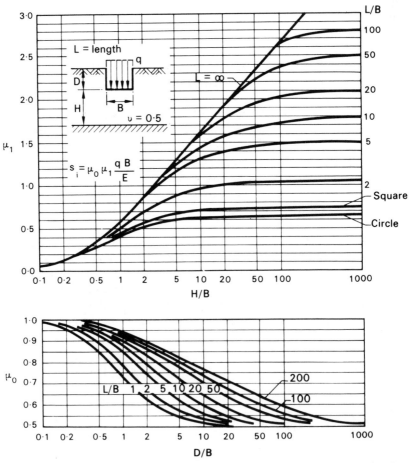

Fig. 5.15   Coefficients for vertical displacement. (Reproduced from N. Janbu, L. Bjerrum and B. Kjaernsli (1956) *Norwegian Geotechnical Institute Publication No. 16*, by permission.)

determined by means of undrained triaxial tests; however, such values are very sensitive to sample disturbance and in the case of unconsolidated-undrained tests would be too low as a result. If the specimen is initially reconsolidated then more realistic values of E_u are obtained. Consolidation may be either isotropic under $\frac{1}{2}$ to $\frac{2}{3}$ of the in-situ effective overburden pressure, or under K_0 conditions to simulate the actual in-situ effective stresses. If possible, however, the value of E_u should be determined from the results of in-situ load tests. It should be recognized, however, that the value obtained from load tests is sensitive to the time interval between excavation and testing, because there will be a gradual change from the undrained condition with time: the greater the time interval between excavation and testing the lower the value of E_u. The value of E_u can be obtained directly if settlement observations are taken during the initial loading of full-scale foundations. For particular clays, correlations can be obtained between E_u and the undrained shear strength parameter c_u.

It has been demonstrated that for certain soils, such as normally consolidated clays, there is a significant departure from linear stress-strain behaviour within the range of working stress, i.e. local yielding will occur within this range, and the immediate settlement will be underestimated. A method of correction for local yield has been given by D'Appolonia, Poulos and Ladd [5.1].

In principle the vertical displacement under fully drained conditions could be estimated using elastic theory if the value of the modulus for this condition (E') and the value of Poisson's ratio for the soil skeleton (ν') could be determined.

Example 5.4

A foundation $4\,\mathrm{m} \times 2\,\mathrm{m}$, carrying a uniform pressure of $150\,\mathrm{kN/m^2}$, is located at a depth of 1 m in a layer of clay 5 m thick for which the value of E_u is $40\,\mathrm{MN/m^2}$. The layer is underlain by a second clay layer 8 m thick for which the value of E_u is $75\,\mathrm{MN/m^2}$. A hard stratum lies below the second layer. Determine the average immediate settlement under the foundation.

Now, $D/B = 0.5$ and $L/B = 2$, therefore from Fig. 5.15, $\mu_0 = 0.90$.

(1) Considering the upper clay layer, with $E_u = 40\,\mathrm{MN/m^2}$:

$$H/B = 4/2 = 2 \quad \text{and} \quad L/B = 2$$
$$\therefore \quad \mu_1 = 0.70$$

Hence, from Equation 5.31,

$$s_{i_1} = 0.90 \times 0.70 \times \frac{150 \times 2}{40} = 4.7\,\mathrm{mm}$$

(2) Considering the two layers combined, with $E_u = 75\,\text{MN/m}^2$:

$$H/B = 12/2 = 6 \quad \text{and} \quad L/B = 2$$

$$\therefore \quad \mu_1 = 0.90$$

$$s_{i_2} = 0.90 \times 0.90 \times \frac{150 \times 2}{75} = 3.2\,\text{mm}$$

(3) Considering the upper layer, with $E_u = 75\,\text{MN/m}^2$:

$$H/B = 2 \quad \text{and} \quad L/B = 2$$

$$\therefore \quad \mu_1 = 0.70$$

$$s_{i_3} = 0.90 \times 0.70 \times \frac{150 \times 2}{75} = 2.5\,\text{mm}$$

Hence, using the principle of superposition, the settlement of the foundation is given by:

$$s_i = s_{i_1} + s_{i_2} - s_{i_3}$$
$$= 4.7 + 3.2 - 2.5 = 5.4\,\text{mm}$$

Problems

5.1 Calculate the vertical stress in a soil mass at a depth of 5 m vertically below a point load of 5000 kN acting near the surface. Plot the variation of vertical stress with radial distance (up to 10 m) at a depth of 5 m.

5.2 Three point loads, 10 000 kN, 7500 kN and 9000 kN, act in line 5 m apart near the surface of a soil mass. Calculate the vertical stress at a depth of 4 m vertically below the centre (7500 kN) load.

5.3 Determine the vertical stress at a depth of 3 m below the centre of a shallow foundation 2 m × 2 m carrying a uniform pressure of 250 kN/m². Plot the variation of vertical stress with depth (up to 10 m) below the centre of the foundation.

5.4 A shallow foundation 25 m × 18 m carries a uniform pressure of 175 kN/m². Determine the vertical stress at a point 12 m below the mid-point of one of the longer sides (a) using influence factors, (b) by means of Newmark's chart.

5.5 A line load of 150 kN/m acts 2 m behind the back surface of an earth-retaining structure 4 m high. Calculate the total thrust, and plot the distribution of pressure, on the structure due to the line load.

5.6 A foundation 4 m × 2 m carries a uniform pressure of 200 kN/m² at a depth of 1 m in a layer of saturated clay 11 m deep and underlain by a hard stratum. If E_u for the clay is 45 MN/m², determine the average value of immediate settlement under the foundation.

References

5.1 D'Appolonia, D. J., Poulos, H. G. and Ladd, C. C. (1971): 'Initial Settlement of Structures on Clay', *Journal ASCE*, Vol. 97, No. SM10.

5.2 Desai, C. S. and Abel, J. F. (1972): *Introduction to the Finite Element Method*, Van Nostrand Reinhold, New York.

5.3 Drucker, D. C. (1950): 'Some Implications of Work Hardening and Ideal Plasticity', *Q. Appl. Math.*, Vol. 7.

5.4 Fadum, R. E. (1948): 'Influence Values for Estimating Stresses in Elastic Foundations', *Proceedings 2nd International Conference SMFE, Rotterdam*, Vol. 3.

5.5 Gibson, R. E. and Anderson, W. F. (1961): 'In-Situ Measurement of Soil Properties with the Pressuremeter', *Civil Engineering and Public Works Review*, Vol. 56.

5.6 Hughes, J. M. O., Wroth, C. P. and Windle, D. (1977): 'Pressuremeter Tests in Sands', *Geotechnique*, Vol. 27, No. 4.

5.7 Janbu, N., Bjerrum, L. and Kjaernsli, B. (1956): *Norwegian Geotechnical Institute Publication* No. 16.

5.8 Kondner, R. L. (1963): 'Hyperbolic Stress-Strain Response: Cohesive Soils', *Journal ASCE*, Vol. 89, No. SM1.

5.9 McKinlay, D. G. and Anderson, W. F. (1975): 'Determination of the Modulus of Deformation of a Till Using a Pressuremeter', *Ground Engineering*, Vol. 8, No. 6.

5.10 Mair, R. J. and Wood, D. M. (1986): 'A Review of the Use of Pressuremeters for In-Situ Testing', *CIRIA Report*.

5.11 Marsland, A. and Randolph, M. F. (1977): 'Comparisons of the Results from Pressuremeter Tests and Large In-Situ Plate Tests in London Clay', *Geotechnique*, Vol. 27, No. 2.

5.12 Newmark, N. M. (1942): *Influence Charts for Computation of Stresses in Elastic Foundations*, University of Illinois Bulletin No. 338.

5.13 Palmer, A. C. (1972): 'Undrained Plane Strain Expansion of a Cylindrical Cavity in Clay: a Simple Interpretation of the Pressuremeter Test', *Geotechnique*, Vol. 22, No. 3.

5.14 Poulos, H. G. and Davis, E. H. (1974): *Elastic Solutions for Soil and Rock Mechanics*, John Wiley and Sons, New York.

5.15 Scott, R. F. (1963): *Principles of Soil Mechanics*, Addison-Wesley, Reading, Massachusetts.

5.16 Terzaghi, K. (1943): *Theoretical Soil Mechanics*, John Wiley and Sons, New York.

5.17 Timoshenko, S. and Goodier, J. N. (1970): *Theory of Elasticity* (3rd edition), McGraw-Hill, New York.

5.18 Windle, D. and Wroth, C. P. (1977): 'The Use of a Self-Boring Pressuremeter to Determine the Undrained Properties of Clay', *Ground Engineering*, Vol. 10, No. 6.

5.19 Wroth, C. P. (1984): 'The Interpretation of In-Situ Soil Tests', *Geotechnique*, Vol. 34, No. 4.

CHAPTER 6

Lateral Earth Pressure

6.1 Introduction

This chapter deals with the magnitude and distribution of lateral pressure between a soil mass and an adjoining retaining structure. Conditions of plane strain are assumed, i.e. strains in the longitudinal direction of the structure are assumed to be zero. The rigorous treatment of this type of problem, with both stresses and displacements being considered, would involve a knowledge of appropriate equations defining the stress-strain relationship for the soil and the solution of the equations of equilibrium and compatibility for the given boundary conditions. The rigorous analysis of earth pressure problems is rarely possible. However, it is the failure condition of the retained soil mass which is of primary interest and in this context, provided a consideration of displacements is not required, it is possible to use the concept of plastic collapse. Earth pressure problems can thus be considered as problems in plasticity.

It is assumed that the behaviour of the soil can be represented by the idealized stress-strain relationship shown in Fig. 6.1, in which the soil after yielding behaves as a perfectly plastic material with unrestricted plastic flow taking place at constant stress, i.e. strains after yielding are entirely plastic. The use of this relationship implies that yielding and shear failure both occur at the same state of stress. A soil mass is said to be in a state of plastic equilibrium if the shear stress at every point within the mass reaches the state of stress represented by point Y.

Plastic collapse occurs after the state of plastic equilibrium has been reached in *part* of a soil mass, resulting in the formation of an unstable

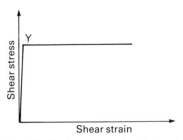

Fig. 6.1   Idealized stress-strain relationship.

mechanism: that part of the soil mass slips relative to the rest of the mass. The applied load system, including body forces, for this condition is referred to as the collapse load. Determination of the collapse load using plasticity theory is complex and would require that the equilibrium equations, the yield criterion and the flow rule were satisfied within the plastic zone: the compatibility condition would not be involved unless specific deformation conditions were imposed. However, plasticity theory also provides the means of avoiding complex analyses. The limit theorems of plasticity can be used to calculate lower and upper bounds to the true collapse load. In certain cases the theorems produce the same result, which would then be the exact value of the collapse load.

The limit theorems can be stated as follows.

Lower bound theorem. If a state of stress can be found which at no point exceeds the failure criterion for the soil and which is in equilibrium with a system of external loads (which includes the self-weight of the soil), then collapse cannot occur: the external load system thus constitutes a lower bound to the true collapse load.

Upper bound theorem. If a mechanism of plastic collapse is postulated and if, in an increment of displacement, the rate of work done by a system of external loads is equal to the rate of dissipation of energy by the internal stresses then collapse must occur: the external load system thus constitutes an upper bound to the true collapse load.

In the lower bound approach the conditions of equilibrium and yield are satisfied with no consideration of the mode of deformation: the Mohr-Coulomb failure criterion is also taken to be the yield criterion. In the upper bound approach a mechanism of plastic collapse is formed by choosing a slip surface and the work done by the external forces is equated to the internal dissipation of energy without consideration of equilibrium. The chosen collapse mechanism is not necessarily the true mechanism but it must be kinematically admissible, i.e. the motion of the sliding soil mass must be compatible with its continuity and with any boundary restrictions. It can be shown that for undrained conditions the slip surface, in section, should consist of a straight line or a circular arc (or a combination of the two): for drained conditions the slip surface should consist of a straight line or a logarithmic spiral (or a combination of the two). This satisfies the requirement that slip surfaces must intersect each other at an angle of $(90° + \phi)$.

6.2 Rankine's Theory of Earth Pressure

Rankine's theory (1857) considers the state of stress in a soil mass when the condition of plastic equilibrium has been reached, i.e. when shear failure is on the point of occurring throughout the mass. The theory satisfies the conditions of a lower bound plasticity solution. The Mohr circle represent-

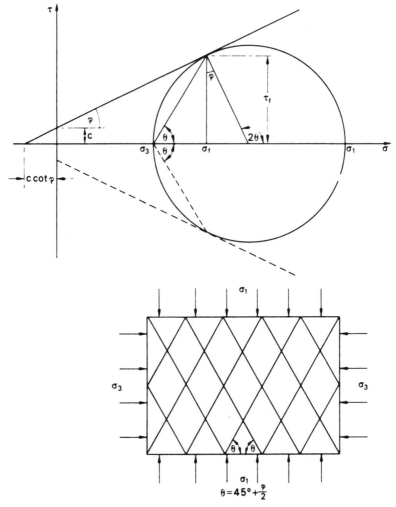

Fig. 6.2   State of plastic equilibrium

ing the state of stress at failure in a two-dimensional element is shown in
Fig. 6.2, the relevant shear strength parameters being denoted by c and ϕ.
Shear failure occurs along a plane at an angle of $(45° + \phi/2)$ to the major
principal plane. If the soil mass as a whole is stressed such that the principal
stresses at every point are in the same directions then, theoretically, there
will be a network of failure planes (known as a slip line field) equally
inclined to the principal planes, as shown in Fig. 6.2. It should be
appreciated that the state of plastic equilibrium can be developed only if
sufficient deformation of the soil mass can take place.

Consider now a semi-infinite mass of soil with a horizontal surface and

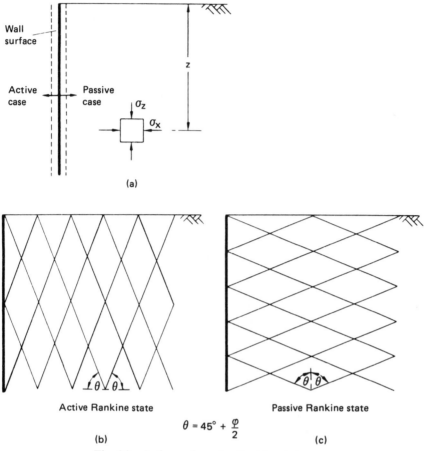

Fig. 6.3   Active and passive Rankine states.

having a vertical boundary formed by a *smooth* wall surface extending to semi-infinite depth, as represented in Fig. 6.3a. The soil is assumed to be homogeneous and isotropic. A soil element at any depth z is subjected to a vertical stress σ_z and a horizontal stress σ_x and, since there can be no lateral transfer of weight if the surface is horizontal, no shear stresses exist on horizontal and vertical planes. The vertical and horizontal stresses, therefore, are principal stresses.

If there is now a movement of the wall away from the soil, the value of σ_x decreases as the soil dilates or expands outwards, the decrease in σ_x being an unknown function of the lateral strain in the soil. If the expansion is large enough the value of σ_x decreases to a minimum value such that a state of plastic equilibrium develops. Since this state is developed by a decrease in the horizontal stress σ_x, this must be the minor principal stress (σ_3). The vertical stress σ_z is then the major principal stress (σ_1).

The stress σ_1 ($= \sigma_z$) is the overburden pressure at depth z and is a fixed value for any depth. The value of σ_3 ($= \sigma_x$) is determined when a Mohr circle through the point representing σ_1 touches the failure envelope for the soil. The relationship between σ_1 and σ_3 when the soil reaches a state of plastic equilibrium can be derived from this Mohr circle. Rankine's original derivation assumed a zero value of the shear strength parameter c but a general derivation with c greater than zero is given below.

Referring to Fig. 6.2,

$$\sin \phi = \frac{\frac{1}{2}(\sigma_1 - \sigma_3)}{\frac{1}{2}(\sigma_1 + \sigma_3 + 2c \cot \phi)}$$

$$\therefore \sigma_3(1 + \sin \phi) = \sigma_1(1 - \sin \phi) - 2c \cos \phi$$

$$\therefore \sigma_3 = \sigma_1 \left(\frac{1 - \sin \phi}{1 + \sin \phi} \right) - 2c \frac{\sqrt{(1 - \sin^2 \phi)}}{1 + \sin \phi}$$

$$\therefore \sigma_3 = \sigma_1 \left(\frac{1 - \sin \phi}{1 + \sin \phi} \right) - 2c \sqrt{\left(\frac{1 - \sin \phi}{1 + \sin \phi} \right)} \tag{6.1}$$

Alternatively, $\tan^2 [45° - (\phi/2)]$ can be substituted for $(1 - \sin \phi)/(1 + \sin \phi)$.

As stated, σ_1 is the overburden pressure at depth z, i.e.

$$\sigma_1 = \gamma z$$

The horizontal stress for the above condition is defined as the *active pressure* (p_A), being due directly to the self-weight of the soil. If

$$K_A = \frac{1 - \sin \phi}{1 + \sin \phi}$$

is defined as the active pressure coefficient then Equation 6.1 can be written as

$$p_A = K_A \gamma z - 2c \sqrt{K_A} \tag{6.2}$$

When the horizontal stress becomes equal to the active pressure the soil is said to be in the *active Rankine state*, there being two sets of failure planes each inclined at $(45° + \phi/2)$ to the *horizontal* (the direction of the major principal plane) as shown in Fig. 6.3b.

In the above derivation a movement of the wall away from the soil was considered. If, on the other hand, the wall is moved against the soil mass, there will be lateral compression of the soil and the value of σ_x will increase until a state of plastic equilibrium is reached. For this condition σ_x becomes a maximum value and is the major principal stress σ_1. The stress σ_z, equal to the overburden pressure, is then the minor principal stress, i.e.

$$\sigma_3 = \gamma z$$

The maximum value σ_1 is reached when the Mohr circle through the point representing the fixed value σ_3 touches the failure envelope for the soil. In this case the horizontal stress is defined as the *passive pressure* (p_P) representing the maximum inherent resistance of the soil to lateral compression. Rearranging Equation 6.1:

$$\sigma_1 = \sigma_3 \left(\frac{1 + \sin \phi}{1 - \sin \phi} \right) + 2c \sqrt{\left(\frac{1 + \sin \phi}{1 - \sin \phi} \right)} \tag{6.3}$$

If

$$K_P = \frac{1 + \sin \phi}{1 - \sin \phi}$$

is defined as the passive pressure coefficient, Equation 6.3 can be written as:

$$p_P = K_P \gamma z + 2c \sqrt{K_P} \tag{6.4}$$

When the horizontal stress becomes equal to the passive pressure the soil is said to be in the *passive Rankine state*, there being two sets of failure planes each inclined at $(45° + \phi/2)$ to the *vertical* (the direction of the major principal plane) as shown in Fig. 6.3c.

Inspection of Equations 6.2 and 6.4 shows that the active and passive pressures increase linearly with depth as represented in Fig. 6.4. When $c = 0$, triangular distributions are obtained in each case.

When c is greater than zero, the value of p_A is zero at a particular depth z_0. From Equation 6.2, with $p_A = 0$,

$$z_0 = \frac{2c}{\gamma \sqrt{K_A}} \tag{6.5}$$

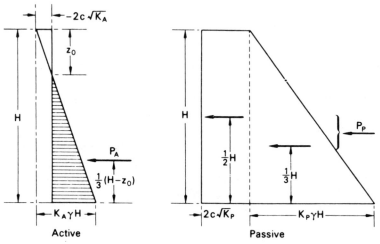

Active   Passive

Fig. 6.4   Active and passive pressure distributions.

This means that in the active case the soil is in a state of tension between the surface and depth z_0. In practice, however, this tension cannot be relied upon to act on the wall, since cracks are likely to develop within the tension zone and the part of the pressure distribution diagram above depth z_0 should be neglected.

The force per unit length of wall due to the active pressure distribution is referred to as the *total active thrust* (P_A). For a vertical wall surface of height H:

$$P_A = \int_{z_0}^{H} p_A dz$$

$$= \tfrac{1}{2}K_A\gamma(H^2 - z_0^2) - 2c(\sqrt{K_A})(H - z_0) \tag{6.6}$$

$$= \tfrac{1}{2}K_A\gamma(H - z_0)^2 \tag{6.6a}$$

The force P_A acts at a distance of $\tfrac{1}{3}(H - z_0)$ above the bottom of the wall surface.

The force due to the passive pressure distribution is referred to as the *total passive resistance* (P_P). For a vertical wall surface of height H,

$$P_P = \int_{0}^{H} p_P dz$$

$$= \tfrac{1}{2}K_P\gamma H^2 + 2c(\sqrt{K_P})H \tag{6.7}$$

The two components of P_P act at distances of $H/3$ and $H/2$ respectively above the bottom of the wall surface.

If a uniformly distributed surcharge pressure of q per unit area acts over the entire surface of the soil mass, the vertical stress σ_z at any depth is increased to $(\gamma z + q)$, resulting in an *additional* pressure of $K_A q$ in the active case or $K_P q$ in the passive case, both distributions being constant with depth as shown in Fig. 6.5. The corresponding forces on a vertical wall surface of height H are $K_A q H$ and $K_P q H$ respectively, each acting at mid-height. In the case of two layers of soil having different shear strengths the weight of the upper layer can be considered as a surcharge acting on the lower layer. There will be a discontinuity in the pressure diagram at the boundary between the two layers due to the different values of shear strength parameters.

If the soil below the water table is in the fully drained condition, the active and passive pressures must be expressed in terms of the effective weight of the soil and the effective stress parameters c' and ϕ'. For example, if the water table is at the surface and if no seepage is taking place, the active pressure at depth z is given by

$$p_A = K_A\gamma'z - 2c'\sqrt{K_A}$$

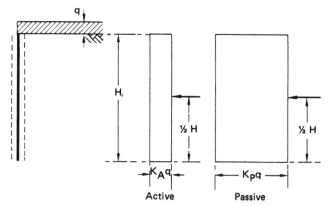

Fig. 6.5   Additional pressures due to surcharge.

where

$$K_A = \frac{1 - \sin \phi'}{1 + \sin \phi'}$$

Corresponding equations apply in the passive case. The hydrostatic pressure $\gamma_w z$ due to the water in the soil pores must be considered *in addition* to the active or passive pressure.

For the undrained condition in a fully saturated soil the active and passive pressures are calculated using the parameters c_u and ϕ_u with the total unit weight γ_{sat} (i.e. the water in the soil pores is not considered separately).

Example 6.1

(a) Calculate the total active thrust on a vertical wall 5 m high retaining a sand of unit weight of 17 kN/m³ for which $\phi' = 35°$: the surface of the sand is horizontal and the water table is below the bottom of the wall.
(b) Determine the thrust on the wall if the water table rises to a level 2 m below the surface of the sand. The saturated unit weight of the sand is 20 kN/m³.

(a) $K_A = \dfrac{1 - \sin 35°}{1 + \sin 35°} = 0.27$

$P_A = \tfrac{1}{2}K_A \gamma H^2 = \tfrac{1}{2} \times 0.27 \times 17 \times 5^2 = 57.5 \, \text{kN/m}$

(b) The pressure distribution on the wall is now as shown in Fig. 6.6, including hydrostatic pressure on the lower 3 m of the wall. The components of the thrust are:

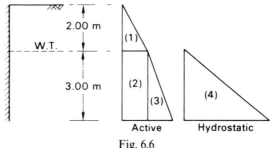

Fig. 6.6

$(1)\ \frac{1}{2} \times 0\cdot27 \times 17 \times 2^2$ $=\ 9\cdot2\,\mathrm{kN/m}$

$(2)\ 0\cdot27 \times 17 \times 2 \times 3$ $=27\cdot6$

$(3)\ \frac{1}{2} \times 0\cdot27 \times (20 - 9\cdot8) \times 3^2 = 12\cdot4$

$(4)\ \frac{1}{2} \times 9\cdot8 \times 3^2$ $=44\cdot1$

Total thrust $=\overline{93\cdot3\,\mathrm{kN/m}}$

Example 6.2

The soil conditions adjacent to a sheet pile wall are given in Fig. 6.7, a surcharge pressure of $50\,\mathrm{kN/m^2}$ being carried on the surface behind the wall. For soil 1, a sand above the water table, $c' = 0$, $\phi' = 38°$ and $\gamma = 18\,\mathrm{kN/m^3}$. For soil 2, a saturated clay, $c' = 10\,\mathrm{kN/m^2}$, $\phi' = 28°$ and $\gamma_{sat} = 20\,\mathrm{kN/m^3}$. Plot the distributions of active pressure behind the wall and passive pressure in front of the wall.

For soil 1,

$$K_A = \frac{1 - \sin 38°}{1 + \sin 38°} = 0\cdot24 \qquad K_P = \frac{1}{0\cdot24} = 4\cdot17$$

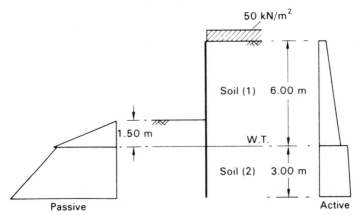

Fig. 6.7

Table 6.1

Soil	Depth (m)	Pressure (kN/m²)	

Active pressure

1	0	0.24×50	$= 12.0$
1	6	$(0.24 \times 50) + (0.24 \times 18 \times 6) = 12.0 + 25.9$	$= 37.9$
2	6	$0.36[50 + (18 \times 6)] - (2 \times 10 \times \sqrt{0.36}) = 56.9 - 12.0$	$= 44.9$
2	9	$0.36[50 + (18 \times 6)] - (2 \times 10 \times \sqrt{0.36}) + (0.36 \times 10.2 \times 3) = 56.9 - 12.0 + 11.0$	$= 55.9$

Passive pressure

1	0	0	
1	1.5	$4.17 \times 18 \times 1.5$	$= 112.6$
2	1.5	$(2.78 \times 18 \times 1.5) + (2 \times 10 \times \sqrt{2.78}) = 75.1 + 33.3$	$= 108.4$
2	4.5	$(2.78 \times 18 \times 1.5) + (2 \times 10 \times \sqrt{2.78}) + (2.78 \times 10.2 \times 3) = 75.1 + 33.3 + 85.1$	$= 193.5$

For soil 2,

$$K_A = \frac{1 - \sin 28°}{1 + \sin 28°} = 0.36 \qquad K_P = \frac{1}{0.36} = 2.78$$

The pressures in soil 1 are calculated using $K_A = 0.24$, $K_p = 4.17$ and $\gamma = 18 \, \text{kN/m}^3$. Soil 1 is then considered as a surcharge of $(18 \times 6) \, \text{kN/m}^2$ on soil 2, in addition to the surface surcharge. The pressures in soil 2 are calculated using $K_A = 0.36$, $K_P = 2.78$ and $\gamma' = (20 - 9.8) = 10.2 \, \text{kN/m}^3$. (See Table 6.1.) The active and passive pressure distributions are shown in Fig. 6.7. In addition there is equal hydrostatic pressure on each side of the wall below the water table.

Sloping Soil Surface

The Rankine theory will now be applied to cases in which the soil surface slopes at a constant angle β to the horizontal. It is assumed that the active and passive pressures act in a direction parallel to the sloping surface. Consider a rhombic element of soil, with sides vertical and at angle β to the horizontal, at depth z in a semi-infinite mass. The vertical stress and the active or passive pressure are each inclined at β to the appropriate sides of the element, as shown in Fig. 6.8a. Since these stresses are not normal to their respective planes they are not principal stresses.

In the active case the vertical stress at depth z on a plane inclined at angle β to the horizontal is given by

$$\sigma_z = \gamma z \cos \beta$$

and is represented by the distance OA on the stress diagram, Fig. 6.8b. If lateral expansion of the soil is sufficient to induce the state of plastic equilibrium, the Mohr circle representing the state of stress in the element must pass through point A (such that the greater part of the circle lies on the side of A towards the origin) and touch the failure envelope for the soil. The active pressure p_A is then represented by OB (numerically equal to OB') on the diagram. When $c = 0$ the relationship between p_A and σ_z, giving the active pressure coefficient, can be derived from the diagram.

$$K_A = \frac{p_A}{\sigma_z} = \frac{OB}{OA} = \frac{OB'}{OA} = \frac{OD - AD}{OD + AD}$$

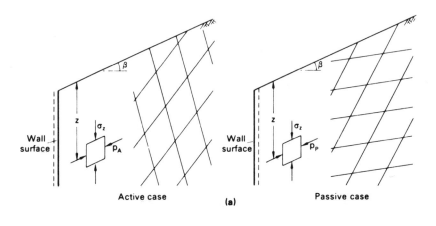

Active case   (a)   Passive case

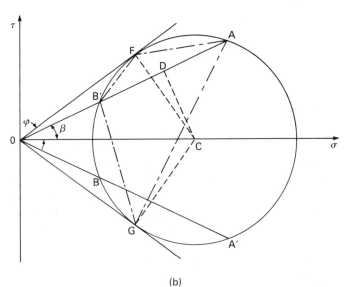

(b)

Fig. 6.8   Active and passive states for sloping surface.

Now,

$$OD = OC \cos \beta$$
$$AD = \sqrt{(OC^2 \sin^2 \phi - OC^2 \sin^2 \beta)}$$

therefore

$$K_A = \frac{\cos \beta - \sqrt{(\cos^2 \beta - \cos^2 \phi)}}{\cos \beta + \sqrt{(\cos^2 \beta - \cos^2 \phi)}} \qquad (6.8)$$

Thus the active pressure, acting parallel to the slope, is given by

$$p_A = K_A \gamma z \cos \beta \qquad (6.9)$$

and the total active thrust on a vertical wall surface of height H is

$$P_A = \tfrac{1}{2} K_A \gamma H^2 \cos \beta \qquad (6.10)$$

In the passive case, the vertical stress σ_z is represented by the distance OB' in Fig. 6.8b. The Mohr circle representing the state of stress in the element, after a state of plastic equilibrium has been induced by lateral compression of the soil, must pass through B' (such that the greater part of the circle lies on the side of B' away from the origin) and touch the failure envelope. The passive pressure p_P is then represented by OA' (numerically equal to OA) and when $c = 0$ the passive pressure coefficient (equal to p_P/σ_z) is given by

$$K_P = \frac{\cos \beta + \sqrt{(\cos^2 \beta - \cos^2 \phi)}}{\cos \beta - \sqrt{(\cos^2 \beta - \cos^2 \phi)}} \qquad (6.11)$$

Then the passive pressure, acting parallel to the slope, is given by

$$p_P = K_P \gamma z \cos \beta \qquad (6.12)$$

and the total passive resistance on a vertical wall surface of height H is

$$P_P = \tfrac{1}{2} K_P \gamma H^2 \cos \beta \qquad (6.13)$$

The active and passive pressures can, of course, be obtained graphically from Fig. 6.8b. The above formulae apply only when the shear strength parameter c is zero: when c is greater than zero the graphical procedure should be used.

The directions of the two sets of failure planes can be obtained from Fig. 6.8b. In the active case the coordinates of point A represent the state of stress on a plane inclined at angle β to the horizontal, therefore point B' is the origin of planes. (A line drawn from the origin of planes intersects the circumference of the circle at a point whose coordinates represent the state of stress on a plane parallel to that line.) The state of stress on a vertical plane is represented by the coordinates of point B. Then the failure planes, which are shown in Fig. 6.8a, are parallel to B'F and B'G (F and G lying on the failure envelope). In the passive case the coordinates of point B' represent the state of stress on a plane inclined at angle β to the horizontal, therefore point A is the origin of planes: the state of stress on a vertical plane is represented by the coordinates of point A'. Then the failure planes in the passive case are parallel to AF and AG.

Example 6.3

A vertical wall 6 m high, above the water table, retains a 20° soil slope, the retained soil having a unit weight of 18 kN/m³: the appropriate shear strength parameters are $c' = 0$ and $\phi' = 40°$. Determine the total active thrust on the wall and the directions of the two sets of failure planes relative to the horizontal.

In this case the total active thrust can be obtained by calculation. Using Equation 6.8,

$$K_A = \frac{\cos 20° - \sqrt{(\cos^2 20° - \cos^2 40°)}}{\cos 20° + \sqrt{(\cos^2 20° - \cos^2 40°)}} = 0.265$$

Then

$$P_A = \tfrac{1}{2}K_A\gamma H^2 \cos \beta$$
$$= \tfrac{1}{2} \times 0.265 \times 18 \times 6^2 \times 0.940 = 81 \text{ kN/m}$$

The result can also be determined using a stress diagram (Fig. 6.9). Draw the failure envelope on the τ/σ plot and a straight line through the origin at 20° to the horizontal. At a depth of 6 m,

$$\sigma_z = \gamma z \cos \beta = 18 \times 6 \times 0.940 = 102 \text{ kN/m}^2$$

and this stress is set off to scale (distance OA) along the 20° line. The Mohr circle is then drawn as in Fig. 6.9 and the active pressure (distance OB or OB') is scaled from the diagram, i.e.

$$p_A = 27 \text{ kN/m}^2$$

Then

$$P_A = \tfrac{1}{2}p_A H = \tfrac{1}{2} \times 27 \times 6 = 81 \text{ kN/m}$$

The failure planes are parallel to B'F and B'G in Fig. 6.9. The directions of these lines are measured as 59° and 71°, respectively, to the horizontal (adding up to $90° + \phi$).

Earth Pressure At-Rest

It has been shown that active pressure is associated with lateral expansion of the soil and is a minimum value; passive pressure is associated with lateral compression of the soil and is a maximum value. If the lateral strain

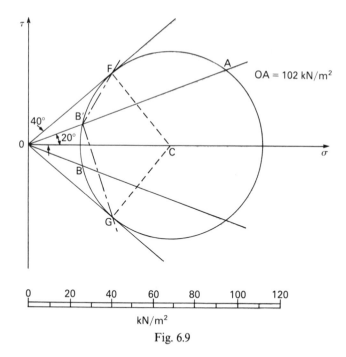

Fig. 6.9

in the soil is zero the corresponding lateral pressure is called the *earth pressure at-rest* and is usually expressed in terms of effective stress by the equation

$$p_0 = K_0 \gamma' z \tag{6.14}$$

where K_0 is defined as the coefficient of earth pressure at-rest, in terms of effective stress.

Since the at-rest condition does not involve failure of the soil (it represents a state of 'elastic' equilibrium) the Mohr circle representing the vertical and horizontal stresses does not touch the failure envelope and the horizontal stress cannot be evaluated. The value of K_0, however, can be determined experimentally by means of a triaxial test in which the axial stress and the all-round pressure are increased simultaneously such that the lateral strain in the specimen is maintained at zero. For soft clays, methods of measuring lateral pressure in situ have been developed by Bjerrum and Anderson [6.2] and, using the pressuremeter, by Wroth and Hughes [6.25].

Generally, for any condition intermediate to the active and passive states the value of the lateral stress is unknown. The range of possible conditions can only be determined experimentally and Fig. 6.10 shows the form of the relationship between strain and the lateral pressure coefficient. The strain required to mobilize the passive pressure is considerably greater

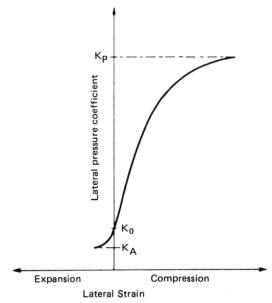

Fig. 6.10   Relationship between lateral strain and lateral pressure coefficient.

than that required to mobilize the active pressure. The effective stress paths for the development of the active and passive states from the K_0 condition (for $K_0 < 1$) are AB and AC respectively in Fig. 6.11.

For normally consolidated soils the value of K_0 can be related approximately to the effective stress parameter ϕ' by the following formula proposed by Jaky:

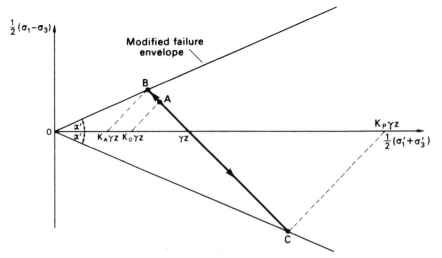

Fig. 6.11   Effective stress paths for active and passive states.

$$K_0 = 1 - \sin \phi' \tag{6.15a}$$

For overconsolidated soils the value of K_0 depends on the stress history and can be greater than unity, a proportion of the at-rest pressure developed during initial consolidation being retained in the soil when the effective vertical stress is subsequently reduced. Mayne and Kulhawy [6.10] proposed the following correlation for overconsolidated soils during expansion (but not recompression):

$$K_0 = (1 - \sin \phi')(OCR)^{\sin\phi'} \tag{6.15b}$$

OCR being the overconsolidation ratio. A typical relationship between K_0 and overconsolidation ratio, determined in the triaxial apparatus, is shown in Fig. 6.12 and some typical values of K_0 for different soils are given in Table 6.2.

Application of Rankine Theory to Retaining Walls

In Rankine's theory the state of stress in a semi-infinite mass of soil is considered, the entire mass being subjected to lateral expansion or compression. The movement of a retaining wall of finite depth, however, cannot produce the active or passive state in the soil mass as a whole. The active state, for example, can be developed only within a wedge of soil

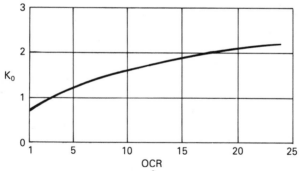

Fig. 6.12   Typical relationship between K_0 and overconsolidation ratio.

Table 6.2   Coefficient of Earth Pressure At-Rest

Soil	K_0
Dense sand	0·35
Loose sand	0·6
Normally consolidated clays (Norway)	0·5–0·6
Clay, OCR = 3·5 (London)	1·0
Clay, OCR = 20 (London)	2·8

between the wall and a failure plane passing through the lower end of the wall and at an angle of $(45° + \phi/2)$ to the horizontal, as shown in Fig. 6.13a: the remainder of the soil mass is in a state of 'elastic' equilibrium. A certain minimum value of lateral strain is necessary to develop the active state within the above wedge. A uniform strain within the wedge can be produced by a rotational movement (A'B) of the wall, away from the soil, about its lower end and a deformation of this type, of sufficient magnitude, constitutes the minimum deformation condition for the development of the active state. If the deformation of the wall does not satisfy the minimum deformation condition the soil adjacent to the wall remains in a state of 'elastic' equilibrium and the lateral pressure will be between the at-rest and active values. If the wall deforms by rotation about its upper end the condition for the development of the active state is not satisfied since adequate strain in the soil near the surface cannot take place. In the passive case the minimum deformation condition is a rotational movement of the wall, about its lower end, into the soil. If this movement is of sufficient magnitude the passive state is developed within a wedge of soil between the wall and a failure plane at $(45° + \phi/2)$ to the vertical, as shown in Fig. 6.13b.

In practice the magnitude of wall deformation is unknown. The soil behind a wall is normally backfilled after the wall has been constructed and some degree of outward wall deformation will take place during backfilling. Since increased deformation results in a reduction of lateral pressure towards the active value, a retaining wall need be designed only to withstand the active pressure, provided free deformation is possible.

In the Rankine theory it is assumed that the wall surface is smooth whereas in practice considerable friction may be developed between the wall and the adjacent soil, depending on the wall material. In principle, the theory results either in an overestimation of active pressure and an underestimation of passive pressure (i.e. lower bounds to the respective 'collapse loads') or in exact values of active and passive pressure.

Shear strength parameters in terms of either total stress or effective stress

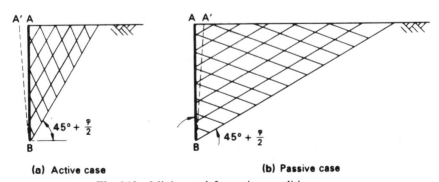

(a) Active case   (b) Passive case

Fig. 6.13   Minimum deformation conditions.

may be used in the theory, according to the drainage conditions of the problem. In practice, most backfills are of granular material, the shear strength parameter c' being zero.

6.3 Coulomb's Theory of Earth Pressure

Coulomb's theory (1776) involves consideration of the stability, as a whole, of the wedge of soil between a retaining wall and a trial failure plane. The *force* between the wedge and the wall surface is determined by considering the equilibrium of forces acting on the wedge when it is on the point of sliding either up or down the failure plane, i.e. when the wedge is in a condition of *limiting equilibrium*. Friction between the wall and the adjacent soil is taken into acount. The angle of friction between the soil and the wall material, denoted by δ, can be determined in the laboratory by means of a direct shear test. At any point on the wall surface a shearing resistance per unit area of $p_n \tan \delta$ will be developed, where p_n is the normal pressure on the wall at that point. A constant component of shearing resistance or 'wall adhesion', c_w, can also be assumed if appropriate in the case of clays. A minimum deformation condition for the wall, either away from or into the soil, is assumed, as in the Rankine theory, such that a state of plastic equilibrium develops throughout the wedge of soil between the wall and a failure surface between the heel of the wall and the soil surface. Due to wall friction the shape of the failure surface is curved near the bottom of the wall in both the active and passive cases, as indicated in Fig. 6.14, but in the Coulomb theory the failure surfaces are assumed to be plane in each case. In the active case the curvature is slight and the error involved in assuming a plane surface is relatively small. This is also true in the passive case for values of δ less than $\phi/3$, but for higher values of δ the error becomes relatively large.

The Coulomb theory is now interpreted as an upper bound plasticity solution, collapse of the soil mass above the chosen failure plane occurring as the wall moves away from or into the soil. Thus, in general, the theory

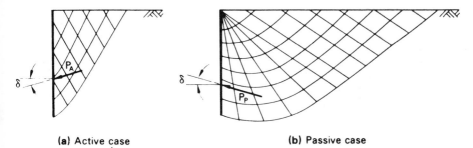

(a) Active case   (b) Passive case

Fig. 6.14   Curvature due to wall friction.

underestimates the total active thrust and overestimates the total passive resistance (i.e. upper bounds to the true collapse loads). When $\delta = 0$, the Coulomb theory gives results which are identical to those of the Rankine theory for the case of a vertical wall and a horizontal soil surface, i.e. the solution for this case is exact because the upper and lower bound results coincide.

Active Case

Fig. 6.15a shows the forces acting on the soil wedge between a wall surface AB, inclined at angle α to the horizontal, and a trial failure plane BC, at angle θ to the horizontal. The soil surface AC is inclined at angle β to the horizontal. The shear strength parameter c will be taken as zero, as will be the case for most backfills. For the failure condition the soil wedge is in equilibrium under its own weight (W), the *reaction* to the force (P) between the soil and the wall, and the reaction (R) on the failure plane. Because the soil wedge tends to move down the plane BC at failure, the reaction P acts at angle δ below the normal to the wall. (If the wall were to settle more than the backfill the reaction P would act at angle δ above the normal.) At failure, when the shear strength of the soil has been fully mobilised, the direction of R is at angle ϕ below the normal to the failure plane (R being the resultant of the normal and shear forces on the failure plane). The directions of all three forces, and the magnitude of W, are known, therefore the triangle of forces (Fig. 6.15b) can be drawn and the magnitude of P determined for the trial in question.

A number of trial failure planes would have to be selected to obtain the maximum value of P, which would be the total active thrust on the wall. However, using the sine rule, P can be expressed in terms of W and the

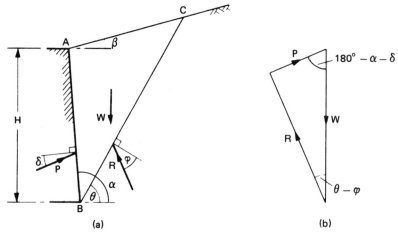

(a)   (b)

Fig. 6.15   Coulomb theory: active case with $c = 0$.

angles in the triangle of forces. Then the maximum value of P, corresponding to a particular value of θ, is given by $\partial P/\partial\theta = 0$. This leads to the following solution for P_A:

$$P_A = \tfrac{1}{2}K_A\gamma H^2 \tag{6.16}$$

where

$$K_A = \left(\frac{\sin(\alpha - \phi)/\sin\alpha}{\sqrt{[\sin(\alpha + \delta)]} + \sqrt{\left[\dfrac{\sin(\phi + \delta)\sin(\phi - \beta)}{\sin(\alpha - \beta)} \right]}} \right)^2 \tag{6.17}$$

The point of application of the total active thrust is not given by the Coulomb theory but is assumed to act at a distance of $H/3$ above the base of the wall.

Passive Case

In the passive case the reaction P acts at angle δ above the normal to the wall surface (or δ below the normal if the wall were to settle more than the adjacent soil) and the reaction R at angle ϕ above the normal to the failure plane. In the triangle of forces the angle between W and P is $(180° - \alpha + \delta)$ and the angle between W and R is $(\theta + \phi)$. The total passive resistance, equal to the minimum value of P, is given by:

$$P_P = \tfrac{1}{2}K_P\gamma H^2 \tag{6.18}$$

where

$$K_P = \left(\frac{\sin(\alpha + \phi)/\sin\alpha}{\sqrt{[\sin(\alpha - \delta)]} - \sqrt{\left[\dfrac{\sin(\phi + \delta)\sin(\phi + \beta)}{\sin(\alpha - \beta)} \right]}} \right)^2 \tag{6.19}$$

When the value of δ is greater than $\phi/3$ the curvature of the failure surface should be taken into account otherwise the total passive resistance will be significantly overestimated, representing an error on the unsafe side. An example of an analysis in which curvature is considered is given in Fig. 6.16. The surface of the soil is horizontal and the shear strength parameter c is equal to zero. In section the failure surface is assumed to consist of a circular arc BC (centre 0, radius r) and a straight line CE which is a tangent to the circular arc. When the wall deformation is such that the total passive resistance is fully mobilized, the soil within the triangle ACE is in the passive Rankine state, the angles EAC and AEC both being $(45° - \phi/2)$. The horizontal force (Q) on the vertical plane DC, therefore, is the Rankine passive value given by Equation 6.7, acting horizontally at a distance of DC/3 above point C.

It is necessary to analyse only the stability of the soil mass ABCD and the forces to be considered are:

1. the weight (W) of ABCD acting through the centroid of the section;
2. the force Q on DC;
3. the reaction (P) to the force between the soil and the wall, acting at angle δ above the normal and at a distance of AB/3 above B;
4. the reaction (R) on the failure surface BC.

When the shear strength along the circular arc BC is fully mobilized the reaction R is *assumed* to act at angle ϕ to the normal. The line of action of R is then tangential to a circle, centre 0, radius $r \sin \phi$: this circle is referred to as the *ϕ-circle*. It has been shown that the line of action of R is actually tangential to a circle, centre 0, of radius slightly greater than $r \sin \phi$ but the error in the value of force P due to the above assumption is on the safe side.

The values of forces W and Q are known and their resultant (S) is determined graphically as in Fig. 6.16. For equilibrium the lines of action of forces S, P and R must intersect. Therefore the line of action of R must pass through the intersection of S and P and be tangential to the ϕ-circle: thus the direction of the force R is fixed. The polygon of forces can now be completed and the value of P determined. The analysis must be repeated for a number of failure surfaces to obtain the minimum value of P.

It is possible to introduce values of the parameters c and c_w into the analysis and to take into account the influence of a surcharge pressure q on the soil surface. In an alternative analysis the curve BC is assumed to be a logarithmic spiral (the correct shape, theoretically, for $\phi > 0$).

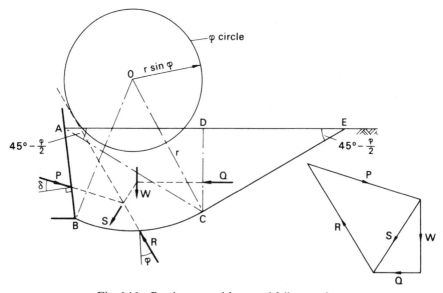

Fig. 6.16   Passive case with curved failure surface.

Earth Pressure Coefficients

Tabulated values of earth pressure coefficients, for the case of a vertical wall and a horizontal soil surface, are given in Civil Engineering Code of Practice No. 2 (*Earth Retaining Structures*) [6.6]. In general the *horizontal components* of total active thrust and total passive resistance can be expressed as

$$P_{An} = \tfrac{1}{2}K_A\gamma H^2 - 2K_{Ac}cH \tag{6.20}$$

and

$$P_{Pn} = \tfrac{1}{2}K_P\gamma H^2 + 2K_{Pc}cH \tag{6.21}$$

where the coefficients K depend on the values of ϕ, c, δ and c_w. The coefficients published in the above Code appear in Table 6.3. It should be noted that this Code is under revision.

The selection of an appropriate value of ϕ is of prime importance in the passive case. The difficulty is that strains vary significantly throughout the soil mass and in particular along the failure surface. The effect of strain, which is governed by the mode of wall deformation, is neglected both in the failure criterion and in analysis. In the earth pressure theories a constant value of ϕ is assumed throughout the soil above the failure surface whereas in fact the mobilised value of ϕ varies. In the case of dense sands the average value of ϕ along the failure surface corresponds to a point beyond the peak value on the stress-strain curve (e.g. Fig. 4.9a): the use of the peak value of ϕ would result, therefore, in an overestimation of passive resistance. It should be noted, however, that peak values of ϕ obtained from triaxial tests are normally less than corresponding values in plane strain, the latter being relevant in retaining wall problems. In the case of loose sands the wall deformation required to mobilize the maximum value of ϕ would be unacceptably large in practice. In both cases, therefore, conservative values of ϕ should be used in design: experimental work by Rowe and Peaker [6.18] confirms this point.

6.4 Other Solutions

Active and passive coefficients were derived by Caquot and Kerisel [6.5] by the integration of the differential equations governing the conditions of limiting equilibrium in a soil mass. Charts of active and passive coefficients for ranges of values of ϕ, δ and surface slope β, based on the work of Caquot and Kerisel, have been produced by Padfield and Mair [6.12].

Sokolovski [6.19] obtained solutions to the active and passive problems by combining the equations of equilibrium (Equations 5.1) with the failure criterion for the soil (Equation 4.6b). The solution, obtained by numerical integration, yields two families of curves on which the maximum value of the stress ratio τ/σ is reached and it is then possible to derive the lateral

Table 6.3   Earth Pressure Coefficients

(a) Shear strength parameter $c = 0$

	δ	ϕ 25°	30°	35°	40°	45°
K_A	0°	0·41	0·33	0·27	0·22	0·17
	10°	0·37	0·31	0·25	0·20	0·16
	20°	0·34	0·28	0·23	0·19	0·15
	30°	—	0·26	0·21	0.17	0.14
K_p	0°	2·5	3·0	3·7	4·6	
	10°	3·1	4·0	4·8	6·5	
	20°	3·7	4·9	6·0	8·8	
	30°	—	5·8	7·3	11·4	

(b) Shear strength parameter $c > 0$

	δ	$\dfrac{c_w}{c}$	ϕ 0°	5°	10°	15°	20°	25°
K_A	0	all	1·00	0·85	0·70	0·59	0·48	0·40
	ϕ	values	1·00	0·78	0·64	0·50	0·40	0·32
K_{Ac}	0	0	2·00	1·83	1·68	1·54	1·40	1·29
	0	1·0	2·83	2·60	2·38	2·16	1·96	1·76
	ϕ	0·5	2·45	2·10	1·82	1·55	1·32	1·15
	ϕ	1·0	2·83	2·47	2·13	1·85	1·59	1·41
K_P	0	all	1·0	1·2	1·4	1·7	2·1	2·5
	ϕ	values	1·0	1·3	1·6	2·2	2·9	3·9
K_{Pc}	0	0	2·0	2·2	2·4	2·6	2·8	3·1
	0	0·5	2·4	2·6	2·9	3·2	3·5	3·8
	0	1·0	2·6	2·9	3·2	3·6	4·0	4·4
	ϕ	0·5	2·4	2·8	3·3	3·8	4·5	5·5
	ϕ	1·0	2·6	2·9	3·4	3·9	4·7	5·7

Active case: if $c < 50\,\text{kN/m}^2$ then $c_w = c$;
if $c > 50\,\text{kN/m}^2$ then $c_w = 50\,\text{kN/m}^2$.
Passive case: if $c < 50\,\text{kN/m}^2$ then $c_w = c/2$;
if $c > 50\,\text{kN/m}^2$ then $c_w = 25\,\text{kN/m}^2$.
If wall tends to move downwards relative to ground then values of c_w may be taken as those for active case. (Data extracted from CP2, *Earth Retaining Structures*, and reproduced by permission of the Institution of Structural Engineers.)

Table 6.4 Sokolovski's Values of Active and Passive Coefficients

ϕ	δ	$\alpha = 90°$		$\alpha = 100°$		$\alpha = 110°$	
		K_A	K_P	K_A	K_P	K_A	K_P
	0°	0·70	1·42	0·72	1·31	0·73	1·18
10°	5°	0·67	1·56	0·70	1·43	0·70	1·29
	10°	0·65	1·66	0·68	1·52	0·70	1·35
	0°	0·49	2·04	0·54	1·77	0·58	1·51
20°	10°	0·45	2·55	0·50	2·19	0·54	1·83
	20°	0·44	3·04	0·50	2·57	0·54	2·13
	0°	0·33	3·00	0·40	2·39	0·46	1·90
30°	15°	0·30	4·62	0·37	3·62	0·43	2·79
	30°	0·31	6·55	0·38	5·03	0·45	3·80
	0°	0·22	4·60	0·29	3·37	0·35	2·50
40°	20°	0·20	9·69	0·27	6·77	0·34	4·70
	40°	0·22	18·2	0·29	12·3	0·38	8·23

stresses on the wall surface. The solution is of the lower bound type and again there is no detailed consideration of deformation. For the case of a horizontal soil surface, values of K_A and K_P for ranges of values of ϕ, δ and wall inclination angle α are given in Table 6.4. For the general case, Sokolovski's values are closer to the exact values than those obtained from the theories of Rankine and Coulomb.

DESIGN OF EARTH-RETAINING STRUCTURES

6.5 Gravity and Cantilever Walls

The stability of a gravity retaining wall (Fig. 6.17a) is due to the self-weight of the wall, perhaps aided by passive resistance developed in front of the wall. Walls of this type are uneconomic because the wall material (masonry or mass concrete) is used only for its dead weight. Cantilever walls of reinforced concrete (Fig. 6.17b) are more economic because the backfill itself is employed to provide most of the required dead weight. Both types of wall are liable to rotational or translational movements and the Rankine or Coulomb theories are used for the calculation of lateral pressure.

The retaining wall as a whole must satisfy two basic conditions: (1) the base pressure at the toe of the wall must not exceed the allowable bearing capacity of the soil (see Chapter 8); (2) the factor of safety against sliding

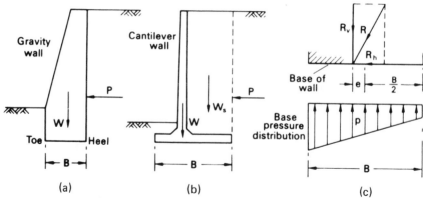

Fig. 6.17   Gravity and cantilever retaining walls.

between the base and the underlying soil must be adequate, a value of at least 1·5 usually being specified. Passive resistance in front of the wall should not be relied upon unless it is certain that the soil will always remain firm and undisturbed, an assumption that can seldom be made.

The first step in the design is to determine all the forces acting on the wall, from which the horizontal and vertical components (R_h and R_v respectively) of the resultant force R acting on the base of the wall are obtained. The position of the force R (Fig. 6.17c) is then determined by dividing the algebraic sum of the moments of all forces about any point on the base by the vertical component R_v. To ensure that the base pressure remains compressive over the entire base width, the resultant R must act within the middle third of the base, i.e. the eccentricity (e) of the base resultant must not exceed $B/6$, where B is the width of the base. If the middle third rule is observed, adequate safety against overturning of the wall will also be ensured. If a linear distribution of pressure (p) under the base is assumed, the maximum and minimum base pressures can be calculated from the expression:

$$p = \frac{R_v}{B}\left(1 \pm \frac{6e}{B}\right) \tag{6.22}$$

The factor of safety against sliding (F_s), ignoring any passive resistance in front of the wall, is given by

$$F_s = \frac{R_v \tan \delta}{R_h} \tag{6.23}$$

where δ is the angle of friction between the base and the underlying soil. If an adequate value of F_s cannot be achieved a key may be incorporated in the base

If a wall is constructed on a compressible soil such as a fully saturated clay, non-uniform base pressure will result in progressive tilting of the wall due to consolidation of the soil. A wall constructed on compressible soil should be dimensioned so that the resultant R acts close to the mid-point of the base.

Clay backfills should be avoided if at all possible since climatic changes are likely to cause successive swelling and shrinkage of the soil. Swelling causes unpredictable pressures on, and movements of, the wall: subsequent shrinkage may result in the formation of cracks in the soil surface.

Some form of filter of coarse permeable material is desirable behind a retaining wall to prevent the development of high pore water pressures within the backfill, the water percolating into the filter draining out through weep-holes in the wall.

Certain categories of retaining walls are essentially unyielding, for example foundation walls supported by the floor system of the building and bridge abutments restrained by the deck structure. In such cases the at-rest value of lateral pressure, or pressures between the at-rest and active values, should be used in design. Sowers *et al.* [6.20] showed that compaction of a backfill against an unyielding wall can result in residual lateral pressures considerably higher than the corresponding values for uncompacted soil.

Example 6.4

Determine the maximum and minimum pressures under the base of the cantilever retaining wall detailed in Fig. 6.18 and the factor of safety against sliding. The appropriate shear strength parameters for the soil are $c' = 0$ and $\phi' = 40°$: the unit weight $\gamma = 17\,\text{kN/m}^3$, the water table being below the base of the wall. Take $\delta = 30°$ on the base of the wall.

To determine the position of the base reaction, the moments of all forces

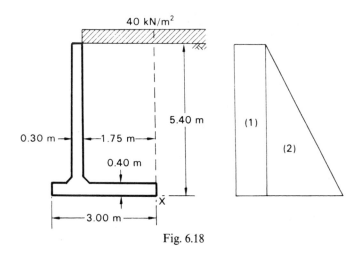

40 kN/m²

5.40 m

(1)

(2)

0.30 m

1.75 m

0.40 m

3.00 m

X

Fig. 6.18

about the heel of the wall (X) are calculated (Table 6.5). The unit weight of concrete is taken to be $23 \cdot 5 \, kN/m^3$. The active pressure is calculated on the vertical through the heel of the wall. No shear stresses act on this vertical, therefore the Rankine theory ($\delta = 0$) is used to calculate the active pressure: the pressure distribution is shown in Fig. 6.18.

For $\phi' = 40°$ (and $\delta = 0$), $K_A = 0 \cdot 22$.

Table 6.5

Force per m (kN)			Arm (m)	Moment per m (kNm)
(1)	$0 \cdot 22 \times 40 \times 5 \cdot 40$	$= 47.5$	$2 \cdot 70$	$128 \cdot 2$
(2)	$\frac{1}{2} \times 0 \cdot 22 \times 17 \times 5 \cdot 40^2$	$= 54 \cdot 6$	$1 \cdot 80$	$98 \cdot 3$
		$R_h = 102 \cdot 1$		
(Stem)	$5 \cdot 00 \times 0 \cdot 30 \times 23 \cdot 5$	$= 35 \cdot 3$	$1 \cdot 90$	$67 \cdot 0$
(Base)	$3 \cdot 00 \times 0 \cdot 40 \times 23 \cdot 5$	$= 28 \cdot 2$	$1 \cdot 50$	$42 \cdot 3$
(Soil)	$5 \cdot 00 \times 1 \cdot 75 \times 17$	$= 148 \cdot 8$	$0 \cdot 875$	$130 \cdot 2$
(Load)	$1 \cdot 75 \times 40$	$= 70 \cdot 0$	$0 \cdot 875$	$61 \cdot 3$
		$R_v = 282 \cdot 3$		$M = 527 \cdot 3$

Lever arm of base resultant:

$$\frac{M}{R_v} = \frac{527 \cdot 3}{282 \cdot 3} = 1 \cdot 86 \, m$$

i.e. the resultant acts within the middle third of the base.

Eccentricity of base reaction:

$$e = 1 \cdot 86 - 1 \cdot 50 = 0 \cdot 36 \, m$$

The maximum and minimum base pressures are given by

$$p = \frac{R_v}{B} \left(1 \pm \frac{6e}{B} \right)$$

$$= \frac{282 \cdot 3}{3} \left(1 \pm \frac{6 \times 0 \cdot 36}{3} \right) = 94(1 \pm 0 \cdot 72)$$

$$= 162 \, kN/m^2 \quad \text{and} \quad 26 \, kN/m^2$$

The factor of safety against sliding is given by

$$F = \frac{R_v \tan \delta}{R_h}$$

$$= \frac{282 \cdot 3 \tan 30°}{102 \cdot 1} = 1 \cdot 6$$

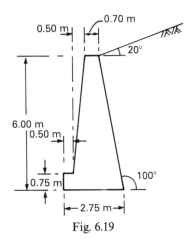

Fig. 6.19

Example 6.5

Details of a gravity retaining wall are shown in Fig. 6.19, the unit weight of the wall material being $23.5\,\text{kN/m}^3$. The unit weight of the backfill is $18\,\text{kN/m}^3$ and the appropriate shear strength parameters are $c' = 0$ and $\phi' = 38°$; the value of δ between the wall and the backfill, and between the wall and the foundation soil, is $25°$. Determine the maximum and minimum pressures under the base of the wall and the factor of safety against sliding.

As the back of the wall and the soil surface are both inclined, the value of K_A will be calculated from Equation 6.17. The values of the angles in this equation are: $\alpha = 100°$, $\phi = 38°$, $\delta = 25°$ and $\beta = 20°$. Hence

$$K_A = \left[\frac{\sin 62°/\sin 100°}{\sqrt{\sin 125°} + \sqrt{(\sin 63° \sin 18°/\sin 80°)}}\right]^2$$

$$= 0.39$$

Then, from Equation 6.16,

$$P_A = \tfrac{1}{2} \times 0.39 \times 18 \times 6^2 = 126 \text{ kN/m}$$

acting at $\tfrac{1}{3}$ height and at $25°$ above the normal, or $35°$ above the horizontal. Consider moments about the heel of the wall (Table 6.6).

Lever arm of base resultant:

$$\frac{M}{R_v} = \frac{520.4}{293.4} = 1.77\,\text{m} \quad \text{(within middle third)}$$

Eccentricity of base reaction:

$$e = 1.77 - 2.75/2 = 0.40\,\text{m}$$

Table 6.6

Force per m (kN)			Arm (m)	Moment per m (kNm)
$P_A \cos 35°$		$= 103·2$	2·00	206·4
	R_h	$= 103·2$		
$P_A \sin 35°$		$= 72·3$	0·35	25·3
(1) $\frac{1}{2} × 1·05 × 6 × 23·5$		$= 74·0$	0·70	51·8
(2) $0·70 × 6 × 23·5$		$= 98·7$	1·40	138·2
(3) $\frac{1}{2} × 0·5 × 5·25 × 23·5$		$= 30·8$	1·92	59·1
(4) $1 × 0·75 × 23·5$		$= 17·6$	2·25	39·6
	R_v	$= 293·4$		$M = 520·4$

The maximum and minimum base pressures are given by

$$p = \frac{R_v}{B}\left(1 \pm \frac{6e}{B}\right)$$

$$= \frac{293.4}{2.75}(1 \pm 0.87)$$

$$= 200 \, \text{kN/m}^2 \quad \text{and} \quad 14 \, \text{kN/m}^2$$

The factor of safety against sliding is given by

$$F = \frac{R_v \tan \delta}{R_h}$$

$$= \frac{293.4 \tan 25°}{103.2} = 1.33$$

Example 6.6

Fig. 6.20a gives details of a retaining wall with a vertical drain adjacent to the back surface. Determine the total thrust on the wall when the backfill becomes fully saturated due to continuous rainfall, with steady seepage towards the drain. Assume a failure plane at 55° to the horizontal. The relevant parameters for the backfill are $c' = 0$, $\phi' = 38°$, $\delta = 15°$ and $\gamma_{sat} = 20 \, \text{kN/m}^3$.

Determine also the thrust on the wall (a) if the vertical drain were replaced by an inclined drain below the failure plane, (b) if there were no drainage system behind the wall.

The flow net for seepage towards the vertical drain is shown in Fig. 6.20a. Since the permeability of the drain must be considerably greater than that of the backfill, the drain remains unsaturated and the pore pressure at every point within the drain is zero (atmospheric). Thus, at every point on the

boundary between the drain and the backfill, total head is equal to
elevation head. The equipotentials, therefore, must intersect this boundary
at points spaced at equal vertical intervals Δh: the boundary itself is neither
a flow line nor an equipotential.

The combination of total weight and boundary water force is used. The

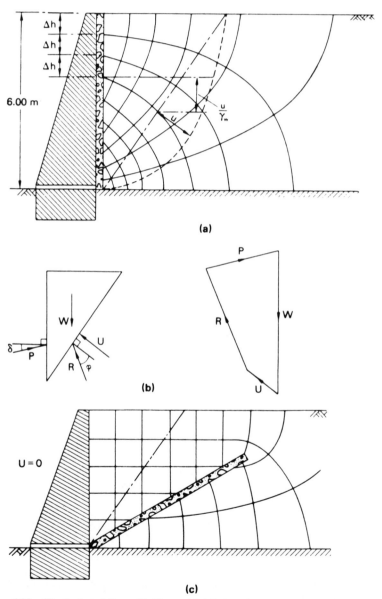

Fig. 6.20 (Reproduced from K. Terzaghi (1943) *Theoretical Soil Mechanics*,
John Wiley & Sons Inc., by permission.)

values of pore water pressure at the points of intersection of the equipotentials with the failure plane are evaluated and plotted normal to the plane. The boundary water force (U), acting normal to the plane, is equal to the area of the pressure diagram, thus:

$$U = 55\,\text{kN/m}$$

The water forces on the other two boundaries of the soil wedge are zero.
The total weight (W) of the soil wedge is now calculated, i.e.

$$W = 252\,\text{kN/m}$$

The forces acting on the wedge are shown in Fig. 6.20b. Since the directions of the four forces are known, together with the values of W and U, the force polygon can be drawn, from which

$$P_A = 108\,\text{kN/m}$$

or

$$P_{An} = P_A \cos \delta = 105\,\text{kN/m}$$

Other failure surfaces would have to be chosen in order that the maximum value of total active thrust can be determined.

For the inclined drain shown in Fig. 6.20c the flow lines and equipotentials above the drain are vertical and horizontal, respectively. Thus at every point on the failure plane the pore water pressure is zero. This form of drain is preferable to the vertical drain. In this case:

$$P_{An} = \tfrac{1}{2} K_A \gamma_{\text{sat}} H^2$$

For $\phi' = 38°$ and $\delta = 15°$, $K_A = 0\cdot21$. Therefore

$$P_{An} = \tfrac{1}{2} \times 0\cdot21 \times 20 \times 6^2 = 76\,\text{kN/m}$$

For the case of no drainage system behind the wall, the pore water is static, therefore

$$P_{An} = \tfrac{1}{2} K_A \gamma' H^2 + \tfrac{1}{2}\gamma_w H^2$$
$$= (\tfrac{1}{2} \times 0\cdot21 \times 10\cdot2 \times 6^2) + (\tfrac{1}{2} \times 9\cdot8 \times 6^2)$$
$$= 39 + 176 = 215\,\text{kN/m}$$

6.6 Cantilever Sheet Pile Walls

Walls of this type are used only when the retained height of soil is relatively small. In sands and gravels these walls may be used as permanent structures, but in general they are used for temporary support. The stability of a

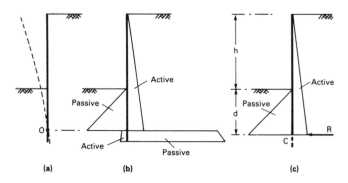

Fig. 6.21   Cantilever sheet pile wall.

cantilever sheet pile wall is due entirely to passive resistance developed below the lower soil surface. The mode of failure is by rotation about a point O near the lower end of the wall as shown in Fig. 6.21a. Consequently, passive resistance acts in front of the wall above O and behind the wall below O, as shown in Fig. 6.21b, thus providing a fixing moment: however this pressure distribution is an idealization as there is unlikely to be a sudden change in passive pressure from front to back at point O.

Design is generally based on the simplification shown in Fig. 6.21c, it being assumed that the net passive resistance below point O is represented by a concentrated force R acting at a point C, slightly below O, at depth d below the lower soil surface. The depth d can be determined by equating moments about C to zero, a factor of safety F being applied to the stabilizing moment i.e. by dividing the available passive resistance (P_P) in front of the wall by F. The value of d is then increased arbitrarily by 20% to allow for the simplification involved in the method, i.e. the required depth of embedment is $1.2d$. However it is desirable to evaluate R by equating horizontal forces to zero and to check that the net passive resistance available over the additional 20% embedment depth is equal to or greater than R.

6.7   Anchored or Propped Sheet Pile Walls

Additional support to sheet pile walls can be provided by a row of tie-backs or props near the top of the wall. Tie-backs are normally high tensile steel cables or rods anchored in the soil some distance behind the wall. Walls of this type are used extensively in waterfront construction and in the support of deep excavations. There are two basic modes of construction. Excavated walls are constructed by driving a row of sheet piling on the required line followed by excavation or dredging to the required depth in front of the piling. Backfilled walls are constructed by partial driving in level ground and backfilling to the required height behind the piling. Stability is due to the passive resistance developed in front of the wall together with the forces in the ties or props. The method of analysis used for sheet pile walls is also applicable to diaphragm walls (Section 6.9).

Free Earth Support Method

In this method of analysis it is assumed that the depth of embedment below excavation level is insufficient to produce fixity at the lower end of the wall. The wall is thus free to rotate at its lower end, the bending moment diagram being of the form shown in Fig. 6.22b. The lateral pressure distributions are assumed to be those given by either the Rankine or Coulomb theory. The mode of failure is rotation about the point of application of the tie or prop (A) and in design it is essential to ensure that the available stabilizing moments exceed the disturbing moments by an adequate margin. Referring to Fig. 6.22, the required depth of embedment (d) can be determined from the condition that the algebraic sum of the factored moments about A must be zero. This condition yields a cubic equation in d which can be solved by the substitution of trial values. Having determined the value of d, the force T in the tie or prop, per unit length of wall, can be calculated from the condition that the algebraic sum of the horizontal forces must be zero. Finally, the bending moment diagram can be drawn, the maximum moment governing the pile section. In the case of quay walls it is recommended that the depth of embedment be increased arbitrarily by 20% to guard against excess dredging, scour, or the presence of weak pockets of soil.

However, the value of d obtained by the method depends on how the factor of safety is introduced into the calculations and on how the lateral pressure distributions are considered, i.e. on whether or not the distributions on the two sides of the wall are combined. The factor of safety can thus be defined in different ways.

Traditionally the factor of safety has been defined in terms of the passive resistance available in front of the wall, the stabilizing moment being that due to passive resistance: this factor is designated F_p. The active and passive

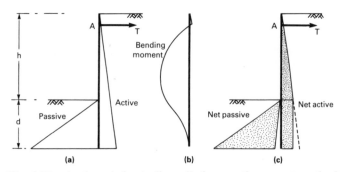

Fig. 6.22   Anchored sheet pile wall: free earth support method.

pressure distributions are considered separately as illustrated in Fig. 6.22a. The passive resistance which must be mobilized to produce a condition of limiting equilibrium (P_{p_m}) is equal to the available passive resistance (the Rankine or Coulomb value P_p) divided by the factor of safety (F_p).

The factor of safety could also be defined in terms of shear strength viz. the ratio of the available shear strength divided by the average shear strength required to produce a condition of limiting equilibrium: this factor is designated F_s. The shear strength parameters normally represent the greatest uncertainty in the analysis. The same factor is usually applied to both parameters, active and passive pressures being calculated using the mobilized shear strength parameters c/F_s and $\tan^{-1}(\tan \phi/F_s)$.

Burland, Potts and Walsh [6.4] presented evidence to show that there was a lack of consistency between F_p and F_s over the practical ranges of wall geometry and shear strength parameters, particularly in the case of clays under undrained conditions. Using an analogy with the bearing capacity of a footing, Burland, Potts and Walsh proposed that the factor of safety should be defined in terms of the *net* available passive resistance in front of the wall, as illustrated in Fig. 6.22c: this factor is designated F_r. It was shown that this definition of factor of safety was consistent with that in terms of shear strength.

The value of factor of safety used in design will depend on how the shear strength parameters are selected, for example whether conservative or worst credible values are used. Guidance on the choice of factor of safety for walls in stiff clays is given by Padfield and Mair [6.12].

It should be realized that full passive resistance is only developed under conditions of limiting equilibrium ($F = 1$). Under working conditions, with a factor of safety greater than unity, experimental and analytical work has indicated that the distribution of lateral pressure is likely to be of the form shown in Fig. 6.23: it should be noted that passive resistance is likely to be fully mobilized close to the surface in front of the wall. The extra depth of embedment required to produce a factor of safety greater than unity results in a partial fixing moment at the lower end of the wall and consequently a lower maximum bending moment than the limiting equilibrium value. In

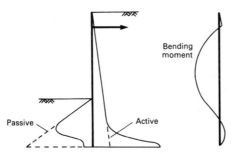

Fig. 6.23 Anchored sheet pile wall: pressure distribution under working conditions.

view of the uncertainty regarding the actual pressure distribution under working conditions there is a case for calculating bending moments, and the tie or prop force, for the condition of limiting equilibrium. The calculated value of tie or prop force should then be increased by 25% to allow for possible redistribution of pressure due to arching (see below). The Burland–Potts–Walsh method can also be used in the case of cantilever sheet pile walls (Section 6.6). Bending moments should also be calculated for $F = 1$ in the case of cantilever walls.

Effect of Flexibility and K_0

The behaviour of an anchored wall is also influenced by its degree of flexibility or stiffness. In the case of *flexible* sheet pile walls, experimental and analytical results indicate that redistributions of lateral pressure take place. The pressures on the most yielding parts of the wall (between the tie and excavation level) are reduced and those on the relatively unyielding parts (in the vicinity of the tie and below excavation level) are increased with respect to the theoretical values, as illustrated in Fig. 6.24. These redistributions of lateral pressure are the result of the phenomenon known as *arching*. No such redistributions take place in the case of stiff walls, such as concrete diaphragm walls.

Arching was defined by Terzaghi [6.21] in the following way. "If one part of the support of a soil mass yields while the remainder stays in place, the soil adjoining the yielding part moves out of its original position between adjacent stationary soil masses. The relative movement within the soil is opposed by shearing resistance within the zone of contact between the yielding and stationary masses. Since the shearing resistance tends to keep the yielding mass in its original position, the pressure on the yielding part of

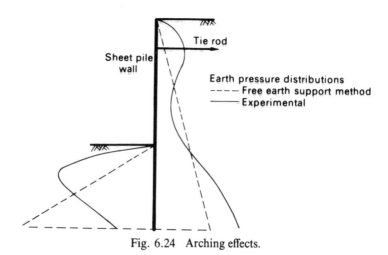

Fig. 6.24   Arching effects.

the support is reduced and the pressure on the stationary parts is increased. This transfer of pressure from a yielding part to adjacent non-yielding parts of a soil mass is called the arching effect. Arching also occurs when one part of a support yields more than the adjacent parts."

The conditions for arching are present in anchored sheet pile walls when they deflect. If yield of the anchor takes place, arching effects are reduced to an extent depending on the amount of yielding. On the passive side of the wall the pressure is increased just below excavation level as a result of larger deflections *into* the soil. In the case of backfilled walls, arching is only partly effective until the fill is above tie level. Arching effects are much greater in sands than in silts or clays and are greater in dense sands than in loose sands.

Redistributions of earth pressure result in lower bending moments than those obtained from the free earth support method of analysis, the greater the flexibility of the wall the greater the moment reduction. Rowe [6.16; 6.17] proposed the use of moment reduction coefficients, to be applied to the results of free earth support analyses, based on the flexibility of the wall. Wall flexibility is represented by the parameter $\rho = H^4/EI$, where H is the overall height of the wall and EI is the flexural rigidity. The tie force is also influenced by earth pressure redistribution and factors are also given for the adjustment of the free earth support value of this force.

Rowe's moment reduction factors should only be used if a factored passive resistance $(F > 1)$ has been used for the calculation of bending moments. If bending moments have been calculated for the limiting equilibrium condition $(F = 1)$, Rowe's factors should not be used.

Potts and Fourie [6.14; 6.15] analysed a propped cantilever wall in clay by means of the finite element method, incorporating an elastic–perfectly plastic stress-strain relationship. The results indicated that the required depth of embedment was in agreement with the value obtained by the free earth support method. However the results also showed that in general the behaviour of the wall depended on the wall stiffness (confirming Rowe's earlier findings), the initial value of K_0 (the coefficient of earth pressure at-rest) for the soil and the method of construction (i.e. backfilling or excavation).

In particular, maximum bending moment and prop force increased as wall stiffness increased. For backfilled walls and for excavated walls in soils having a low K_0 value (of the order of 0.5), maximum bending moment and prop force were both lower than those obtained using the free earth support method. However for stiff walls, such as diaphragm walls, formed by excavation in soils having a high K_0 value (in the range 1–2), such as over-consolidated clays, maximum bending moment and prop force were both significantly higher than those obtained using the free earth support method. For the particular (excavated) wall and material properties considered by Potts and Fourie, the patterns of variation shown in Fig. 6.25 were obtained for a factor of safety (F_r) of 2.0. In this figure M_{fe} and T_{fe} denote the maximum bending moment and prop force, respectively,

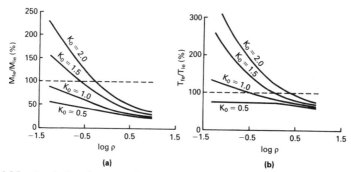

Fig. 6.25 Analysis of propped cantilever wall in clay by the finite element method. (Reproduced from D. M. Potts and A. B. Fourie (1985), *Geotechnique*, Vol. 35, No. 3, by permission of Thomas Telford Ltd.).

obtained from the finite element analysis, and M_{le} and T_{le} denote the corresponding values obtained from a limiting equilibrium (free earth support) analysis.

Pore Water Pressure Distribution

Sheet pile and diaphragm walls are normally analysed in terms of effective stress. Care is therefore required in deciding on the appropriate distribution of pore water pressure. Several different situations are illustrated in Fig. 6.26.

If the water table levels are the same on both sides of the wall the pore water pressure distributions will be hydrostatic and will balance (Fig. 6.26a): they can thus be eliminated from the calculations.

If the water table levels are different and if steady seepage conditions have developed the distributions on the two sides of the wall will be unbalanced. The net distribution, on the back of the wall, could be determined from the flow net, as illustrated in Example 2.1. However, for certain conditions an approximate distribution, ABC in Fig. 6.26b, can be obtained by assuming that the total head is dissipated uniformly along the back and front wall surfaces between the two water table levels. The maximum net pressure occurs opposite the lower water table level and, referring to Fig. 6.26b, is given by:

$$u_C = \frac{2ba}{2b + a}\gamma_w$$

In general, the approximate method will underestimate net water pressure especially if the bottom of the wall is relatively close to the lower boundary of the flow region (i.e. if there are large differences in the sizes of curvilinear squares in the flow net).

In Fig. 6.26c a depth of water is shown in front of the wall, the water level being below that of the water table behind the wall. In this case the approxi-

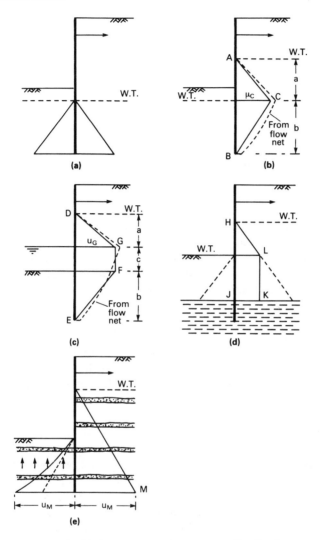

Fig. 6.26 Various pore water pressure distributions.

mate distribution DEFG should be used in appropriate cases, the net pressure at G being given by:

$$u_G = \frac{(2b + c)a}{2b + c + a}\gamma_w$$

A wall constructed mainly in a soil of relatively high permeability but penetrating a layer of clay of low permeability, is shown in Fig. 6.26d. If undrained conditions apply within the clay the pore water pressure in the overlying soil would be hydrostatic and the net pressure distribution would be HJKL as shown.

A wall constructed in a clay which contains thin layers or partings of fine sand or silt is shown in Fig. 6.26e. In this case it should be assumed that the sand or silt allows water at hydrostatic pressure to reach the back surface of the wall. This implies pressure in excess of hydrostatic, and consequent upward seepage, in front of the wall.

For short term situations for walls in clay (e.g. during and immediately after excavation), the possibility exists of tension cracks developing or fissures opening. If such cracks or fissures fill with water, hydrostatic pressure should be assumed over the depth in question: the water in the cracks or fissures would also result in softening of the clay. Softening would also occur near the soil surface in front of the wall as a result of stress relief on excavation. An effective stress analysis would ensure a safe design in the event of rapid softening of the clay taking place or if work were delayed during the temporary stage of construction: however a relatively low factor of safety could be used in such cases. Padfield and Mair [6.12] give details of a mixed total and effective stress design method as an alternative for short term situations in clay, i.e. effective stress conditions within the zones liable to soften and total stress conditions below.

Seepage Pressure

Under conditions of steady seepage, use of the approximation that total head is dissipated uniformly along the wall has the consequent advantage that the seepage pressure is constant. For the conditions shown in Fig. 6.26b, for example, the seepage pressure at any depth is:

$$j = \frac{a}{2b + a}\gamma_w$$

The effective unit weight of the soil therefore would be increased to $(\gamma' + j)$ behind the wall, where seepage is downwards, and reduced to $(\gamma' - j)$ in front of the wall, where seepage is upwards: these values should be used in the calculation of active and passive pressures respectively. Thus under conditions of steady seepage, active pressures are increased and passive pressures are decreased relative to the corresponding static state values.

Tie-back Anchorage

Tie rods are normally anchored in beams, plates or concrete blocks some distance behind the wall (Fig. 6.27). The tie rod force is resisted by the passive resistance mobilized by the anchor, reduced by the active pressure, a factor of safety of not less than 2 being employed to ensure that the anchor yield is not excessive. If the height (b) of the anchor is not less than half the depth (d_a) from the surface to the bottom of the anchor, the anchor can be assumed to develop passive resistance over the depth d_a. The anchor must be situated beyond the plane YZ (Fig. 6.27a) to ensure that the passive

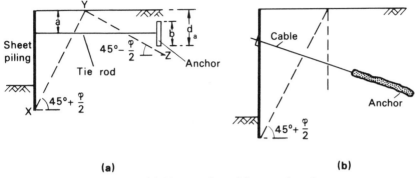

(a)   (b)

Fig. 6.27   (a) Plate anchor, (b) ground anchor.

wedge of the anchor does not encroach on the active wedge behind the wall: the lower end (X) of the active wedge should be taken at the bottom of the wall if the free earth support method is used in design.

If T = tie rod force per unit length of wall, s = spacing of tie rods, F = factor of safety, and l = length of anchor per tie rod, then:

$$Ts = \frac{\gamma d_a^2 l}{2F}(K_P - K_A) \tag{6.24}$$

If an individual anchor is used for each tie rod, the shearing resistance on the sides of the passive wedge produces additional anchor resistance. Tensioned cables anchored in a mass of cement grout or grouted soil (Fig. 6.27b) can also be used to support sheet pile walls: these are called ground anchors and are described in Section 8.8.

Example 6.7

The sides of an excavation 2·5 m deep in sand are to be supported by a canti-lever sheet pile wall, the water table being 1·0 m below the bottom of the excavation. The unit weight of the sand above the water table is 17 kN/m³ and below the water table the saturated unit weight is 20 kN/m³. If $c' = 0$, $\phi' = 35°$ and $\delta = 0$, determine the required depth of embedment of the piling to give a factor of safety of 2·0 with respect to passive resistance.

For $\phi' = 35°$ and $\delta = 0$, $K_A = 0·27$ and $K_P = 3·7$.

Below the water table the effective unit weight of the soil is $(20 - 9·8) = 10·2$ kN/m³.

The earth pressure diagrams are shown in Fig. 6.28. The distributions of hydrostatic pressure on the two sides of the wall balance and can be eliminated from the calculations. The procedure is to equate moments about C, the point of application of the force representing the net passive resistance below the point of rotation. The forces, lever arms and moments

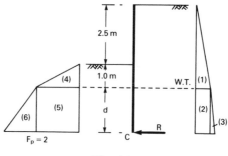

Fig. 6.28

are set out in Table 6.7, the specified factor of safety being applied to forces (4), (5) and (6).

Table 6.7

Force per m (kN)			Arm (m)	Moment per m (kNm)
(1) $\frac{1}{2} \times 0.27 \times 17 \times 3.5^2$	=	28·11	$d + 3.5/3$	$28\cdot11d + 32\cdot79$
(2) $0.27 \times 17 \times 3.5 \times d$	=	$16\cdot06d$	$d/2$	$8\cdot03d^2$
(3) $\frac{1}{2} \times 0.27 \times 10.2 \times d^2$	=	$1\cdot38d^2$	$d/3$	$0\cdot46d^3$
(4) $-\frac{1}{2} \times 3.7 \times 17 \times 1^2 \times \frac{1}{2}$	=	$-15\cdot72$	$d + 1/3$	$-15\cdot72d - 5\cdot24$
(5) $-3.7 \times 17 \times 1 \times d \times \frac{1}{2}$	=	$-31\cdot45d$	$d/2$	$-15\cdot72d^2$
(6) $-\frac{1}{2} \times 3.7 \times 10.2 \times d^2 \times \frac{1}{2}$	=	$-9\cdot43d^2$	$d/3$	$-3\cdot14d^3$

Equating the algebraic sum of the moments about C to zero produces the following equation:

$$-2\cdot68d^3 - 7\cdot69d^2 + 12\cdot39d + 27\cdot55 = 0$$
$$\therefore \qquad d^3 + 2\cdot87d^2 - 4\cdot62d = 10\cdot28$$

By trial the solution is:

$$d = 2\cdot00\,\text{m}$$

Then the required depth of embedment $= 1\cdot2(2\cdot0 + 1\cdot0) = 3\cdot6\,\text{m}$.

The force R should be evaluated and compared with the net passive resistance available over the additional 20% embedment depth. Thus for $d = 2\cdot00\,\text{m}$:

$$R = -(28\cdot11 + 32\cdot12 + 5\cdot51 - 15\cdot72 - 62\cdot90 - 37\cdot72)$$
$$= 50\cdot6\,\text{kN}$$

Passive pressure acts on the back of the wall between depths of 5·5 and 6·1 m. At a depth of 5·8 m the net passive pressure is given by:

$$p_P - p_A = (3\cdot7 \times 17 \times 3\cdot5) - (0\cdot27 \times 17 \times 1\cdot0)$$
$$+ \{(3\cdot7 - 0\cdot27) \times 10\cdot2 \times 2\cdot3\}$$
$$= 220\cdot1 - 4\cdot6 + 80\cdot5$$
$$= 296 \text{ kN/m}^2$$

Net passive resistance available over the additional embedded depth is:

$$296(6\cdot1 - 5\cdot5)$$
$$= 177\cdot6 \text{ kN} \qquad (>R \text{ therefore satisfactory})$$

Example 6.8

A quay wall is to be constructed using anchored sheet piling as shown in Fig. 6.29. Above the water table the unit weight of the soil is 17 kN/m³ and below the water table the saturated unit weight is 20 kN/m³. The relevant shear strength parameters are $c' = 0$ and $\phi' = 36°$. For a factor of safety (F_p) of 2·0 with respect to passive resistance, determine the required depth of embedment and the force in each tie if these are spaced at 2 m centres. Design a continuous anchor to support the ties.

For $\phi' = 36°$ (and $\delta = 0$), $K_A = 0\cdot26$ and $K_P = 3\cdot85$.

Below the water table the effective unit weight of the soil is 10·2 kN/m³.

The earth pressure diagrams are shown in Fig. 6.29. The distributions of hydrostatic pressure on the two sides of the wall balance and can be eliminated from the calculations. The procedure is to equate moments about A, the point of application of the tie force (T per unit length), to zero. The forces and their lever arms about A are set out in Table 6.8, the factor of safety being applied to force (4).

Equating the algebraic sum of the moments about A to zero yields the following equation:

$$-5\cdot66d^3 - 44\cdot7d^2 + 253\cdot0d + 714\cdot6 = 0$$
$$\therefore \qquad d^3 + 7\cdot9d^2 - 44\cdot7d = 126\cdot3$$

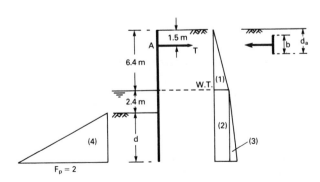

Fig. 6.29

Table 6.8

Force per m (kN)		Arm (m)
(1) $\frac{1}{2} \times 0\cdot26 \times 17 \times 6\cdot4^2$	$= \quad 90\cdot5$	$2\cdot77$
(2) $0\cdot26 \times 17 \times 6\cdot4 \times (d + 2\cdot4)$	$= \quad 28\cdot3d + 67\cdot9$	$d/2 + 6\cdot1$
(3) $\frac{1}{2} \times 0\cdot26 \times 10\cdot2 \times (d + 2\cdot4)^2$	$= \quad 1\cdot33d^2 + 6\cdot36d + 7\cdot64$	$2d/3 + 6\cdot5$
(4) $-\frac{1}{2} \times 3\cdot85 \times 10\cdot2 \times d^2 \times \frac{1}{2}$	$= \quad -9\cdot82d^2$	$2d/3 + 7\cdot3$
Tie	$= \quad -T$	0

By trial the solution is:

$$d = 5\cdot24 \text{ m}$$

For a quay wall this value would be increased by 20%, giving a depth of embedment of $6\cdot29$ m.

The algebraic sum of the forces in Table 6.8 must equate to zero, thus for $d = 5\cdot24$ m:

$$90\cdot5 + 216\cdot2 + 77\cdot5 - 269\cdot6 - T = 0$$
$$\therefore \quad T = 114\cdot6 \text{ kN}$$

Hence the force in each tie is $(2 \times 114\cdot6) = 229$ kN.

For a continuous anchor, $s = l$ in Equation 6.24. Therefore, using a factor of safety of $2\cdot0$:

$$d_a^2 = \frac{2FT}{\gamma(K_P - K_A)}$$

$$= \frac{2 \times 2 \times 114\cdot6}{17(3\cdot85 - 0\cdot26)} = 7\cdot51 \text{ m}^2$$

$$\therefore \quad d_a = 2\cdot74 \text{ m}$$

Hence $b = 2\cdot48$ m for an anchor centred $1\cdot50$ m below the surface.

Example 6.9

A propped cantilever wall supporting the sides of an excavation in stiff clay is shown in Fig. 6.30. The saturated unit weight of the clay (above and below the water table) is 20 kN/m^3. The appropriate shear strength parameters are $c' = 0$ and $\phi' = 29°$. On the back of the wall $\delta = \frac{2}{3}\phi'$ and on the front $\delta = \frac{1}{2}\phi'$. Using the Burland–Potts–Walsh method and assuming conditions of steady seepage, determine the required depth of embedment to give a factor of safety of $2\cdot0$ with respect to net passive resistance. Determine the force in each prop and draw the shearing force and bending moment diagrams for the wall.

The earth pressure coefficients are $K_A = 0\cdot30$ and $K_P = 4\cdot2$.

The distributions of earth pressure and net pore water pressure (assuming

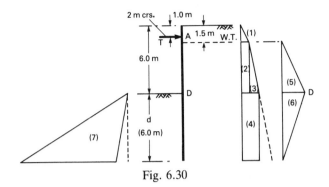

Fig. 6.30

uniform decrease of total head around the wall) are shown in Fig. 6.30: the diagram (7) represents the net available passive resistance. The procedure is to equate moments about A to zero. However if the forces, lever arms and moments were expressed in terms of the unknown embedment depth d, complex algebraic expressions would result and it is preferable to assume a series of trial values of d and calculate the corresponding values of factor of safety F_r: (a computer program could be written). The depth of embedment for $F_r = 2 \cdot 0$ can then be obtained by interpolation.

Following this procedure, a trial value of $d = 6 \cdot 0$ m will be selected. Then the maximum net water pressure, at level D is:

$$u_D = \frac{12 \cdot 0}{16 \cdot 5} \times 4 \cdot 5 \times 9 \cdot 8 = 32 \cdot 1 \text{ kN/m}^2$$

and the average seepage pressure is:

$$j = \frac{4 \cdot 5}{16 \cdot 5} \times 9 \cdot 8 = 2 \cdot 7 \text{ kN/m}^3$$

Thus, below the water table, active forces are calculated using:

$$(\gamma' + j) = 10 \cdot 2 + 2 \cdot 7 = 12 \cdot 9 \text{ kN/m}^3$$

and passive forces are calculated using:

$$(\gamma' - j) = 10 \cdot 2 - 2 \cdot 7 = 7 \cdot 5 \text{ kN/m}^3$$

The calculations are set out in Table 6.9.

Hence the factor of safety is:

$$F_r = \frac{4475 \cdot 7}{2441 \cdot 1} = 1 \cdot 83$$

The calculations are repeated for three other values of d, the results being as follows:

$d(m)$	3·9	5·1	6·0	6·9
F_r	0·87	1·40	1·83	2·33

Table 6.9

Force per m (kN)		Arm (m)	Moment per m (kNm)
(1) $\frac{1}{2} \times 0\cdot30 \times 20 \times 1\cdot5^2$	= $\quad$ 6·8	0	0
(2) $0\cdot30 \times 20 \times 1\cdot5 \times 4\cdot5$	= $\quad$ 40·5	2·75	111·4
(3) $\frac{1}{2} \times 0\cdot30 \times 12\cdot9 \times 4\cdot5^2$	= $\quad$ 39·1	3·5	136·8
(4) $0\cdot30\{(20 \times 1\cdot5) + (12\cdot9 \times$			
$\quad 4\cdot5)\}6\cdot0$	= $\quad$ 158·2	8·0	1266·0
(5) $\frac{1}{2} \times 32\cdot1 \times 4\cdot5$	= $\quad$ 72·2	3·5	252·8
(6) $\frac{1}{2} \times 32\cdot1 \times 6\cdot0$	= $\quad$ 96·3	7·0	674·1
			2441·1
(7) $-\frac{1}{2}\{(4\cdot2 \times 7\cdot5) -$			
$\quad (0\cdot30 \times 12\cdot9)\}6\cdot0^2$	= $-497\cdot3$	9·0	$-4475\cdot7$

Referring to Fig. 6.31 it is apparent that for $F_r = 2\cdot0$ the required depth of embedment is 6·3 m.

The prop load, shearing forces and bending moments will be calculated for the condition of limiting equilibrium, i.e. for $F_r = 1\cdot0$: the corresponding value of d is 4·2 m. For this value of d:

$$u_D = \frac{8\cdot4}{12\cdot9} \times 4\cdot5 \times 9\cdot8 = 28\cdot7 \text{ kN/m}^2$$

$$j = \frac{4\cdot5}{12\cdot9} \times 9\cdot8 = 3\cdot4 \text{ kN/m}^3$$

$$\gamma' + j = 10\cdot2 + 3\cdot4 = 13\cdot6 \text{ kN/m}^3$$

$$\gamma' - j = 10\cdot2 - 3\cdot4 = 6\cdot8 \text{ kN/m}^3$$

The forces on the wall are calculated for $d = 4\cdot2$ m, as shown in Table 6.10.

Table 6.10

Force per m (kN)	
(1) $\frac{1}{2} \times 0\cdot30 \times 20 \times 1\cdot5^2$	= $\quad$ 6·8
(2) $0\cdot30 \times 20 \times 1\cdot5 \times 4\cdot5$	= $\quad$ 40·5
(3) $\frac{1}{2} \times 0\cdot30 \times 13\cdot6 \times 4\cdot5^2$	= $\quad$ 41·3
(4) $0\cdot30\{(20 \times 1\cdot5) + (13\cdot6 \times 4\cdot5)\}4\cdot2$	= $\quad$ 114·9
(5) $\frac{1}{2} \times 28\cdot7 \times 4\cdot5$	= $\quad$ 64·6
(6) $\frac{1}{2} \times 28\cdot7 \times 4\cdot2$	= $\quad$ 60·3
	328·4
(7) $-\frac{1}{2}\{(4\cdot2 \times 6\cdot8) - (0\cdot30 \times 13\cdot6)\}4\cdot2^2$	= $-215\cdot9$
$\therefore$ Prop force (T)	= $\quad$ 112·5

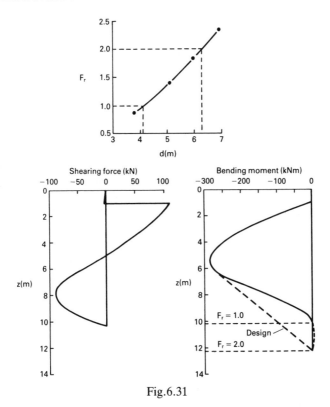

Fig.6.31

Multiplying the calculated value by 1·25 to allow for arching, the prop force for 2 m spacing is:

$$1·25 \times 2 \times 112·5 = 281 \text{ kN}$$

The shearing forces and bending moments, calculated for $d = 4·2$ m, are given in Table 6.11 and are plotted to scale in Fig. 6.31. For the required embedment depth of 6·3 m it is recommended that bending moments given by the dotted line should be used in design.

Table 6.11

Depth (m)	Shearing Force (kN)	Bending Moment (kNm)
0	0	0
1	−3/+109·5	+1
2	+93·6	−103·3
4	+34·6	−248·5
6	−40·7	−278·5
8	−90·2	−126·8
10	−14·3	− 1·5

6.8   Braced Excavations

Sheet piling or timbering is normally used to support the sides of deep, narrow excavations, stability being maintained by means of struts acting across the excavation, as shown in Fig. 6.32a. The piling is usually driven first, the struts being installed in stages as excavation proceeds. When the first row of struts is installed the depth of excavation is small and no significant yielding of the soil mass will have taken place. As the depth of excavation increases, yielding of the soil before strut installation becomes appreciable but the first row of struts prevents yielding near the surface. Deformation of the wall, therefore, will be of the form shown in Fig. 6.32a, being negligible at the top and increasing with depth. Thus the deformation condition of the Rankine theory is not satisfied and the theory cannot be used for this type of wall. Failure of the soil will take place along a surface of the form shown in Fig. 6.32a, only the lower part of the soil wedge within this surface reaching a state of plastic equilibrium, the upper part remaining in a state of elastic equilibrium.

Failure of a braced wall is normally due to the initial failure of one of the struts, resulting in the progressive failure of the whole system. The forces in the individual struts may differ widely because they depend on such random factors as the force with which the struts are wedged home and the time between excavation and installation of struts. The usual design procedure for braced walls is semi-empirical, being based on actual measurements of strut loads in excavations in sands and clays in a number of locations. For example, Fig. 6.32b shows the apparent distributions of earth pressure derived from load measurements in the struts at three sections of a braced excavation in a dense sand. Since it is essential that no individual strut should fail, the pressure distribution assumed in design is taken as the envelope covering all the random distributions obtained from field measurements. Such an envelope should not be thought of as representing the actual distribution of earth pressure with depth but as a hypothetical pressure diagram from which the strut loads can be obtained with some degree of confidence. The pressure envelope proposed by Terzaghi and Peck [6.23] for medium to dense sands is shown in Fig. 6.32c, being a uniform distribution of 0·65 times the Rankine active value.

According to Peck [6.13] the behaviour of a braced excavation in clay depends on the value of the stability number $\gamma H/c_u$, where c_u is the average undrained shear strength of the clay adjacent to the excavation. It should be noted that the value of the stability number increases as the depth of excavation increases in a particular clay. If the stability number is less than 4, most of the clay adjacent to the excavation should be in a state of elastic equilibrium and for this condition Peck proposed that the envelope shown in Fig. 6.32d should be used to estimate the strut loads. If the stability number is greater than 4, plastic zones can be expected to develop near the bottom of the excavation and the envelope shown in Fig. 6.32e should be

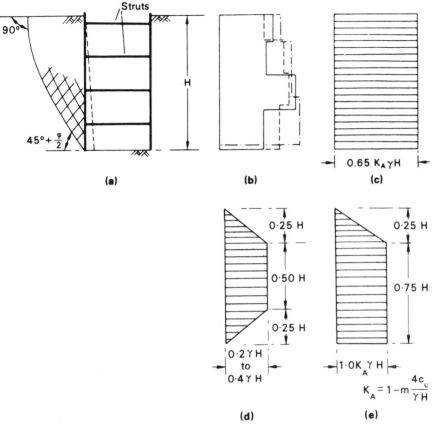

Fig. 6.32 Braced excavation.

used provided the abscissae are greater than those in Fig. 6.32d. If this is not the case, Fig. 6.32d should be used regardless of the value of the stability number. In general the value of m in Fig. 6.32e should be taken as 1·0: however in the case of soft, normally consolidated clays a value of m as low as 0·4 may be appropriate.

In the case of excavations in clay for which the stability number is greater than about 7 there is a possibility that the base of the excavation will fail by heaving (see Section 8.2) and this should be analysed before the strut loads are considered. Due to base heave and the inward deformation of the clay there will be horizontal and vertical movement of the soil outside the excavation. Such movements may result in damage to adjacent structures and services and should be monitored during excavation: advance warning of excessive movement or possible instability can thus be obtained. In general the greater the flexibility of the wall system and the longer the time before struts or anchors are installed, the greater will be the movements outside the excavation.

6.9 Diaphragm Walls

A diaphragm wall is a relatively thin reinforced concrete membrane cast in a trench, the sides of which are supported prior to casting by the hydrostatic pressure of a slurry of bentonite (a montmorillonite clay) in water. When mixed with water, bentonite readily disperses to form a colloidal suspension which exhibits thixotropic properties, i.e. it gels when left undisturbed but becomes fluid when agitated. The trench, the width of which is equal to that of the wall, is excavated progressively in suitable lengths from the ground surface, generally using a power-closing clamshell grab: shallow concrete guide walls are normally constructed as an aid to excavation. The trench is filled with the bentonite slurry as excavation proceeds: excavation thus takes place through the slurry already in place. The excavation process turns the gel into a fluid but the gel becomes re-established when disturbance ceases. The slurry tends to become contaminated with soil and cement in the course of construction but can be cleaned and re-used.

The bentonite particles form a skin of very low permeability, known as the filter cake, on the excavated soil faces. This occurs due to the fact that water filters from the slurry into the soil, leaving a layer of bentonite particles, a few millimetres thick, on the surface of the soil. Consequently, the full hydrostatic pressure of the slurry acts against the sides of the trench enabling stability to be maintained. The filter cake will form only if the fluid pressure in the trench is greater than the pore water pressure in the soil: a high water table level can thus be a considerable impediment to diaphragm wall construction. In soils of low permeability, such as clays, there will be virtually no filtration of water into the soil and therefore no significant filter cake formation will take place. In soils of high permeability, such as sandy gravels, there may be excessive loss of bentonite into the soil, resulting in a layer of bentonite-impregnated soil and poor filter cake formation. However, if a small quantity of fine sand (around 1%) is added to the slurry the sealing mechanism in soils of high permeability can be improved, with a considerable reduction in bentonite loss. Trench stability depends on the presence of an efficient seal on the soil surface: the efficiency of the seal becomes more vital the higher the permeability of the soil.

A slurry having a relatively high density is desirable from the points of view of trench stability, reduction of loss into soils of high permeability and the retention of contaminating particles in suspension. On the other hand a slurry of relatively low density will be displaced more cleanly from the soil surfaces and the reinforcement, and will be more easily pumped and decontaminated. The specification for the slurry must reflect a compromise between these conflicting requirements. Slurry specifications are usually based on density, viscosity, gel strength and pH.

On completion of excavation the reinforcement is positioned and the length of trench is filled with wet concrete using a tremie pipe. The wet concrete (which has a density of approximately twice that of the slurry) displaces the slurry upwards from the bottom of the trench, the tremie being raised in stages as the level of concrete rises. Once the wall (constructed as a series of individual panels keyed together) has been completed and the concrete has achieved adequate strength, the soil on one side of the wall can be excavated. It is usual for ground anchors (Section 8.8) to be installed at appropriate levels, as excavation proceeds, to tie the wall back into the retained soil. The method is very convenient for the construction of deep basements and underpasses, an important advantage being that the wall can be constructed close to adjoining structures: provided that the soil is moderately compact, ground deformations are usually tolerable. Diaphragm walls are often preferred to sheet pile walls because of their relative rigidity and their ability to be incorporated as part of the structure.

The decision whether to use a triangular or a trapezoidal distribution of lateral pressure in the design of a diaphragm wall depends on the anticipated wall deformation. A triangular distribution would probably be indicated in the case of a single row of tie-backs near the top of the wall. In the case of multiple rows of tie-backs over the height of the wall, a trapezoidal distribution might be considered appropriate.

Trench Stability

It is assumed that the full hydrostatic pressure of the slurry acts against the sides of the trench, the bentonite forming a thin, virtually impermeable layer on the surface of the soil. Consider a wedge of soil above a plane inclined at angle α to the horizontal, as shown in Fig. 6.33. When the wedge is on the point of sliding into the trench, i.e. the soil within the wedge is in a condition of limiting equilibrium, the angle α can be assumed to be $(45° + \phi/2)$. The unit weight of the slurry is γ_s and that of the soil is γ. The depth of slurry is nH and the height of the water table above the bottom of the trench is mH, where H is the depth of the trench. The normal and tangential components of the resultant force on the failure plane are N and T respectively.

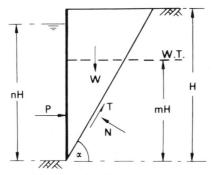

Fig. 6.33 Stability of slurry trench.

Considering force equilibrium,

$$P + T\cos\alpha - N\sin\alpha = 0 \tag{6.25}$$

$$W - T\sin\alpha - N\cos\alpha = 0 \tag{6.26}$$

Now:

$$P = \tfrac{1}{2}\gamma_s(nH)^2$$

and

$$W = \tfrac{1}{2}\gamma H^2 \cot\alpha$$

In the case of a sand $(c' = 0)$ an effective stress analysis is relevant. Hence:

$$T = (N - U)\tan\phi'$$

where U, the boundary water force on the failure plane, is given by

$$U = \tfrac{1}{2}\gamma_w(mH)^2 \operatorname{cosec}\alpha$$

and

$$\alpha = 45° + \phi'/2$$

In the case of a saturated clay $(\phi_u = 0)$ a total stress analysis is relevant. Hence:

$$T = c_u H \operatorname{cosec}\alpha$$

where $\alpha = 45°$.

Using Equations 6.25 and 6.26 the minimum slurry density can be determined for a factor of safety (F) of unity against shear failure (the above expressions assume $F = 1$). Alternatively, for a slurry of given density, the factor of safety can be obtained by using mobilized shear strength parameters $\tan^{-1}(\tan\phi'/F)$ or c_u/F in the above equations. The slurry unit weight required for stability is very sensitive to the level of the water table.

6.10   Reinforced Earth

Reinforced earth consists of a compacted soil mass within which tensile reinforcing elements, usually in the form of horizontal metal strips, are embedded. It should be noted that the technique is the subject of patents by Henri Vidal and the Reinforced Earth Company. The mass is stabilized as a result of frictional forces developed between the soil and the reinforcing, the stresses within the soil mass being transferred to the elements which are thereby placed in tension. The soil used as the fill material should be predominantly coarse-grained and it has been proposed that not more than 10% of the particles should pass the 63 μm BS sieve. It is vital that the fill should be adequately drained to prevent it from becoming saturated. The reinforcing elements are usually of galvanized steel. Data available on the rate of corrosion of galvanized steel in soils indicates that elements of this material are likely to have a minimum service life of 120 years. Other materials used for reinforcement include stainless steel, aluminium alloys, plastics and geotextiles.

In a reinforced earth wall, a facing is attached to the reinforcing elements to prevent the soil from spilling out and to satisfy aesthetic requirements. The facing should be sufficiently flexible to withstand any deformation of the fill. In most cases the facing consists of either precast concrete units, which can move to a limited extent relative to one another, or pliant U-shaped steel sections arranged horizontally. Reinforced earth walls possess considerable inherent flexibility and consequently can withstand relatively large differential settlements. The basic features of a reinforced earth wall are shown in Fig. 6.34. A wall of this type is generally more economic than an equivalent concrete cantilever retaining wall. The reinforced earth principle can also be used in embankments, normally by the use of geotextiles.

The main consideration in design concerns the tensile stresses which must be carried by the reinforcing elements. Tensile failure of one of the elements could result in the progressive collapse of the whole structure. Also important is the frictional resistance between the reinforcement and the soil. Local slipping due to inadequate resistance would result in a redistribution of tensile stress and a gradual deformation of the structure, not necessarily leading to collapse.

The shearing resistance between soil and reinforcement can be determined by means of direct shear tests or full-scale pull-out tests. The resistance depends on the relative density of the soil, the effective vertical stress and the surface texture of the reinforcement. It should be noted that in the direct shear test a dense soil is free to dilate, whereas in an earth structure dilation is restricted. Values of angle of friction obtained from a direct shear test can thus be expected to be lower than those for the same initial density in an earth structure. Direct shear values would have to be adjusted to allow for dilatancy.

Experimental work has indicated that the maximum tensile stress in a reinforcing element occurs not at the face of the structure but at a point within the reinforced soil, the position of this point varying with depth as indicated by the curve AB in Fig. 6.34a. This curve divides the soil mass into an 'active' zone within which shear stresses on reinforcement act outwards towards the face of the structure and a zone of resistance within which the shear stresses act inwards.

Juran and Schlosser [6.9] developed a design method based on the stability analysis of the active zone. The assumed mode of failure is that the reinforcing elements fracture progressively at the points of maximum tensile stress and, consequently, that conditions of plastic equilibrium develop in a thin layer of soil along the path of fracture. The curve of maximum tensile stress therefore defines the potential failure surface. If it is assumed that the

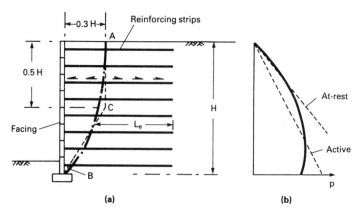

Fig. 6.34   Reinforced earth wall.

soil becomes perfectly plastic, the failure surface will be a logarithmic spiral. The spiral is assumed to pass through the bottom of the wall face and to intersect the surface at right angles, at a point $0\cdot3H$ from the face, as shown in Fig. 6.34a. The distribution of lateral pressure predicted by the method is of the form shown in Fig. 6.34b. The theoretical value of K, the lateral pressure coefficient, is approximately equal to K_0 in the upper part of the reinforced soil, decreasing to approximately K_A near the bottom.

A simplified analysis may be made by assuming that the curve of maximum tensile stress can be represented by the bilinear approximation ACB shown in Fig. 6.34a. Consider a reinforcing element at depth z below the surface of the soil mass. The tensile force in the element due to the transfer of lateral stress from the soil to the element is given by

$$T = K\sigma_z S_x S_z \tag{6.27}$$

where K is the appropriate earth pressure coefficient at depth z, σ_z is the vertical stress, S_x is the horizontal spacing of the elements and S_z is the vertical spacing. The maximum vertical stress within a reinforced earth wall will be greater than the overburden pressure (γz) because of the total active thrust of the backfill beyond the elements. Assuming a trapezoidal distribution of vertical stress (cf. Fig. 6.17c), the maximum value can be estimated as:

$$\sigma_z = \gamma z \left(1 + K_A \frac{z^2}{L^2}\right) \tag{6.28}$$

where L is the length of the reinforcing element at depth z and K_A is the active pressure coefficient for the backfill beyond the elements. Given the permissible tensile strength of the material, the required cross-sectional area of the element can be obtained from Equation 6.27.

The frictional resistance available on the surfaces of the element (only the top and bottom surfaces being considered) is given by

$$R = 2bL_e\gamma z \tan \delta \qquad (6.29)$$

where b = width of element, L_e = effective length of element, i.e. the length of element in the zone of resistance (beyond ACB in Fig. 6.34), and δ = angle of friction between soil and element. Hence the factor of safety against 'bond' failure, which should not be less than 2, is given by the ratio R/T.

The external stability of a reinforced earth structure must also be considered. A reinforced earth wall, although behaving as a relatively flexible structure, should be designed as if it were a gravity wall from the point of view of external stability. The back of the wall should be taken as the vertical plane through the inner end of the lowest reinforcing element. The total active thrust on this plane should be calculated by the Rankine theory. The factor of safety against sliding between the reinforced fill and the foundation soil should not be less than 2, the angle of shearing resistance of the weaker soil being used in the analysis. The pressure distribution on the base must be wholly compressive and must not exceed the allowable bearing capacity of the foundation soil. All potential failure surfaces encompassing the structure should be analysed, as for slopes (see Chapter 9): a minimum factor of safety of 1.5 should be ensured.

Problems

6.1 The backfill behind a retaining wall above the water table consists of a sand of unit weight $17 \, \text{kN/m}^3$, having shear strength parameters $c' = 0$ and $\phi' = 37°$. The height of the wall is 6 m and the surface of the backfill is horizontal. Determine the total active thrust on the wall according to the Rankine theory. If the wall is prevented from yielding, what is the approximate value of the thrust on the wall?

6.2 Plot the distribution of active pressure on the wall surface shown in Fig. 6.35. Calculate the total thrust on the wall (active + hydrostatic) and determine its point of application.

6.3 The front of a retaining wall slopes outwards at an angle of 10° to the vertical. The depth of soil in front of the wall is 2 m, the soil surface being horizontal, and the water table is well below the base of the wall. The following parameters are known for the soil: $c' = 0$, $\phi' = 34°$, $\delta = 15°$ and $\gamma = 18 \, \text{kN/m}^3$. Determine the total passive resistance available in front of the wall (a) according to Coulomb's theory, (b) using Sokolovski's coefficients.

6.4 Details of a cantilever retaining wall are given in Fig. 6.36. Calculate the maximum and minimum pressures under the base if the water table rises behind the wall to a level 3·90 m from the top of the wall. The shear strength parameters for the soil are $c' = 0$ and $\phi' = 38°$. The saturated unit weight of the soil is $20 \, \text{kN/m}^3$ and above the water table

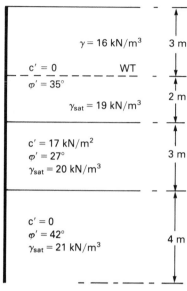

Fig. 6.35

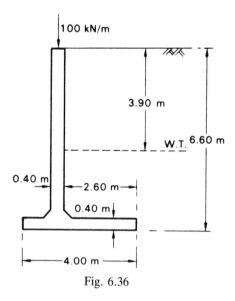

Fig. 6.36

the unit weight is $17\,\text{kN/m}^3$: the unit weight of the concrete is $23\cdot5\,\text{kN/m}^3$. If $\delta = 25°$ on the base of the wall, what is the factor of safety against sliding?

6.5 The sides of an excavation $3\cdot00\,\text{m}$ deep in sand are to be supported by a cantilever sheet pile wall. The water table is $1\cdot5\,\text{m}$ below the bottom

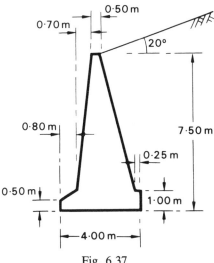

Fig. 6.37

of the excavation. The sand has a saturated unit weight of 20kN/m³, a unit weight of 17 kN/m³ above the water table and $\phi' = 36°$. Determine the depth of penetration of the piling below the bottom of the excavation to give a factor of safety of 2·0 with respect to passive resistance.

6.6 The section through a gravity retaining wall is shown in Fig. 6.37: the unit weight of the wall material is 23·5 kN/m³. The unit weight of the backfill is 19 kN/m³ and the appropriate shear strength parameters are $c' = 0$ and $\phi' = 36°$: the value of δ between the wall and the backfill and between the wall and the foundation soil is 25°. Calculate the maximum and minimum base pressures and the factor of safety against sliding. What would be the factor of safety against sliding if it were assumed that passive resistance is mobilised over a depth of 1·5 m in front of the wall?

6.7 An anchored sheet pile wall is constructed by driving a line of piling into a soil of saturated unit weight 21 kN/m³ and having shear strength parameters $c' = 10$ kN/m² and $\phi' = 27°$. Backfilling to a depth of 8 m is placed behind the piling, the backfill having a saturated unit weight of 20 kN/m³, a unit weight above the water table of 17 kN/m³ and shear strength parameters $c' = 0$ and $\phi' = 35°$. Tie rods are spaced at 2·5 m centres, 1·5 m below the surface of the backfill. The water level in front of the wall and the water table behind the wall are both 5 m below the surface of the backfill. Using the free earth support method, determine the depth of penetration required for a factor of safety of 2·0 with respect to passive resistance and the force in each tie rod.

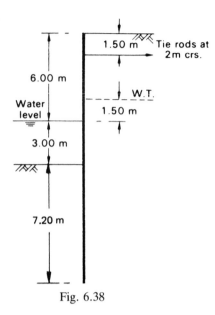

Fig. 6.38

6.8 The soil on both sides of the anchored sheet pile wall detailed in Fig. 6.38 has a saturated unit weight of $21 \, \text{kN/m}^3$, a unit weight above the water table of $18 \, \text{kN/m}^3$ and shear strength parameters $c' = 0$ and $\phi' = 36°$. There is a lag of $1\cdot5 \, \text{m}$ between the water table behind the wall and the tidal level in front. Using the free earth support method, determine the factor of safety with respect to passive resistance and the force in each tie rod.

6.9 Details of a propped cantilever wall are shown in Fig. 6.39. The relevant parameters for the soil are $c' = 0$, $\phi' = 30°$, $\delta = 15°$ (for which $K_A = 0\cdot30$ and $K_p = 4\cdot6$). The saturated unit weight of the soil is $20 \, \text{kN/m}^3$: above the water table the unit weight is $17 \, \text{kN/m}^3$. Determine the factor of safety of the wall according to the Burland–Potts–Walsh method.

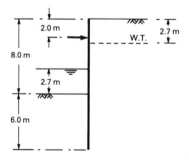

Fig. 6.39

6.10 The struts in a braced excavation 9 m deep in a dense sand are placed at 1·5 m centres vertically and 3·0 m centres horizontally: the bottom of the excavation is above the water table. The unit weight of the sand is 19 kN/m³ and the shear strength parameters are $c' = 0$ and $\phi' = 40°$. What load should each strut be designed to carry?

6.11 A diaphragm wall is to be constructed in a soil having a unit weight of 18 kN/m³ and shear strength parameters $c' = 0$ and $\phi' = 34°$. The depth of the trench is 3·50 m and the water table is 1·85 m above the bottom of the trench. Determine the factor of safety against failure of the trench if the unit weight of the slurry is 10·6 kN/m³ and the depth of slurry in the trench is 3·35 m.

References

6.1 Bishop, A. W. (1958): 'Test Requirements for Measuring the Coefficient of Earth Pressure at Rest', *Proc. Conference on Earth Pressure Problems, Brussels*, Vol. 1.

6.2 Bjerrum, L. and Andersen, K. (1972): 'In-situ Measurement of Lateral Pressures in Clay', *Proc. 5th European Conference SMFE, Madrid*, Vol. 1.

6.3 Brooker, E. W. and Ireland, H. O. (1965): 'Earth Pressures at Rest Related to Stress History', *Canadian Geotechnical Journal*, Vol. 2.

6.4 Burland, J. B., Potts, D. M. and Walsh, N. M. (1981): 'The Overall Stability of Free and Propped Embedded Cantilever Retaining Walls', *Ground Engineering*, Vol. 14, No. 5.

6.5 Caquot, A. and Kerisel, J. (1948): '*Tables for the Calculation of Passive Pressure, Active Pressure and Bearing Capacity of Foundations*', Gauthier-Villars, Paris.

6.6 Civil Engineering Code of Practice No. 2 (1951): *Earth Retaining Structures*, Institution of Structural Engineers, London.

6.7 Drucker, D. C. and Prager, W. (1952): 'Soil Mechanics and Plastic Analysis or Limit Design', *Q. Appl. Math.*, Vol. 10.

6.8 Ingold, T. S. (1982): *Reinforced Earth*, Thomas Telford, London.

6.9 Juran, I. and Schlosser, F. (1978): 'Theoretical Analysis of Failure in Reinforced Earth Structures', *Proc. Symposium on Earth Reinforcement*, ASCE Convention, Pittsburgh.

6.10 Mayne, P. W. and Kulhawy, F. H. (1982): 'K_0 – OCR Relationships in Soil', *Journal ASCE*, Vol. 108, No. GT6.

6.11 Nash, J. K. T. L. and Jones, G. K. (1963): 'The Support of Trenches Using Fluid Mud', *Grouts and Drilling Muds in Engineering Practice*, ICE, London.

6.12 Padfield, C. J. and Mair, R. J. (1984): 'Design of Retaining Walls Embedded in Stiff Clay', *CIRIA Report* 104.

6.13 Peck, R. B. (1969): 'Deep Excavations and Tunnelling in Soft

Ground', *Proc. 7th International Conference SMFE, Mexico* (State of the Art Volume).

6.14 Potts, D. M. and Fourie, A. B. (1984): 'The Behaviour of a Propped Retaining Wall: Results of a Numerical Experiment', *Geotechnique*, Vol. 34, No. 3.

6.15 Potts, D. M. and Fourie, A. B. (1985): 'The Effect of Wall Stiffness on the Behaviour of a Propped Retaining Wall', *Geotechnique*, Vol. 35, No. 3.

6.16 Rowe, P. W. (1952): 'Anchored Sheet Pile Walls', *Proc. Institution of Civil Engineers*, Part 1.

6.17 Rowe, P. W. (1957): 'Sheet Pile Walls in Clay', *Proc. Institution of Civil Engineers*, Part 1.

6.18 Rowe, P. W. and Peaker, K. (1965): 'Passive Earth Pressure Measurements', *Geotechnique*, Vol. 15, No. 1.

6.19 Sokolovski, V. V. (1965): *Statics of Granular Media*, Pergamon, Oxford.

6.20 Sowers, G. F., Robb, A. D., Mullis, C. H. and Glenn, A. J. (1957): 'The Residual Lateral Pressures Produced by Compacting Soils', *Proc. 4th International Conference SMFE, London*, Vol. 2.

6.21 Terzaghi, K. (1943): *Theoretical Soil Mechanics*, John Wiley and Sons, New York.

6.22 Terzaghi, K. (1954): 'Anchored Bulkheads', *Transactions ASCE*, Vol. 119, p. 1243.

6.23 Terzaghi, K. and Peck, R. B. (1967): *Soil Mechanics in Engineering Practice* (2nd Edition), John Wiley and Sons, New York.

6.24 Tschebotarioff. G. P. (1962): 'Retaining Structures', Chapter 5 of *Foundation Engineering* (Ed. G. A. Leonards), McGraw-Hill, New York.

6.25 Wroth, C. P. and Hughes, J. M. O. (1973): 'An Instrument for the In-Situ Measurement of the Properties of Soft Clays', *Proc. 8th International Conference SMFE, Moscow*, Vol. 1(2).

CHAPTER 7

Consolidation Theory

7.1 Introduction

As explained in Chapter 3, consolidation is the gradual reduction in volume of a fully saturated soil of low permeability due to drainage of some of the pore water, the process continuing until the excess pore water pressure set up by an increase in total stress has completely dissipated: the simplest case is that of one-dimensional consolidation, in which a condition of zero lateral strain is implicit. The process of swelling, the reverse of consolidation, is the gradual increase in volume of a soil under negative excess pore water pressure.

Consolidation settlement is the vertical displacement of the surface corresponding to the volume change at any stage of the consolidation process. Consolidation settlement will result, for example, if a structure is built over a layer of saturated clay or if the water table is lowered permanently in a stratum overlying a clay layer. If, on the other hand, an excavation is made in a saturated clay, heaving (the reverse of settlement) will result in the bottom of the excavation due to swelling of the clay. In cases in which significant lateral strain takes place, there will be an immediate settlement due to deformation of the soil under undrained conditions, in addition to consolidation settlement. Immediate settlement can be estimated using the results from elastic theory given in Chapter 5. This chapter is concerned with the prediction of both the magnitude and rate of consolidation settlement.

The progress of consolidation in situ can be monitored by installing piezometers to record the change in pore water pressure with time. The magnitude of settlement can be measured by recording the levels of suitable reference points on a structure or in the ground: precise levelling is essential, working from a benchmark which is not subject to even the slightest settlement. Every opportunity should be taken of obtaining settlement data as it is only through such measurements that the adequacy of theoretical methods can be assessed.

7.2   The Oedometer Test

The characteristics of a soil during one-dimensional consolidation or swelling can be determined by means of the oedometer test. Fig. 7.1 shows diagrammatically a cross-section through an oedometer. The test specimen is in the form of a disc, held inside a metal ring and lying between two porous stones. The upper porous stone, which can move inside the ring with a small clearance, is fixed below a metal loading cap through which pressure can be applied to the specimen. The whole assembly sits in an open cell of water to which the pore water in the specimen has free access. The ring confining the specimen may be either fixed (clamped to the body of the cell) or floating (being free to move vertically): the inside of the ring should have a smooth polished surface to reduce side friction. The confining ring imposes a condition of zero lateral strain on the specimen, the ratio of lateral to vertical effective stress being K_0, the coefficient of earth pressure at rest. The compression of the specimen under pressure is measured by means of a dial gauge operating on the loading cap.

The test procedure has been standardized in BS 1377 [7.4] which specifies that the oedometer shall be of the fixed ring type. The initial pressure will depend on the type of soil, then a sequence of pressures is applied to the specimen, each being double the previous value. Each pressure is normally maintained for a period of 24 hours (in exceptional cases a period of 48 hours may be required), compression readings being observed at suitable intervals during this period. At the end of the increment period, when the excess pore water pressure has completely dissipated, the applied pressure equals the effective vertical stress in the specimen. The results are presented by plotting the thickness (or percentage change in thickness) of the specimen or the void ratio at the end of each increment period against the corresponding effective stress. The effective stress may be plotted to either a natural or a logarithmic scale. If desired, the expansion of the specimen can be measured under successive decreases in applied pressure. However, even if the swelling characteristics of the soil are not required, the expansion of the specimen due to the removal of the final pressure should be measured.

The void ratio at the end of each increment period can be calculated from the dial gauge readings and either the water content or dry weight of the specimen at the end of the test. Referring to the phase diagram in Fig. 7.2, the two methods of calculation are as follows.

(1) Water content measured at end of test $= w_1$
     Void ratio at end of test $= e_1 = w_1 G_s$ (assuming $S_r = 100\%$)
     Thickness of specimen at start of test $= H_0$
     Change in thickness during test $= \Delta H$
     Void ratio at start of test $= e_0 = e_1 + \Delta e$
     where:

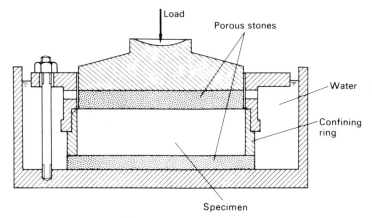

Fig. 7.1   The oedometer.

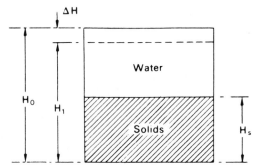

Fig. 7.2   Phase diagram.

$$\frac{\Delta e}{\Delta H} = \frac{1 + e_0}{H_0} \tag{7.1}$$

In the same way Δe can be calculated up to the end of any increment period.

(2) Dry weight measured at end of test $= M_s$ (i.e. mass of solids)
Thickness at end of any increment period $= H_1$
Area of specimen $= A$
Equivalent thickness of solids $= H_s = M_s/AG_s\rho_w$
Void ratio,

$$e_1 = \frac{H_1 - H_s}{H_s} = \frac{H_1}{H_s} - 1 \tag{7.2}$$

Compressibility Characteristics

Typical plots of void ratio (e) after consolidation, against effective stress (σ') for a saturated clay are shown in Fig. 7.3, the plots showing an initial

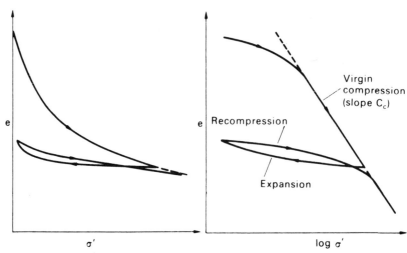

Fig. 7.3   Void ratio-effective stress relationship.

compression followed by expansion and recompression (cf. Fig. 4.10 for isotropic consolidation). The shapes of the curves are related to the stress history of the clay. The e-log σ' relationship for a normally consolidated clay is linear (or very nearly so) and is called the virgin compression line. If a clay is overconsolidated its state will be represented by a point on the expansion or recompression parts of the e-log σ' plot. The recompression curve ultimately joins the virgin compression line: further compression then occurs along the virgin line. During compression, changes in soil structure continuously take place and the clay does not revert to the original structure during expansion. The plots show that a clay in the overconsolidated state will be much less compressible than the same clay in a normally consolidated state.

The compressibility of the clay can be represented by one of the following coefficients.

(1) The coefficient of volume compressibility (m_v), defined as the volume change per unit volume per unit increase in effective stress. The units of m_v are the inverse of pressure (m^2/MN). The volume change may be expressed in terms of either void ratio or specimen thickness. If, for an increase in effective stress from σ'_0 to σ'_1 the void ratio decreases from e_0 to e_1, then:

$$m_v = \frac{1}{1 + e_0}\left(\frac{e_0 - e_1}{\sigma'_1 - \sigma'_0}\right) \tag{7.3}$$

$$= \frac{1}{H_0}\left(\frac{H_0 - H_1}{\sigma'_1 - \sigma'_0}\right) \tag{7.4}$$

The value of m_v for a particular soil is not constant but depends on the stress range over which it is calculated. BS 1377 specifies the use of the m_v coefficient calculated for a stress increment of 100 kN/m^2 in excess of the effective overburden pressure of the in-situ soil at the depth of interest, although the coefficient may also be calculated, if required, for any other stress range.

(2) *The compression index* (C_c) is the slope of the linear portion of the e-log σ' plot and is dimensionless. For any two points on the linear portion of the plot:

$$C_c = \frac{e_0 - e_1}{\log \dfrac{\sigma'_1}{\sigma'_0}} \tag{7.5}$$

The expansion part of the e-log σ' plot can be approximated to a straight line the slope of which is referred to as the *expansion index* C_e.

Preconsolidation Pressure

Casagrande proposed an empirical construction to obtain from the e-log σ' curve for an overconsolidated clay the maximum effective vertical stress that has acted on the clay in the past, referred to as the *preconsolidation pressure* (σ'_c). Fig. 7.4 shows a typical e-log σ' curve for a specimen of clay, initially overconsolidated. The initial curve indicates that the clay is undergoing recompression in the oedometer, having at some stage in its history undergone expansion. Expansion of the clay in situ may, for example, have been due to melting of ice sheets, erosion of overburden or a rise in water table level. The construction for estimating the preconsolidation pressure consists of the following steps.

1. Produce back the straight line part (BC) of the curve.
2. Determine the point (D) of maximum curvature on the recompression part (AB) of the curve.
3. Draw the tangent to the curve at D and bisect the angle between the tangent and the horizontal through D.
4. The vertical through the point of intersection of the bisector and CB produced gives the approximate value of the preconsolidation pressure.

Whenever possible the preconsolidation pressure for an overconsolidated clay should not be exceeded in construction. Compression will not usually be great if the effective vertical stress remains below σ'_c: only if σ'_c is exceeded will compression be large.

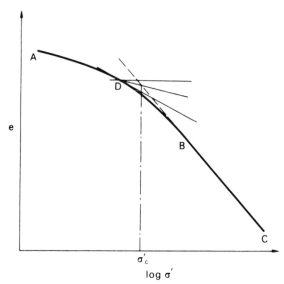

Fig. 7.4   Determination of preconsolidation pressure.

In-Situ e-log σ' Curve

Due to the effects of sampling and preparation the specimen in an oedometer test will be slightly disturbed. It has been shown that an increase in the degree of specimen disturbance results in a slight decrease in the slope of the virgin compression line. It can therefore be expected that the slope of the line representing virgin compression of the in-situ soil will be slightly greater than the slope of the virgin line obtained in a laboratory test.

No appreciable error will be involved in taking the in-situ void ratio as being equal to the void ratio (e_0) at the start of the laboratory test. Schmertmann [7.17] pointed out that the laboratory virgin line may be expected to intersect the in-situ virgin line at a void ratio of approximately 0·42 times the initial void ratio. Thus the in-situ virgin line can be taken as the line EF in Fig. 7.5 where the coordinates of E are $\log \sigma'_c$ and e_0, and F is the point on the laboratory virgin line at a void ratio of 0·42 e_0.

In the case of overconsolidated clays the in-situ condition is represented by the point (G) having coordinates σ'_0 and e_0, where σ'_0 is the present effective overburden pressure. The in-situ recompression curve can be approximated to the straight line GH parallel to the mean slope of the laboratory recompression curve.

Example 7.1

The following compression readings were obtained in an oedometer test on a specimen of saturated clay ($G_s = 2·73$):

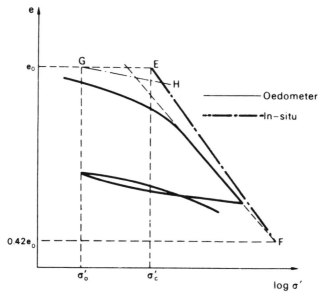

Fig. 7.5   In-situ e-log σ' curve.

Pressure (kN/m²)	0	54	107	214	429	858	1716	3432	0

Dial gauge
after 24 h (mm) 5·000 4·747 4·493 4·108 3·449 2·608 1·676 0·737 1·480

The initial thickness of the specimen was 19·0 mm and at the end of the test the water content was 19·8%. Plot the e-log σ' curve and determine the preconsolidation pressure. Determine the values of m_v for the stress increments 100–200 kN/m² and 1000–1500 kN/m². What is the value of C_c for the latter increment?

Void ratio at end of test $= e_1 = w_1 G_s = 0·198 \times 2·73 = 0·541$
Void ratio at start of test $= e_0 = e_1 + \Delta e$

Now,

$$\frac{\Delta e}{\Delta H} = \frac{1 + e_0}{H_0} = \frac{1 + e_1 + \Delta e}{H_0}$$

i.e.

$$\frac{\Delta e}{3·520} = \frac{1·541 + \Delta e}{19·0}$$

$$\Delta e = 0·350$$

$$e_0 = 0·541 + 0·350 = 0·891$$

In general the relationship between Δe and ΔH is given by:

$$\frac{\Delta e}{\Delta H} = \frac{1 \cdot 891}{19 \cdot 0}$$

i.e. $\Delta e = 0 \cdot 0996 \, \Delta H$, and can be used to obtain the void ratio at the end of each increment period (see Table 7.1). The e-log σ' curve using these values is shown in Fig. 7.6. Using Casagrande's construction the value of the preconsolidation pressure is 325 kN/m².

$$m_v = \frac{1}{1 + e_0} \cdot \frac{e_0 - e_1}{\sigma_1' - \sigma_0'}$$

For $\sigma_0' = 100 \, \text{kN/m}^2$ and $\sigma_1' = 200 \, \text{kN/m}^2$,

$$e_0 = 0 \cdot 845 \quad \text{and} \quad e_1 = 0 \cdot 808$$

therefore

$$m_v = \frac{1}{1 \cdot 845} \times \frac{0 \cdot 037}{100} = 2 \cdot 0 \times 10^{-4} \, \text{m}^2/\text{kN} = 0 \cdot 20 \, \text{m}^2/\text{MN}$$

For $\sigma_0' = 1000 \, \text{kN/m}^2$ and $\sigma_1' = 1500 \, \text{kN/m}^2$,

$$e_0 = 0 \cdot 632 \quad \text{and} \quad e_1 = 0 \cdot 577$$

therefore

$$m_v = \frac{1}{1 \cdot 632} \times \frac{0 \cdot 055}{500} = 6 \cdot 7 \times 10^{-5} \, \text{m}^2/\text{kN} = 0 \cdot 067 \, \text{m}^2/\text{MN}$$

and

$$C_c = \frac{0 \cdot 632 - 0 \cdot 577}{\log \dfrac{1500}{1000}} = \frac{0 \cdot 055}{0 \cdot 176} = 0 \cdot 31$$

Note that C_c will be the same for any stress range on the linear part of the e-log σ' curve; m_v will vary according to the stress range, even for ranges on the linear part of the curve.

7.3   Consolidation Settlement: One-Dimensional Method

In order to estimate consolidation settlement, the value of either the coefficient of volume compressibility or the compression index is required. Consider a layer of saturated clay of thickness H: due to construction the total vertical stress in an elemental layer of thickness dz at depth z is increased by $\Delta\sigma$ (Fig. 7.7). It is assumed that the condition of zero lateral strain applies within the clay layer. After the completion of consolidation an equal increase $\Delta\sigma'$ in effective vertical stress will have taken place corresponding to a stress increase from σ_0' to σ_1' and a reduction in void

Table 7.1

Pressure (kN/m²)	ΔH (mm)	Δe	e
0	0	0	0·891
54	0·253	0·025	0·866
107	0·507	0·050	0·841
214	0·892	0·089	0·802
429	1·551	0·154	0·737
858	2·392	0·238	0·653
1716	3·324	0·331	0·560
3432	4·263	0·424	0·467
0	3·520	0·350	0·541

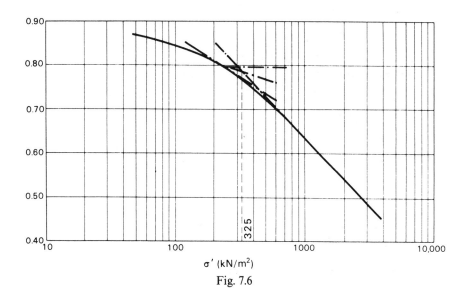

σ' (kN/m²)

Fig. 7.6

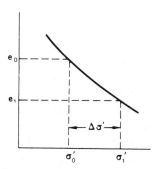

Fig. 7.7 Consolidation settlement.

ratio from e_0 to e_1 on the e-σ' curve. The reduction in volume per unit volume of clay can be written in terms of void ratio:

$$\frac{\Delta V}{V_0} = \frac{e_0 - e_1}{1 + e_0}$$

Since the lateral strain is zero, the reduction in volume per unit volume is equal to the reduction in thickness per unit thickness, i.e. the settlement per unit depth. Therefore, by proportion, the settlement of the layer of thickness dz will be given by:

$$ds_c = \frac{e_0 - e_1}{1 + e_0} dz$$

$$= \left(\frac{e_0 - e_1}{\sigma_1' - \sigma_0'}\right)\left(\frac{\sigma_1' - \sigma_0'}{1 + e_0}\right) dz$$

$$= m_v \Delta\sigma' dz$$

where s_c = consolidation settlement.

The settlement of the layer of thickness H is given by

$$s_c = \int_0^H m_v \Delta\sigma' dz$$

If m_v and $\Delta\sigma'$ are assumed constant with depth, then

$$s_c = m_v \Delta\sigma' H \tag{7.6}$$

or

$$s_c = \frac{e_0 - e_1}{1 + e_0} H \tag{7.7}$$

or, in the case of a normally consolidated clay,

$$s_c = \frac{C_c \log\dfrac{\sigma_1'}{\sigma_0'}}{1 + e_0} H \tag{7.8}$$

In order to take into account the variation of m_v and/or $\Delta\sigma'$ with depth, the graphical procedure shown in Fig. 7.8 can be used to determine s_c. The variations of initial effective vertical stress (σ_0') and effective vertical stress increment $(\Delta\sigma')$ over the depth of the layer are represented in Fig. 7.8a: the variation of m_v is represented in Fig. 7.8b. The curve in Fig. 7.8c represents the variation with depth of the dimensionless product $m_v\Delta\sigma'$ and the area under this curve is the settlement of the layer. Alternatively the layer can be divided into a suitable number of sublayers and the product $m_v\Delta\sigma'$ evaluated at the centre of each sublayer: each product $m_v\Delta\sigma'$ is then multiplied by the appropriate sublayer thickness to give the sublayer

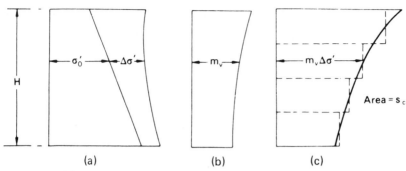

Fig. 7.8   Consolidation settlement: graphical procedure.

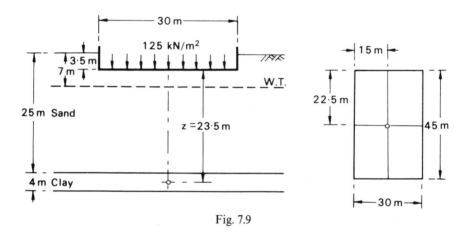

Fig. 7.9

settlement. The settlement of the whole layer is equal to the sum of the sublayer settlements.

Example 7.2

A building is supported on a raft 45 m × 30 m, the net foundation pressure (assumed to be uniformly distributed) being 125 kN/m². The soil profile is as shown in Fig. 7.9. The value of m_v for the clay is 0·35 m²/MN. Determine the final settlement under the centre of the raft due to consolidation of the clay.

The clay layer is thin relative to the dimensions of the raft, therefore it can be assumed that consolidation is approximately one-dimensional. In this case it will be sufficiently accurate to consider the clay layer as a whole. Because the consolidation settlement is to be calculated in terms of m_v, only the effective stress *increment* at mid-depth of the layer is required (the increment being assumed constant over the depth of the layer). Also,

$\Delta\sigma' = \Delta\sigma$ for one-dimensional consolidation and can be evaluated from Fig. 5.10.

At mid-depth of the layer, $z = 23\cdot5$ m. Below the centre of the raft:

$$m = \frac{22\cdot5}{23\cdot5} = 0\cdot96$$

$$n = \frac{15}{23\cdot5} = 0\cdot64$$

$$I_r = 0\cdot140$$
$$\Delta\sigma' = 4 \times 0\cdot140 \times 125 = 70\,\text{kN/m}^2$$
$$s_c = m_v\Delta\sigma'H = 0\cdot35 \times 70 \times 4 = 98\,\text{mm}$$

7.4   Settlement by the Skempton-Bjerrum Method

Predictions of consolidation settlement using the one-dimensional method are based on the results of oedometer tests using representative samples of the clay. Due to the confining ring the net lateral strain in the test specimen is zero and for this condition the initial excess pore water pressure is equal theoretically to the increase in total vertical stress, i.e. the pore pressure coefficient A is equal to unity.

In practice the condition of zero lateral strain is satisfied approximately in the cases of thin clay layers and of layers under loaded areas which are large compared with the layer thickness. In many practical situations, however, significant lateral strain will occur and the initial excess pore water pressure will depend on the in-situ stress conditions and the value of the pore pressure coefficient A (which will not be equal to unity).

In cases in which the lateral strain is not zero, there will be an immediate settlement, under undrained conditions, in addition to the consolidation settlement. Immediate settlement is zero if the lateral strain is zero, as assumed in the one-dimensional method of calculating settlement. In the Skempton-Bjerrum method [7.21] the total settlement (s) of a foundation on clay is given by

$$s = s_i + s_c$$

where s_i = immediate settlement, occurring under undrained conditions, and s_c = consolidation settlement, due to the volume reduction accompanying the gradual dissipation of excess pore water pressure.

The immediate settlement (s_i) can be estimated from the results of elastic theory (Section 5.3). The value of Poisson's ratio (v) relevant to undrained conditions in a fully saturated clay is taken to be $0\cdot5$. The undrained Young's modulus (E_u) must be estimated from the results of laboratory tests, in-situ loading tests or from correlations with undrained shear strength c_u.

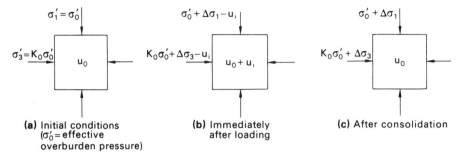

$\sigma_1' = \sigma_0'$ $\sigma_0' + \Delta\sigma_1 - u_i$ $\sigma_0' + \Delta\sigma_1$

$\sigma_3' = K_0\sigma_0'$ u_0 $K_0\sigma_0' + \Delta\sigma_3 - u_i$ $u_0 + u_i$ $K_0\sigma_0' + \Delta\sigma_3$ u_0

(a) Initial conditions   **(b)** Immediately   **(c)** After consolidation
($\sigma_0' =$ effective   after loading
overburden pressure)

Fig. 7.10   In-situ effective stresses.

The initial excess pore water pressure at a point in the clay layer is given by Equation 4.25 (with $B = 1$ for a fully saturated soil), i.e.

$$u_i = \Delta\sigma_3 + A(\Delta\sigma_1 - \Delta\sigma_3)$$

$$= \Delta\sigma_1\left[A + \frac{\Delta\sigma_3}{\Delta\sigma_1}(1 - A)\right] \tag{7.9}$$

where $\Delta\sigma_1$ and $\Delta\sigma_3$ are the total principal stress increments due to surface loading. From Equation 7.9 it is seen that

$$u_i > \Delta\sigma_3$$

if A is positive. Note also that $u_i = \Delta\sigma_1$ if $A = 1$. The value of A depends on the type of clay, the stress levels and the stress system.

The in-situ effective stresses before loading, immediately after loading and after consolidation are represented in Fig. 7.10 and the corresponding Mohr circles (A, B and C respectively) in Fig. 7.11. In Fig. 7.11, *abc* is the effective stress path for in-situ loading and consolidation, *ab* representing an immediate change of stress and *bc* a gradual change of stress as the excess pore water pressure dissipates. Immediately after loading there is a reduction in σ_3' due to u_i being greater than $\Delta\sigma_3$ and lateral expansion will occur. Subsequent consolidation will therefore involve lateral recompression. Circle D in Fig. 7.11 represents the corresponding stresses in the oedometer test after consolidation and *ad* is the corresponding effective stress path for the oedometer test. As the excess pore water pressure dissipates, Poisson's ratio decreases from the undrained value (0·5) to the drained value at the end of consolidation. The decrease in Poisson's ratio does not significantly affect the vertical stress but results in a small decrease in horizontal stress (point *c* would become *c'* in Fig. 7.11): this decrease is neglected in the Skempton-Bjerrum method.

Skempton and Bjerrum proposed that the effect of lateral strain be neglected in the calculation of consolidation settlement (s_c), thus enabling the oedometer test to be maintained as the basis of the method. It was admitted, however, that this simplification could involve errors of up to 20% in vertical settlements. However, the value of excess pore water

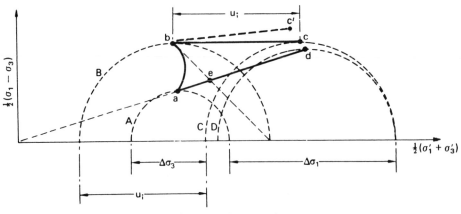

Fig. 7.11   Stress paths.

pressure given by Equation 7.9 is used in the method.

By the one-dimensional method, consolidation settlement (equal to the total settlement) is given as:

$$S_{\text{oed}} = \int_0^H m_v \Delta\sigma_1 \, dz \qquad (\text{i.e. } \Delta\sigma' = \Delta\sigma_1)$$

where H = thickness of clay layer, and $_{\text{oed}}$ = 'based on oedometer test only'. By the Skempton-Bjerrum method, consolidation settlement is expressed in the form:

$$s_c = \int_0^H m_v u_i \, dz$$

$$= \int_0^H m_v \Delta\sigma_1 \left[A + \frac{\Delta\sigma_3}{\Delta\sigma_1}(1 - A) \right] dz$$

A settlement coefficient μ is introduced, such that

$$s_c = \mu s_{\text{oed}} \qquad\qquad\qquad\qquad\qquad\qquad (7.10)$$

where

$$\mu = \frac{\displaystyle\int_0^H m_v \Delta\sigma_1 \left[A + \frac{\Delta\sigma_3}{\Delta\sigma_1}(1 - A) \right] dz}{\displaystyle\int_0^H m_v \Delta\sigma_1 \, dz}$$

If it can be assumed that m_v and A are constant with depth (sublayers can be used in analysis) then μ can be expressed as:

$$\mu = A + (1 - A)\alpha \qquad (7.11)$$

where

$$\alpha = \frac{\displaystyle\int_0^H \Delta\sigma_3 \, dz}{\displaystyle\int_0^H \Delta\sigma_1 \, dz}$$

Taking Poisson's ratio (v) as 0·5 for a saturated clay during loading under undrained conditions, the value of α depends only on the shape of the loaded area and the thickness of the clay layer in relation to the dimensions of the loaded area: thus α can be estimated from elastic theory.

The value of initial excess pore water pressure (u_i) should, in general, correspond to the in-situ stress conditions. The use of a value of pore pressure coefficient A obtained from the results of a triaxial test on a cylindrical clay specimen is strictly applicable only for the condition of axial symmetry, i.e. for the case of settlement under the centre of a circular footing. However, a value of A so obtained will serve as a good approximation for the case of settlement under the centre of a square footing (using a circular footing of the same area). Under a strip footing, however, plane strain conditions apply and the intermediate principal stress increment $\Delta\sigma_2$, in the direction of the longitudinal axis, is equal to $0 \cdot 5(\Delta\sigma_1 + \Delta\sigma_3)$. Scott [7.19] has shown that the value of u_i appropriate in the case of a strip footing can be obtained by using a pore pressure coefficient A_s, where

$$A_s = 0 \cdot 866A + 0 \cdot 211$$

The coefficient A_s replaces A (the coefficient for conditions of axial symmetry) in Equation 7.11 for the case of a strip footing, the expression for α being unchanged.

Values of the settlement coefficient μ, for circular and strip footings, in terms of A and the ratio of layer thickness/breadth of footing (H/B) are given in Fig. 7.12.

Values of μ are typically within the following ranges:

Soft, sensitive clays	1·0 to 1·2
Normally consolidated clays	0·6 to 1·0
Lightly overconsolidated clays	0·4 to 0·7
Heavily overconsolidated clays	0·25 to 0·4

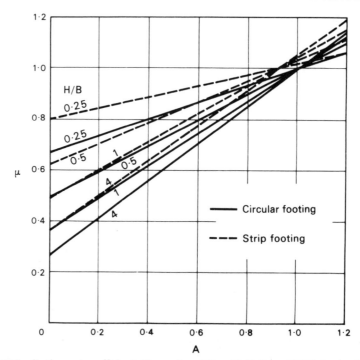

Fig. 7.12   Settlement coefficient. (Reproduced from R. F. Scott (1963) *Principles of Soil Mechanics*, by permission of Addison-Wesley Publishing Company, Inc., Reading, Mass.)

Example 7.3

A footing 6 m square, carrying a net pressure of 160 kN/m², is located at a depth of 2 m in a deposit of stiff clay 17 m thick: a firm stratum lies immediately below the clay. From oedometer tests on specimens of the clay the value of m_v was found to be 0·13 m²/MN and from triaxial tests the value of A was found to be 0·35. The undrained Young's modulus for the clay is estimated to be 55 MN/m². Determine the total settlement under the centre of the footing.

In this case there will be significant lateral strain in the clay beneath the footing (resulting in immediate settlement) and it is appropriate to use the Skempton-Bjerrum method. The section is shown in Fig. 7.13.

(a) Immediate settlement. The influence factors are obtained from Fig. 5.15. Now:

$$H/B = 15/6 = 2·5$$
$$D/B = 2/6 = 0·33$$
$$L/B = 1$$

$\therefore$ $\mu_0 = 0·91$ and $\mu_1 = 0·60$

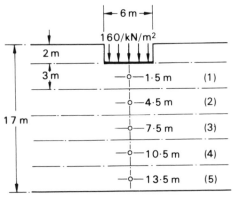

Fig. 7.13

Table 7.2

Layer	z (m)	m, n	I_r	$\Delta\sigma'$ (kN/m²)	s_{oed} (mm)
1	1·5	2·00	0·233	149	58·1
2	4·5	0·67	0·121	78	30·4
3	7·5	0·40	0·060	38	14·8
4	10·5	0·285	0·033	21	8·2
5	13·5	0·222	0·021	13	5·1
					116·6

Hence,

$$s_i = \mu_0 \mu_1 \frac{qB}{E_u}$$

$$= 0·91 \times 0·60 \times \frac{160 \times 6}{55} = 9·5 \,\text{mm}$$

(b) Consolidation settlement. In Table 7.2,

$$\Delta\sigma' = 4 \times 160 \times I_r \quad (\text{kN/m}^2)$$
$$s_{oed} = 0·13 \times \Delta\sigma' \times 3 = 0·39 \,\Delta\sigma' \quad (\text{mm})$$

Now,

$$\frac{H}{B} = \frac{15}{6·77} = 2·2$$

(Equivalent diameter = 6·77 m.)

$$A = 0.35$$

Hence, from Fig. 7.12,

$$\mu = 0.55$$

Then

$$s_c = 0.55 \times 116.6 = 64 \, \text{mm}$$

$$\text{Total settlement} = s_i + s_c$$
$$= 9 + 64$$
$$= 73 \, \text{mm}$$

7.5 The Stress Path Method

In this method it is recognized that soil deformation is dependent on the stress path followed prior to the final state of stress. The stress path for a soil element subjected to undrained loading followed by consolidation (neglecting the decrease in Poisson's ratio) is *abc* in Fig. 7.11, while the stress paths for consolidation only according to the one-dimensional and Skempton-Bjerrum methods are *ad* and *ed* respectively. In the stress path method, due to Lambe [7.12], the actual stress paths for a number of 'average' in-situ elements are estimated and laboratory triaxial tests are run as closely as possible along the same stress paths, beginning at the initial stresses prior to construction. The measured vertical strains (ε_1) during the test are then used to obtain the settlement, i.e. for a layer of thickness H:

$$s = \int_0^H \varepsilon_1 \, dz \tag{7.12}$$

In-situ pore water pressure conditions and partial drainage during the construction period can be simulated if desired. As an example, Fig. 7.14 shows a soil element under a circular storage tank and the effective stress path and corresponding vertical strains for a triaxial specimen in which undrained loading (*ab*), consolidation (*bc*), undrained unloading (*cd*) and swelling (*de*) are simulated.

Although sound in principle, the method depends on the correct selection of typical soil elements and on the test specimens being truly representative of the in-situ material. In addition, the triaxial techniques involved in running the correct stress paths are complex and time-consuming unless computer-controlled equipment is available. Knowledge of the value of K_0 is also required.

Simons and Som [7.20] investigated the effect of stress path on axial and volumetric compressibility and proposed a method of calculating settle-

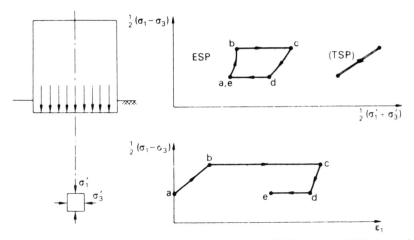

Fig. 7.14 The stress path method. (Reproduced from T. W. Lambe (1976) *Journal ASCE*, Vol. 93 No. SM6, by permission of the American Society of Civil Engineers.)

ment based on the relationship between the ratio of vertical to volumetric strain $(\varepsilon_1/\varepsilon_v)$ and the ratio $\Delta\sigma_3'/\Delta\sigma_1'$.

7.6 Degree of Consolidation

For an element of soil at a particular depth z in a clay layer the progress of the consolidation process under a particular total stress increment can be expressed in terms of void ratio as follows:

$$U_z = \frac{e_0 - e}{e_0 - e_1}$$

where U_z is defined as the degree of consolidation, at a particular instant of time, at depth z $(0 \leqslant U_z \leqslant 1)$, and e_0 = void ratio before the start of consolidation, e_1 = void ratio at the end of consolidation, and e = void ratio, at the time in question, during consolidation.

If the e-σ' curve is assumed to be linear over the stress range in question, as shown in Fig. 7.15, the degree of consolidation can be expressed in terms of σ':

$$U_z = \frac{\sigma' - \sigma_0'}{\sigma_1' - \sigma_0'}$$

Suppose that the total vertical stress in the soil at the depth z is increased from σ_0 to σ_1 and there is *no lateral strain*. Immediately after the increase takes place, although the total stress has increased to σ_1, the effective vertical stress will still be σ_0'; only after the completion of consolidation will

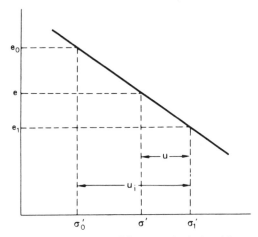

Fig. 7.15   Assumed linear e-σ' relationship.

the effective stress become σ'_1. During consolidation, $\Delta\sigma' = -\Delta u$.

If u_0 = pore water pressure *before* the increase in total stress; u_i (or Δu) = *increase* in pore water pressure above u_0 *immediately after* the increase in total stress; and u = pore water pressure *in excess of* u_0 at a particular time during consolidation under the increase in total stress:

$$\sigma'_1 = \sigma'_0 + u_i = \sigma' + u$$

The degree of consolidation can then be expressed as:

$$U_z = \frac{u_i - u}{u_i} = 1 - \frac{u}{u_i} \tag{7.13}$$

7.7   Terzaghi's Theory of One-Dimensional Consolidation

The assumptions made in the theory are:

1. The soil is homogeneous.
2. The soil is fully saturated.
3. The solid particles and water are incompressible.
4. Compression and flow are one-dimensional (vertical).
5. Strains are small.
6. Darcy's law is valid at all hydraulic gradients.
7. The coefficient of permeability and the coefficient of volume compressibility remain constant throughout the process.
8. There is a unique relationship, independent of time, between void ratio and effective stress.

Regarding assumption 6, there is evidence of deviation from Darcy's law at low hydraulic gradients. Regarding assumption 7, the coefficient of permeability decreases as the void ratio decreases during consolidation. The coefficient of volume compressibility also decreases during consolidation since the e-σ' relationship is non-linear. However for small stress increments assumption 7 is reasonable. The main limitations of Terzaghi's theory (apart from its one-dimensional nature) arise from assumption 8. Experimental results show that the relationship between void ratio and effective stress is not independent of time.

The theory relates the following three quantities.

1. The *excess* pore water pressure (u).
2. The depth (z) below the top of the clay layer.
3. The time (t) from the instantaneous application of a total stress increment.

Consider an element having dimensions dx, dy and dz within a clay layer of thickness $2d$, as shown in Fig. 7.16. An increment of total vertical stress $\Delta\sigma$ is applied to the element.

The flow velocity through the element is given by Darcy's law as

$$v_z = ki_z = -k\frac{\partial h}{\partial z}$$

Since any change in total head (h) is due only to a change in pore water pressure:

$$v_z = -\frac{k}{\gamma_w}\frac{\partial u}{\partial z}$$

The condition of continuity (Equation 2.7) can therefore be expressed as

$$-\frac{k}{\gamma_w}\frac{\partial^2 u}{\partial z^2}dx\,dy\,dz = \frac{dV}{dt} \tag{7.14}$$

The rate of volume change can be expressed in terms of m_v:

$$\frac{dV}{dt} = m_v\frac{\partial\sigma'}{\partial t}dx\,dy\,dz$$

The total stress increment is gradually transferred to the soil skeleton, increasing effective stress, as the excess pore water pressure decreases. Hence the rate of volume change can be expressed as

$$\frac{dV}{dt} = -m_v\frac{\partial u}{\partial t}dx\,dy\,dz \tag{7.15}$$

Combining Equations 7.14 and 7.15,

$$m_v\frac{\partial u}{\partial t} = \frac{k}{\gamma_w}\frac{\partial^2 u}{\partial z^2}$$

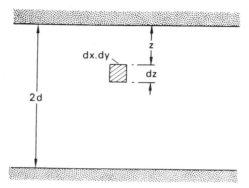

Fig. 7.16   Element within a clay layer.

or

$$\frac{\partial u}{\partial t} = c_v \frac{\partial^2 u}{\partial z^2} \tag{7.16}$$

This is the differential equation of consolidation, in which

$$c_v = \frac{k}{m_v \gamma_w} \tag{7.17}$$

c_v being defined as the *coefficient of consolidation*, suitable units being m²/year. Since k and m_v are assumed constant, c_v is constant during consolidation.

Solution of the Consolidation Equation

The total stress increment is assumed to be applied instantaneously, and at zero time will be carried entirely by the pore water, i.e. the initial value of excess pore water pressure (u_i) is equal to $\Delta\sigma$ and the initial condition is:

$u = u_i$ for $0 \leqslant z \leqslant 2d$ when $t = 0$

The upper and lower boundaries of the clay layer are assumed to be free draining, the permeability of the soil adjacent to each boundary being very high compared to that of the clay. Thus the boundary conditions at any time after the application of $\Delta\sigma$ are:

$u = 0$ for $z = 0$ and $z = 2d$ when $t > 0$

The solution for the excess pore water pressure at depth z after time t is:

$$u = \sum_{n=1}^{n=\infty} \left(\frac{1}{d} \int_0^{2d} u_i \sin\frac{n\pi z}{2d}\, dz \right) \left(\sin\frac{n\pi z}{2d} \right) \exp\left(-\frac{n^2\pi^2 c_v t}{4d^2} \right) \tag{7.18}$$

where d = length of longest drainage path, and u_i = initial excess pore water pressure, in general a function of z.

For the particular case in which u_i is constant throughout the clay layer:

$$u = \sum_{n=1}^{n=\infty} \frac{2u_i}{n\pi}(1 - \cos n\pi)\left(\sin \frac{n\pi z}{2d}\right)\exp\left(-\frac{n^2\pi^2 c_v t}{4d^2}\right) \tag{7.19}$$

When n is even, $(1 - \cos n\pi) = 0$, and when n is odd, $(1 - \cos n\pi) = 2$. Only odd values of n are therefore relevant and it is convenient to make the substitutions:

$$n = 2m + 1$$

and

$$M = \frac{\pi}{2}(2m + 1)$$

It is also convenient to substitute

$$T_v = \frac{c_v t}{d^2} \tag{7.20}$$

a dimensionless number called the *time factor*. Equation 7.19 then becomes

$$u = \sum_{m=0}^{m=\infty} \frac{2u_i}{M}\left(\sin \frac{Mz}{d}\right)\exp(-M^2 T_v) \tag{7.21}$$

The progress of consolidation can be shown by plotting a series of curves of u against z for different values of t. Such curves are called *isochrones* and their form will depend on the initial distribution of excess pore water pressure and the drainage conditions at the boundaries of the clay layer. A layer for which both the upper and lower boundaries are free-draining is described as an *open* layer: a layer for which only one boundary is free-draining is a *half-closed* layer. Examples of isochrones are shown in Fig. 7.17. In part (a) of the figure the initial distribution of u_i is constant and for an open layer of thickness $2d$ the isochrones are symmetrical about the centre line. The upper half of this diagram also represents the case of a half-closed layer of thickness d. The slope of an isochrone at any depth gives the hydraulic gradient and also indicates the direction of flow. In parts (b) and (c) of the figure, with a triangular distribution of u_i, the direction of flow changes over certain parts of the layer. In part (c) the lower boundary is impermeable and for a time swelling takes place in the lower part of the layer.

The degree of consolidation at depth z and time t can be obtained by substituting the value of u (Equation 7.21) in Equation 7.13, giving

$$U_z = 1 - \sum_{m=0}^{m=\infty} \frac{2}{M}\left(\sin \frac{Mz}{d}\right)\exp(-M^2 T_v) \tag{7.22}$$

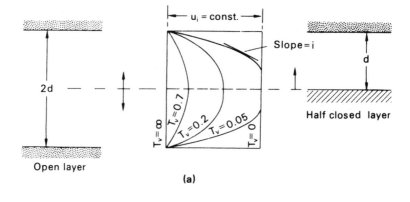

(a)

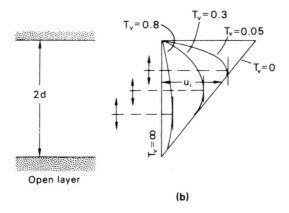

(b)

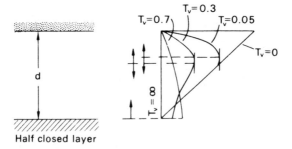

(c)

Fig. 7.17   Isochrones.

In practical problems it is the *average* degree of consolidation (U) over the depth of the layer as a whole that is of interest, the consolidation settlement at time t being given by the product of U and the final settlement. The average degree of consolidation at time t for constant u_i is given by

$$U = 1 - \frac{\dfrac{1}{2d}\displaystyle\int_0^{2d} u\,dz}{u_i}$$

$$= 1 - \sum_{m=0}^{m=\infty} \frac{2}{M^2}\exp(-M^2 T_v) \tag{7.23}$$

The relationship between U and T_v given by Equation 7.23 is represented by curve 1 in Fig. 7.18. Equation 7.23 can be represented almost exactly by the following empirical equations:

for $U < 0\cdot60$, $\quad T_v = \dfrac{\pi}{4}U^2$ $\tag{7.24a}$

for $U > 0\cdot60$, $\quad T_v = -0\cdot933\log(1 - U) - 0\cdot085$ $\tag{7.24b}$

If u_i is not constant the average degree of consolidation is given by

$$U = 1 - \frac{\displaystyle\int_0^{2d} u\,dz}{\displaystyle\int_0^{2d} u_i\,dz} \tag{7.25}$$

where

$$\int_0^{2d} u\,dz = \text{area under isochrone at the time in question}$$

and

$$\int_0^{2d} u_i\,dz = \text{area under initial isochrone}$$

(For a half-closed layer the limits of integration are 0 and d in the above equations.)

The initial variation of excess pore water pressure in a clay layer can usually be approximated in practice to a linear distribution. Curves 1, 2 and 3 in Fig. 7.18 represent the solution of the consolidation equation for the cases shown in Fig. 7.19.

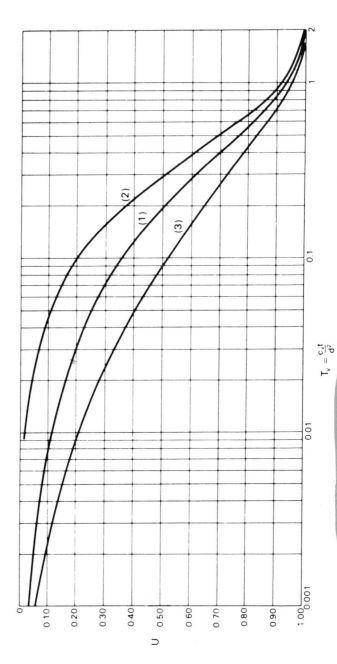

Fig. 7.18 Relationships between average degree of consolidation and time factor.

$$T_v = \frac{c_v t}{d^2}$$

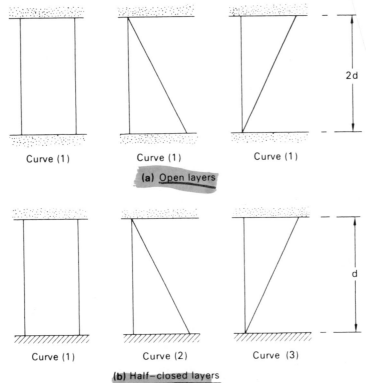

Curve (1)   Curve (1)   Curve (1)

(a) Open layers

Curve (1)   Curve (2)   Curve (3)

(b) Half–closed layers

Fig. 7.19   Initial variations of excess pore water pressure.

7.8   Determination of Coefficient of Consolidation

The value of c_v for a particular pressure increment in the oedometer test can be determined by comparing the characteristics of the experimental and theoretical consolidation curves, the procedure being referred to as *curve fitting*. The characteristics of the curves are brought out clearly if time is plotted to a square root or a logarithmic scale. Once the value of c_v has been determined, the coefficient of permeability can be calculated from Equation 7.17, the oedometer test being a useful method for obtaining the permeability of a clay.

The Log Time Method (due to Casagrande)

The forms of the experimental and theoretical curves are shown in Fig. 7.20. The experimental curve is obtained by plotting the dial gauge readings in the oedometer test against the logarithm of time in minutes. The theoretical curve is given as the plot of the average degree of consolidation against the logarithm of the time factor. The theoretical curve consists of three parts: an initial curve which approximates closely to a parabolic relationship, a part which is linear and a final curve to which the horizontal axis is an asymptote at $U = 1.0$ (or 100%). In the experimental curve the point corresponding to $U = 0$ can be determined by using the fact that the initial part of the curve represents an approximately parabolic relationship between compression and time. Two points on the curve are selected (A and B in Fig. 7.20) for which the values of t are in the ratio of 4:1, and the vertical distance between them is measured. An equal distance set off above the first point fixes the point (a_s) corresponding to $U = 0$. As a check the procedure should be repeated using different pairs of points. The point corresponding to $U = 0$ will not generally correspond to the point (a_0) representing the initial dial gauge reading, the difference being due mainly to the compression of small quantities of air in the soil, the degree of saturation being marginally below 100%: this compression is called *initial compression*. The final part of the experimental curve is linear but not horizontal and the point (a_{100}) corresponding to $U = 100\%$ is taken as the intersection of the two linear parts of the curve. The compression between the a_s and a_{100} points is called *primary consolidation* and represents that part of the process accounted for by Terzaghi's theory. Beyond the point of intersection, compression of the soil continues at a very slow rate for an indefinite period of time and is called *secondary compression*.

The point corresponding to $U = 50\%$ can be located midway between the a_s and a_{100} points and the corresponding time t_{50} obtained. The value of T_v corresponding to $U = 50\%$ is 0.196 and the coefficient of consolidation is given by

$$c_v = \frac{0.196\, d^2}{t_{50}} \tag{7.26}$$

the value of d being taken as half the average thickness of the specimen for the particular pressure increment. BS 1377 states that if the average temperature of the soil in situ is known and differs from the average test temperature, a correction should be applied to the value of c_v, correction factors being given in the standard.

The Root Time Method (due to Taylor)

Fig. 7.21 shows the forms of the experimental and theoretical curves, the dial gauge readings being plotted against the square root of time in minutes and the average degree of consolidation against the square root of time factor. The theoretical curve is linear up to about 60% consolidation and at 90% consolidation the abscissa (AC) is 1.15 times the abscissa (AB) of the production of the linear part of the curve. This characteristic is used to determine the point on the experimental curve corresponding to $U = 90\%$.

The experimental curve usually consists of a short curve representing

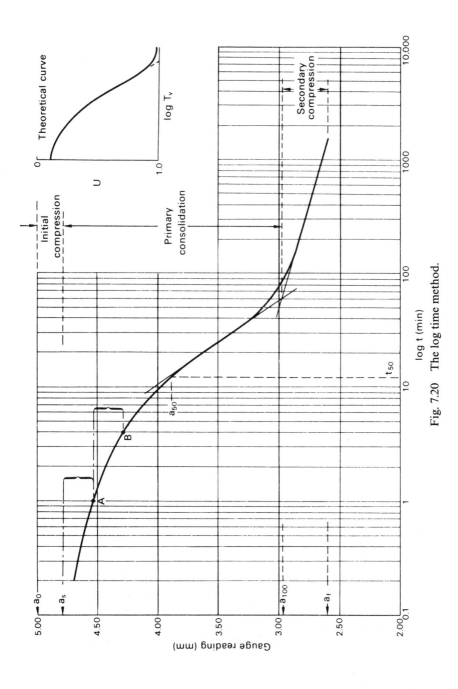

Fig. 7.20   The log time method.

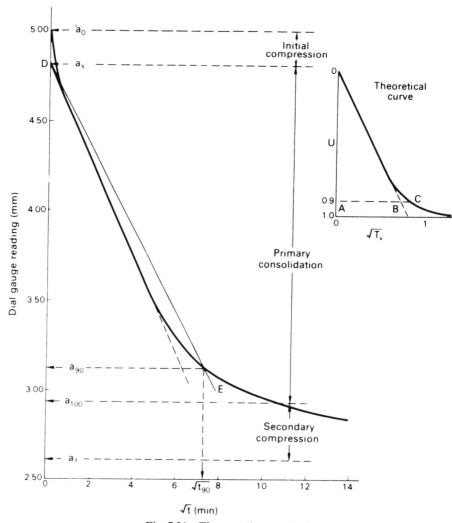

Fig. 7.21   The root time method.

initial compression, a linear part and a second curve. The point (D) corresponding to $U = 0$ is obtained by producing back the linear part of the curve to the ordinate at zero time. A straight line (DE) is then drawn having abscissae 1·15 times the corresponding abscissae on the linear part of the experimental curve. The intersection of the line DE with the experimental curve locates the point (a_{90}) corresponding to $U = 90\%$ and the corresponding value $\sqrt{t_{90}}$ can be obtained. The value of T_v corresponding to $U = 90\%$ is 0·848 and the coefficient of consolidation is given by

$$c_v = \frac{0 \cdot 848 d^2}{t_{90}} \tag{7.27}$$

If required, the point (a_{100}) on the experimental curve corresponding to $U = 100\%$, the limit of primary consolidation, can be obtained by proportion. As in the log time plot the curve extends beyond the 100% point into the secondary compression range. The root time method requires compression readings covering a much shorter period of time compared with the log time method, which requires the accurate definition of the second linear part of the curve well into the secondary compression range. On the other hand, a straight line portion is not always obtained on the root time plot and in such cases the log time method should be used.

Other methods of determining c_v have been proposed by Naylor and Doran [7.14], Scott [7.18] and Cour [7.6].

The Compression Ratios

The relative magnitudes of the initial compression, the compression due to primary consolidation and the secondary compression can be expressed by the following ratios (reference Figs. 7.20 and 7.21).

$$\text{Initial compression ratio: } r_0 = \frac{a_0 - a_s}{a_0 - a_f} \tag{7.28}$$

$$\text{Primary compression ratio (log time): } r_p = \frac{a_s - a_{100}}{a_0 - a_f} \tag{7.29}$$

$$\text{Primary compression ratio (root time): } r_p = \frac{10(a_s - a_{90})}{9(a_0 - a_f)} \tag{7.30}$$

$$\text{Secondary compression ratio: } r_s = 1 - (r_0 + r_p) \tag{7.31}$$

In-Situ Value of c_v

Settlement observations have indicated that the rates of settlement of full-scale structures are generally much greater than those predicted using values of c_v obtained from the results of oedometer tests on small specimens (e.g. 75 mm diameter $\times$ 20 mm thick). Rowe [7.15] has shown that such discrepancies are due to the influence of the clay macro-fabric on drainage behaviour. Features such as laminations, layers of silt and fine sand, silt-filled fissures, organic inclusions and root-holes, if they reach a major permeable stratum, have the effect of increasing the overall permeability of the clay mass. In general the macro-fabric of a clay is not represented accurately in a small oedometer specimen and the permeability of such a specimen will be lower than the mass permeability.

In cases where fabric effects are significant, more realistic values of c_v can be obtained by means of the hydraulic oedometer developed by Rowe and Barden [7.16] and manufactured for a range of specimen sizes. Specimens 250 mm in diameter by 100 mm thick are considered large enough to represent the natural macro-fabric of most clays: values of c_v obtained from tests on specimens of this size have been shown to be consistent with observed settlement rates.

Details of a hydraulic oedometer are shown in Fig. 7.22. Vertical pressure is applied to the specimen by means of water pressure acting across a convoluted rubber jack. The system used to apply the pressure must be capable of compensating for pressure changes due to leakage and specimen volume change. Compression of the specimen is measured by means of a central spindle passing through a sealed housing in the top plate of the oedometer. Drainage from the specimen can be either vertical or radial. Pore water pressure can be measured during the test and back pressure may be applied to the specimen. The apparatus can also be used for flow tests, from which the coefficient of permeability can be determined directly.

Piezometers can be used for the in-situ determination of c_v but the method requires the use of three-dimensional consolidation theory. The most satisfactory procedure is to maintain a constant head (above or below the ambient pore water pressure in the clay) at the piezometer tip and measure the rate of flow into or out of the system. If the rate of flow is measured at various times the value of c_v (and of the coefficient of permeability k) can be deduced. See papers by Gibson [7.8; 7.9] and Wilkinson [7.24] for details.

Another method of determining c_v is to combine laboratory values of m_v (which from experience are known to be more reliable than laboratory values of c_v) with in-situ measurements of k, using Equation 7.17.

Secondary Compression

In the Terzaghi theory it is implied by assumption 8 that a change in void ratio is due entirely to a change in effective stress brought about by the dissipation of excess pore water pressure, with permeability alone governing the time dependency of the process. However, experimental results show that compression does not cease when the excess pore water pressure has dissipated to zero but continues at a gradually decreasing rate under constant effective stress. Secondary compression is thought to be due to the gradual readjustment of the clay particles into a more stable configuration following the structural disturbance caused by the decrease in void ratio, especially if the clay is laterally confined. An additional factor is the gradual lateral displacements which take place in thick clay layers subjected to shear stresses. The rate of secondary compression is thought to be controlled by the highly viscous film of adsorbed water surrounding the clay mineral particles in the soil. A very slow viscous flow of adsorbed water takes place from the zones of film contact, allowing the solid particles to move closer together. The viscosity of the film increases as the particles move closer, resulting in a decrease in the rate of compression of the soil. It is presumed that primary consolidation and secondary compression

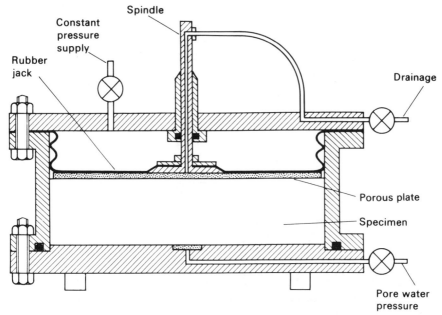

Fig. 7.22   Hydraulic oedometer.

proceed simultaneously from the time of loading.

The rate of secondary compression in the oedometer test can be defined by the slope (C_α) of the final part of the compression/log time curve, measured as the unit compression over one decade on the log time scale. The magnitude of secondary compression in a given time is generally greater in normally consolidated clays than in overconsolidated clays. For certain highly plastic clays and organic clays the secondary compression part of the compression/log time curve may completely mask the primary consolidation part. For a particular soil the magnitude of secondary compression over a given time, as a percentage of the total compression, increases as the ratio of pressure increment to initial pressure decreases: the magnitude of secondary compression also increases as the thickness of the oedometer specimen decreases and as temperature increases. Thus the secondary compression characteristics of an oedometer specimen cannot normally be extrapolated to the case of a full-scale foundation.

In a small number of normally consolidated clays it has been found that secondary compression forms the greater part of the total compression under applied pressure. Bjerrum [7.3] showed that such clays have gradually developed a reserve resistance against further compression as a result of the considerable decrease in void ratio which has occurred, under constant effective stress, over the hundreds or thousands of years since sedimentation. These clays, although normally consolidated, exhibit a quasi-preconsolidation pressure. It has been shown that provided any additional applied pressure is less than approximately 50% of the difference between the quasi-preconsolidation pressure and the effective overburden pressure the resultant settlement will be relatively small.

Example 7.4

The following compression readings were taken during an oedometer test on a saturated clay specimen ($G_s = 2 \cdot 73$) when the applied pressure was increased from 214 to 429 kN/m²:

Time (min)	0	$\frac{1}{4}$	$\frac{1}{2}$	1	$2\frac{1}{4}$	4	9	16	25
Gauge (mm)	5·00	4·67	4·62	4·53	4·41	4·28	4·01	3·75	3·49

Time (min)	36	49	64	81	100	200	400	1440
Gauge (mm)	3·28	3·15	3·06	3·00	2·96	2·84	2·76	2·61

After 1440 min the thickness of the specimen was 13·60 mm and the water content 35·9%. Determine the coefficient of consolidation from both the log time and the root time plots and the values of the three compression ratios. Determine also the value of the coefficient of permeability.

Total change in thickness during increment $= 5 \cdot 00 - 2 \cdot 61 = 2 \cdot 39$ mm

Average thickness during increment $= 13 \cdot 60 + 2 \cdot 39/2 = 14 \cdot 80$ mm

Length of drainage path, $d = 14 \cdot 80/2 = 7 \cdot 40$ mm

From the log time plot (Fig. 7.20),

$$t_{50} = 12 \cdot 5 \, \text{min}$$

$$c_v = \frac{0 \cdot 196 d^2}{t_{50}} = \frac{0 \cdot 196 \times 7 \cdot 40^2}{12 \cdot 5} \times \frac{1440 \times 365}{10^6} = 0 \cdot 45 \, \text{m}^2/\text{year}$$

$$r_0 = \frac{5 \cdot 00 - 4 \cdot 79}{5 \cdot 00 - 2 \cdot 61} = 0 \cdot 088$$

$$r_p = \frac{4 \cdot 79 - 2 \cdot 98}{5 \cdot 00 - 2 \cdot 61} = 0 \cdot 757$$

$$r_s = 1 - (0 \cdot 088 + 0 \cdot 757) = 0 \cdot 155$$

From the root time plot (Fig. 7.21) $\sqrt{t_{90}} = 7 \cdot 30$, therefore

$$t_{90} = 53 \cdot 3 \, \text{min}$$

$$c_v = \frac{0 \cdot 848 d^2}{t_{90}} = \frac{0 \cdot 848 \times 7 \cdot 40^2}{53 \cdot 3} \times \frac{1440 \times 365}{10^6} = 0 \cdot 46 \, \text{m}^2/\text{year}$$

$$r_0 = \frac{5 \cdot 00 - 4 \cdot 81}{5 \cdot 00 - 2 \cdot 61} = 0 \cdot 080$$

$$r_p = \frac{10(4\cdot81 - 3\cdot12)}{9(5\cdot00 - 2\cdot61)} = 0\cdot785$$

$$r_s = 1 - (0\cdot080 + 0\cdot785) = 0\cdot135$$

In order to determine the permeability, the value of m_v must be calculated.

Final void ratio: $e_1 = w_1 G_s = 0\cdot359 \times 2\cdot73 = 0\cdot98$

Initial void ratio: $e_0 = e_1 + \Delta e$

Now,

$$\frac{\Delta e}{\Delta H} = \frac{1 + e_0}{H_0}$$

i.e.

$$\frac{\Delta e}{2\cdot39} = \frac{1\cdot98 + \Delta e}{15\cdot99}$$

Therefore

$$\Delta e = 0\cdot35 \quad \text{and} \quad e_0 = 1\cdot33.$$

Now,

$$m_v = \frac{1}{1 + e_0} \cdot \frac{e_0 - e_1}{\sigma_1' - \sigma_0'}$$

$$= \frac{1}{2\cdot33} \times \frac{0\cdot35}{215} = 7\cdot0 \times 10^{-4}\,\text{m}^2/\text{kN}$$

$$= 0\cdot70\,\text{m}^2/\text{MN}$$

Coefficient of permeability:

$$k = c_v m_v \gamma_w$$

$$= \frac{0\cdot45 \times 0\cdot70 \times 9\cdot8}{60 \times 1440 \times 365 \times 10^3}$$

$$= 1\cdot0 \times 10^{-10}\,\text{m/s}$$

7.9 Correction for Construction Period

In practice, structural loads are applied to the soil not instantaneously but over a period of time. Initially there is usually a reduction in net load due to excavation, resulting in swelling of the clay: settlement will not begin until the applied load exceeds the weight of the excavated soil. Terzaghi proposed an empirical method of correcting the instantaneous time/settlement curve to allow for the construction period.

The net load (P') is the gross load less the weight of soil excavated and the effective construction period (t_c) is measured from the time when P' is zero. It is assumed that the net load is applied uniformly over the time t_c (Fig. 7.23) and that the degree of consolidation at time t_c is the same as if the load P' had been acting as a constant load for the period $t_c/2$. Thus the settlement at any time during the construction period is equal to that occurring for instantaneous loading at half that time; however, since the load acting is not the total load, the value of settlement so obtained must be reduced in the proportion of that load to the total load.

For the period subsequent to the completion of construction, the settlement curve will be the instantaneous curve offset by half the effective construction period. Thus at any time after the end of construction the corrected time corresponding to any value of settlement is equal to the time from the start of loading less half the effective construction period. After a

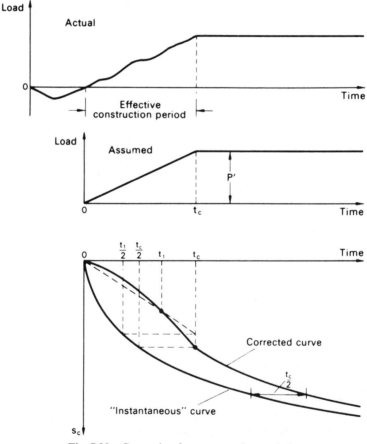

Fig. 7.23   Correction for construction period.

long period of time the magnitude of settlement is not appreciably affected by the construction time.

Example 7.5

A layer of clay 8 m thick lies between two layers of sand. The upper sand layer extends from ground level to a depth of 4 m, the water table being at a depth of 2 m. The lower sand layer is under artesian pressure, the piezometric level being 6 m above ground level. For the clay $m_v = 0.94\,\text{m}^2/\text{MN}$ and $c_v = 1.4\,\text{m}^2/\text{year}$. As a result of pumping from the artesian layer the piezometric level falls by 3 m over a period of 2 years. Draw the time/settlement curve due to consolidation of the clay for a period of 5 years from the start of pumping.

In this case, consolidation is due only to the difference in the static and steady-state pore water pressures: there is no change in total vertical stress. The effective vertical stress remains unchanged at the top of the clay layer but will be increased by $3\gamma_w$ at the bottom of the layer due to the decrease in pore water pressure in the adjacent artesian layer. The distribution of $\Delta\sigma'$ is shown in Fig. 7.24. The problem is one-dimensional since the increase in effective vertical stress is the same over the entire area in question. In calculating the consolidation settlement it is necessary to consider only the value of $\Delta\sigma'$ at the centre of the layer. Note that in order to obtain the value of m_v it would have been necessary to calculate the initial and final values of effective vertical stress in the clay.

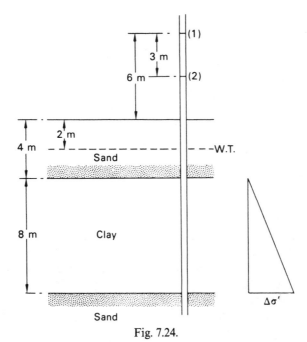

Fig. 7.24.

Table 7.3

U	T_v	t (years)	s_c (mm)
0·10	0·008	0·09	11
0·20	0·031	0·35	22
0·30	0·070	0·79	33
0·40	0·126	1·42	44
0·50	0·196	2·21	55
0·60	0·285	3·22	66
0·73	0·437	5·00	80

At the centre of the clay layer, $\Delta\sigma' = 1\cdot5\gamma_w = 14\cdot7\,\text{kN/m}^2$. The *final* consolidation settlement is given by:

$$s_{cf} = m_v\Delta\sigma'H$$
$$= 0\cdot94 \times 14\cdot7 \times 8$$
$$= 110\,\text{mm}$$

The clay layer is open, therefore $d = 4\,\text{m}$. For $t = 5$ years,

$$T_v = \frac{c_v t}{d^2}$$
$$= \frac{1\cdot4 \times 5}{4^2}$$
$$= 0\cdot437$$

From curve 1, Fig. 7.18, the corresponding value of U is 0·73. To obtain the time/settlement relationship a series of values of U is selected up to 0·73 and the corresponding times calculated from the time factor equation: the corresponding values of settlement (s_c) are given by the product of U and s_{cf}. (See Table 7.3.) The plot of s_c against t gives the 'instantaneous' curve. Terzaghi's method of correction for the 2 year period over which pumping takes place is then carried out as shown in Fig. 7.25.

Example 7.6

An 8 m depth of sand overlies a 6 m layer of clay, below which is an impermeable stratum (Fig. 7.26); the water table is 2 m below the surface of the sand. Over a period of 1 year a 3 m depth of fill (unit weight 20 kN/m³) is to be dumped on the surface over an extensive area. The saturated unit weight of the sand is 19 kN/m³ and that of the clay 20 kN/m³; above the water table the unit weight of the sand is 17 kN/m³. For the clay, the relationship between void ratio and effective stress (units kN/m²) can be represented by the equation

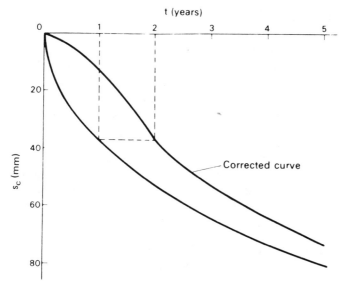

Fig. 7.25

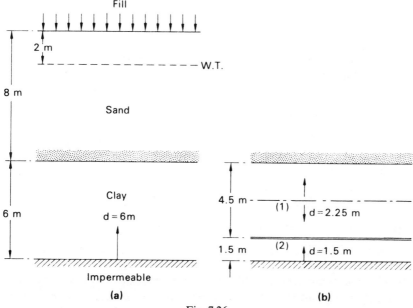

Fig. 7.26

$$e = 0.88 - 0.32 \log \frac{\sigma'}{100}$$

and the coefficient of consolidation is $1.26\,\text{m}^2/\text{year}$.

(a) Calculate the final settlement of the area due to consolidation of the clay and the settlement after a period of 3 years from the start of dumping.

(b) If a very thin layer of sand, freely draining, existed $1.5\,\text{m}$ above the bottom of the clay layer, what would be the values of the final and 3 year settlements?

(a) Since the fill covers a wide area, the problem can be considered to be one-dimensional. The consolidation settlement will be calculated in terms of C_c, considering the clay layer as a whole, therefore the initial and final values of effective vertical stress at the centre of the clay layer are required.

$$\sigma'_0 = (17 \times 2) + (9.2 \times 6) + (10.2 \times 3) = 119.8\,\text{kN/m}^2$$

$$e_0 = 0.88 - 0.32 \log 1.198 = 0.88 - 0.025 = 0.855$$

$$\sigma'_1 = 119.8 + (3 \times 20) = 179.8\,\text{kN/m}^2$$

$$\log \frac{179.8}{119.8} = 0.176$$

The final settlement is calculated from Equation 7.8:

$$S_{cf} = \frac{0.32 \times 0.176 \times 6000}{1.855} = 182\,\text{mm}$$

In the calculation of the degree of consolidation 3 years after the start of dumping, the corrected value of time to allow for the 1 year dumping period is

$$t = 3 - \tfrac{1}{2} = 2.5\,\text{years}$$

The layer is half-closed, therefore $d = 6\,\text{m}$. Then

$$T_v = \frac{c_v t}{d^2} = \frac{1.26 \times 2.5}{6^2}$$

$$= 0.0875$$

From curve 1, Fig. 7.18, $U = 0.335$. Settlement after 3 years:

$$S_c = 0.335 \times 182 = 61\,\text{mm}$$

(b) The final settlement will still be $182\,\text{mm}$ (ignoring the thickness of the drainage layer): only the rate of settlement will be affected. From the point of view of drainage there is now an open layer of thickness $4.5\,\text{m}$ ($d = 2.25\,\text{m}$) above a half-closed layer of thickness $1.5\,\text{m}$ ($d = 1.5\,\text{m}$): these layers are numbered 1 and 2 respectively.

By proportion,

$$T_{v_1} = 0.0875 \times \frac{6^2}{2.25^2} = 0.622$$

$$\therefore \quad U_1 = 0.825$$

and

$$T_{v_2} = 0.0875 \times \frac{6^2}{1.5^2} = 1.40$$

$$\therefore \quad U_2 = 0.97$$

Now for each layer, $s_c = U s_{cf}$, which is proportional to UH. Hence if $\bar{U}$ is the overall degree of consolidation for the two layers combined:

$$4.5 U_1 + 1.5 U_2 = 6.0 \bar{U}$$

i.e. $(4.5 \times 0.825) + (1.5 \times 0.97) = 6.0 \bar{U}$. Hence

$$\bar{U} = 0.86$$

and the 3 year settlement is:

$$s_c = 0.86 \times 182 = 157 \, \text{mm}$$

7.10   Numerical Solution

The one-dimensional consolidation equation can be solved numerically by the method of finite differences. The method has the advantage that any pattern of initial excess pore water pressure can be adopted and it is possible to consider problems in which the load is applied gradually over a period of time. The errors associated with the method are negligible and the solution is easily programmed for the computer.

The method is based on a depth-time grid as shown in Fig. 7.27. The depth of the clay layer is divided into m equal parts of thickness Δz and any specified period of time is divided into n equal intervals Δt. Any point on the grid can be identified by the subscripts i and j, the depth position of the point being denoted by $i \, (0 \leqslant i \leqslant m)$ and the elapsed time by $j \, (0 \leqslant j \leqslant n)$. The value of excess pore water pressure at any depth after any time is therefore denoted by $u_{i,j}$.

The following finite difference approximations can be derived from Taylor's theorem:

$$\frac{\partial u}{\partial t} = \frac{1}{\Delta t} (u_{i,j+1} - u_{i,j})$$

$$\frac{\partial^2 u}{\partial z^2} = \frac{1}{(\Delta z)^2} (u_{i-1,j} + u_{i+1,j} - 2u_{i,j})$$

Substituting these values in Equation 7.16 yields the finite difference approximation of the one-dimensional consolidation equation:

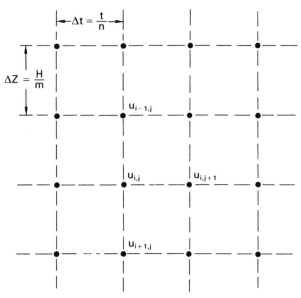

Fig. 7.27   Depth-time grid.

$$u_{i,j+1} = u_{i,j} + \frac{c_v \Delta t}{(\Delta z)^2}(u_{i-1,j} + u_{i+1,j} - 2u_{i,j}) \tag{7.32}$$

It is convenient to write

$$\beta = \frac{c_v \Delta t}{(\Delta z)^2} \tag{7.33}$$

this term being called the *operator* of Equation 7.32. It has been shown that for convergence the value of the operator must not exceed $\frac{1}{2}$. The errors due to neglecting higher order derivatives in Taylor's theorem are reduced to a minimum when the value of the operator is $\frac{1}{6}$.

It is usual to specify the number of equal parts m into which the depth of the layer is to be divided and as the value of β is limited a restriction is thus placed on the value of Δt. For any specified period of time t in the case of an *open* layer:

$$T_v = \frac{c_v(n\Delta t)}{(\frac{1}{2}m\Delta z)^2}$$

$$= 4\frac{n}{m^2}\beta \tag{7.34}$$

In the case of a *half-closed* layer the denominator becomes $(m\Delta z)^2$ and

$$T_v = \frac{n}{m^2}\beta \tag{7.35}$$

A value of n must therefore be chosen such that the value of β in Equation 7.34 or 7.35 does not exceed $\frac{1}{2}$.

Equation 7.32 does not apply to points on an impermeable boundary. There can be no flow across an impermeable boundary, a condition represented by the equation:

$$\frac{\partial u}{\partial z} = 0$$

which can be represented by the finite difference approximation:

$$\frac{1}{2\Delta z}(u_{i-1,j} - u_{i+1,j}) = 0$$

the impermeable boundary being at a depth position denoted by subscript i, i.e.

$$u_{i-1,j} = u_{i+1,j}$$

For all points *on* an impermeable boundary, Equation 7.32 becomes

$$u_{i,j+1} = u_{i,j} + \frac{c_v \Delta t}{(\Delta z)^2}(2u_{i-1,j} - 2u_{i,j}) \tag{7.36}$$

The degree of consolidation at any time t can be obtained by determining the areas under the initial isochrone and the isochrone at time t as in Equation 7.25.

Example 7.7

A half-closed clay layer (free-draining at the upper boundary) is 10 m thick and the value of c_v is 7·9 m²/year. The initial distribution of excess pore water pressure is as follows:

Depth (m)	0	2	4	6	8	10
Pressure (kN/m²)	60	54	41	29	19	15

Obtain the values of excess pore water pressure after consolidation has been in progress for 1 year.

The layer is half-closed, therefore $d = 10$ m. For $t = 1$ year,

$$T_v = \frac{c_v t}{d^2} = \frac{7·9 \times 1}{10^2} = 0·079$$

The layer is divided into five equal parts, i.e. $m = 5$. Now,

$$T_v = \frac{n}{m^2}\beta$$

therefore

$$n\beta = 0·079 \times 5^5 = 1·98, \text{ say } 2·0$$

(This makes the actual value of $T_v = 0·080$ and $t = 1·01$ years.) The value of

Table 7.4

i \ j	0	1	2	3	4	5	6	7	8	9	10
0	0	0	0	0	0	0	0	0	0	0	0
1	540	406	326	273	235	207	185	167	153	141	131
2	410	412	387	357	329	304	282	263	246	232	219
3	290	294	299	300	296	290	283	275	267	260	253
4	190	202	213	224	233	240	245	249	251	252	252
5	150	166	180	194	206	217	226	234	240	244	247

n will be taken as 10 (i.e. $\Delta t = 1/10$ year), making $\beta = 0\cdot2$. The finite difference equation then becomes:

$$u_{i, j+1} = u_{i, j} + 0\cdot2(u_{i-1, j} + u_{i+1, j} - 2u_{i, j})$$

but on the impermeable boundary:

$$u_{i, j+1} = u_{i, j} + 0\cdot2(2u_{i-1, j} - 2u_{i, j})$$

On the permeable boundary, $u = 0$ for all values of t, assuming the initial pressure of $60\,\text{kN/m}^2$ instantaneously becomes zero.

The computation is set out in Table 7.4, all pressures having been multiplied by 10.

7.11 Vertical Drains

The slow rate of consolidation in saturated clays of low permeability may be accelerated by means of vertical drains which shorten the drainage path within the clay. Consolidation is then due mainly to horizontal radial drainage, resulting in the faster dissipation of excess pore water pressure; vertical drainage becomes of minor importance. In theory the final magnitude of consolidation settlement is the same, only the *rate* of settlement being affected.

In the case of an embankment constructed over a highly compressible clay layer (Fig. 7.28), vertical drains installed in the clay would enable the embankment to be brought into service much sooner and there would be a quicker increase in the shear strength of the clay. A degree of consolidation of the order of 80% would be desirable at the end of construction. Any advantages, of course, must be set against the additional cost of the installation.

The traditional method of installing vertical drains is by driving boreholes through the clay layer and backfilling with a suitably graded sand. Typical diameters are 200–400 mm and drains have been installed to

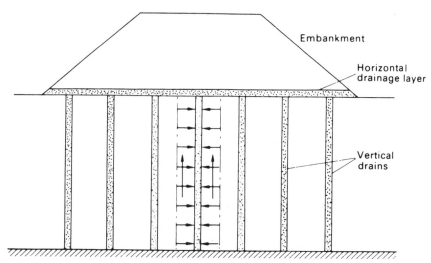

Fig. 7.28 Vertical drains.

depths of over 30 m. The sand should be capable of allowing the efficient drainage of water without permitting fine soil particles to be washed in. Prefabricated drains are also used and are generally cheaper than backfilled drains for a given area of treatment. One such type is the 'prepackaged' drain consisting of a filter stocking, generally of woven polypropylene, filled pneumatically with sand, a typical diameter being 65 mm. This type of drain is very flexible and is usually unaffected by any lateral ground movements. Another type of prefabricated drain is the *band* drain, consisting of a flat plastic core with drainage channels, surrounded by a thin layer of filter fabric: the fabric must have sufficient strength to prevent it from being squeezed into the channels. The main function of the fabric is to prevent the passage of fine soil particles which might clog the channels in the core. Typical dimensions of a band drain are 100 mm by 4 mm and the equivalent diameter is generally assumed to be the perimeter divided by π. Prefabricated drains are installed either by insertion into pre-bored holes or by placing them inside a mandrel or casing which is then driven or vibrated into the ground.

As the object is to reduce the length of the drainage path, the spacing of the drains is the most important design consideration. The drains are usually spaced in either a square or a triangular pattern. The spacing of the drains must obviously be less than the thickness of the clay layer and there is no point in using vertical drains in relatively thin clay layers. It is important for a successful design that the coefficients of consolidation in both the horizontal and vertical directions (c_h and c_v respectively) are known fairly accurately. The ratio c_h/c_v is normally between 1 and 2, the higher the ratio the more beneficial the drain installation will be. The values of the coefficients for the clay adjacent to the drains may be significantly reduced due to remoulding during installation (especially if a mandrel is used), an effect known as *smear*. The smear effect can be taken into account by assuming a reduced value of c_h or by using a reduced drain diameter. Another design complication is that large diameter sand drains tend to act as weak piles, reducing the vertical stress increment in the clay to an unknown degree and resulting in lower values of excess pore water pressure and therefore of consolidation settlement. This effect is minimal in the case of prefabricated drains, due to their flexibility. Experience has shown that vertical drains are not successful in soils having a high secondary compression ratio, such as highly plastic clays and peat: the rate of secondary compression cannot be controlled by vertical drains.

In polar coordinates the three-dimensional form of the consolidation equation, with different soil properties in the horizontal and vertical directions, is

$$\frac{\partial u}{\partial t} = c_h \left(\frac{\partial^2 u}{\partial r^2} + \frac{1}{r}\frac{\partial u}{\partial r} \right) + c_v \frac{\partial^2 u}{\partial z^2} \tag{7.37}$$

The vertical prismatic blocks of soil surrounding the drains are replaced by cylindrical blocks, of radius R, having the same cross-sectional area (Fig. 7.29). The solution to Equation 7.37 can be written in two parts:

$$U_v = f(T_v)$$

and

$$U_r = f(T_r)$$

where U_v = average degree of consolidation due to vertical drainage only; U_r = average degree of consolidation due to horizontal (radial) drainage only;

$$T_v = \frac{c_v t}{d^2} \tag{7.38}$$

= time factor for consolidation due to vertical drainage only

$$T_r = \frac{c_h t}{4R^2} \tag{7.39}$$

= time factor for consolidation due to radial drainage only

The expression for T_r confirms the fact that the closer the spacing of the drains the quicker the consolidation process due to radial drainage proceeds. The solution for radial drainage, due to Barron, is given in Fig. 7.30, the U_r/T_r relationship depending on the ratio $n = R/r_d$, where R is

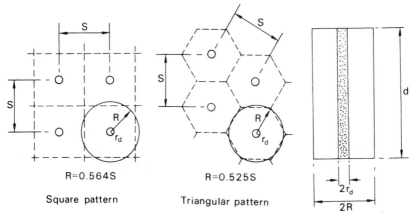

Fig. 7.29 Cylindrical blocks.

the radius of the equivalent cylindrical block and r_d is the radius of the drain. It can also be shown that

$$(1 - U) = (1 - U_v)(1 - U_r) \tag{7.40}$$

where U is the average degree of consolidation under combined vertical and radial drainage.

Example 7.8

An embankment is to be constructed over a layer of clay 10 m thick, with an impermeable lower boundary. Construction of the embankment will increase the total vertical stress in the clay layer by 65 kN/m². For the clay, $c_v = 4·7 \, \text{m}^2/\text{year}$, $c_h = 7·9 \, \text{m}^2/\text{year}$ and $m_v = 0·25 \, \text{m}^2/\text{MN}$. The design requirement is that all but 25 mm of the settlement due to consolidation of the clay layer will have taken place after 6 months. Determine the spacing, in a square pattern, of 400 mm diameter sand drains to achieve the above requirement.

$$\text{Final settlement} = m_v \Delta \sigma' H = 0·25 \times 65 \times 10$$
$$= 162 \, \text{mm}$$

For $t = 6$ months,

$$U = \frac{162 - 25}{162} = 0·85$$

Diameter of sand drains is 0·4 m, i.e. $r_d = 0·2$ m.
Radius of cyclindrical block:

$$R = nr_d = 0·2n$$

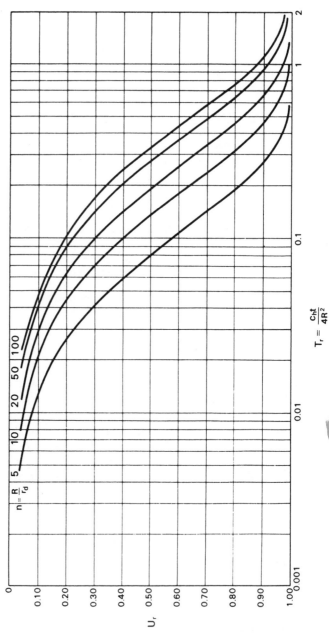

Fig. 7.30   Solution for radial consolidation.

The layer is half-closed, therefore $d = 10\,\text{m}$.

$$T_v = \frac{c_v t}{d^2} = \frac{4\cdot7 \times 0\cdot5}{10^2} = 0\cdot0235$$

From curve 1, Fig. 7.18: $U_v = 0\cdot17$

$$T_r = \frac{c_h t}{4R^2} = \frac{7\cdot9 \times 0\cdot5}{4 \times 0\cdot2^2 \times n^2} = \frac{24\cdot7}{n^2}$$

i.e.

$$n = \sqrt{\left(\frac{24\cdot7}{T_r}\right)}$$

Now $(1 - U) = (1 - U_v)(1 - U_r)$, therefore

$$0\cdot15 = 0\cdot83\,(1 - U_r)$$
$$U_r = 0\cdot82$$

A trial and error solution is necessary to obtain the value of n. Starting with a value of n corresponding to one of the curves in Fig. 7.30 the value of T_r for $U_r = 0\cdot82$ is obtained from that curve. Using this value of T_r the value of $\sqrt{(24\cdot7/T_r)}$ is calculated and plotted against the selected value of n.

n	T_r	$\sqrt{(24\cdot7/T_r)}$
5	0·20	11·1
10	0·33	8·6
15	0·42	7·7

From Fig. 7.31 it is seen that $n = 9$. Therefore

$$R = 0\cdot2 \times 9 = 1\cdot8\,\text{m}$$

Spacing of drains in a square pattern is given by

$$S = \frac{R}{0\cdot564} = \frac{1\cdot8}{0\cdot564} = 3\cdot2\,\text{m}$$

Problems

7.1 In an oedometer test on a specimen of saturated clay ($G_s = 2\cdot72$) the applied pressure was increased from 107 to $214\,\text{kN/m}^2$ and the following compression readings recorded:

Time (min)	0	$\frac{1}{4}$	$\frac{1}{2}$	1	$2\frac{1}{4}$	4	$6\frac{1}{4}$	9	16
Gauge (mm)	7·82	7·42	7·32	7·21	6·99	6·78	6·61	6·49	6·37

Time (min)	25	36	49	64	81	100	300	1440
Gauge (mm)	6·29	6·24	6·21	6·18	6·16	6·15	6·10	6·02

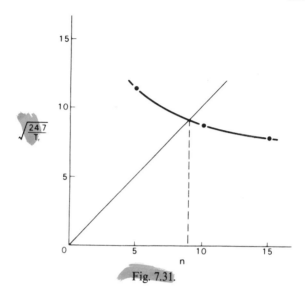

Fig. 7.31.

After 1440 min the thickness of the specimen was 15·30 mm and the water content 23·2%. Determine the values of the coefficient of consolidation and the compression ratios from (a) the root time plot and (b) the log time plot. Determine also the values of the coefficient of volume compressibility and the coefficient of permeability.

7.2 The following results were obtained from an oedometer test on a specimen of saturated clay:

Pressure (kN/m²)	27	54	107	214	429	214	107	54
Void ratio		1·243	1·217	1·144	1·068	0·994	1·001	1·012	1·024

A layer of this clay 8 m thick lies below a 4 m depth of sand, the water table being at the surface. The saturated unit weight for both soils is 19 kN/m³. A 4 m depth of fill of unit weight 21 kN/m³ is placed on the sand over an extensive area. Determine the final settlement due to consolidation of the clay. If the fill were to be removed some time after the completion of consolidation, what heave would eventually take place due to swelling of the clay?

7.3 In an oedometer test a specimen of saturated clay 19 mm thick reaches 50% consolidation in 20 min. How long would it take a layer of this clay 5 m thick to reach the same degree of consolidation under the same stress and drainage conditions? How long would it take the layer to reach 30% consolidation?

7.4 Assuming the fill in Problem 7.2 is dumped very rapidly, what would be the value of excess pore water pressure at the centre of the clay layer after a period of 3 years? The layer is open and the value of c_v is $2.4 \, \text{m}^2/\text{year}$.

7.5 An open clay layer is 6 m thick, the value of c_v being $1.0 \, \text{m}^2/\text{year}$. The initial distribution of excess pore water pressure varies linearly from $60 \, \text{kN/m}^2$ at the top of the layer to zero at the bottom. Using the finite difference approximation of the one-dimensional consolidation equation, plot the isochrone after consolidation has been in progress for a period of 3 years and from the isochrone determine the average degree of consolidation in the layer.

7.6 A 10 m depth of sand overlies an 8 m layer of clay, below which is a further depth of sand. For the clay, $m_v = 0.83 \, \text{m}^2/\text{MN}$ and $c_v = 4.4 \, \text{m}^2/\text{year}$. The water table is at surface level but is to be lowered permanently by 4 m, the initial lowering taking place over a period of 40 weeks. Calculate the final settlement due to consolidation of the clay, assuming no change in the weight of the sand, and the settlement 2 years after the start of lowering.

7.7 A raft foundation 60 m × 40 m carrying a net pressure of $145 \, \text{kN/m}^2$ is located at a depth of 4.5 m below the surface in a deposit of dense sandy gravel 22 m deep: the water table is at a depth of 7 m. Below the sandy gravel is a layer of clay 5 m thick which, in turn, is underlain by dense sand. The value of m_v for the clay is $0.22 \, \text{m}^2/\text{MN}$. Determine the settlement below the centre of the raft, the corner of the raft and the centre of each edge of the raft, due to consolidation of the clay.

7.8 An oil storage tank 35 m in diameter is located 2 m below the surface of a deposit of clay 32 m thick, the water table being at the surface: the net foundation pressure is $105 \, \text{kN/m}^2$. A firm stratum underlies the clay. The average value of m_v for the clay is $0.14 \, \text{m}^2/\text{MN}$ and that of pore pressure coefficient A is 0.65. The undrained value of Young's modulus is estimated to be $40 \, \text{MN/m}^2$. Determine the total settlement under the centre of the tank.

7.9 A half-closed clay layer is 8 m thick and it can be assumed that $c_v = c_h$. Vertical sand drains 300 mm in diameter, spaced at 3 m centres in a square pattern, are to be used to increase the rate of consolidation of the clay under the increased vertical stress due to the construction of an embankment. Without sand drains the degree of consolidation at the time the embankment is due to come into use has been calculated as 25%. What degree of consolidation would be reached with the sand drains at the same time?

7.10 A layer of saturated clay is 10 m thick, the lower boundary being impermeable; an embankment is to be constructed above the clay. Determine the time required for 90% consolidation of the clay layer. If 300 mm diameter sand drains at 4 m centres in a square pattern were installed in the clay, in what time would the same overall degree of consolidation be reached? The coefficients of consolidation in the vertical and horizontal directions respectively are $9.6 \, \text{m}^2/\text{year}$ and $14.0 \, \text{m}^2/\text{year}$.

References

7.1 Atkinson, M. S. and Eldred, P. J. L. (1981): 'Consolidation of Soil using Vertical Drains', *Geotechnique*, Vol. 31, No. 1.

7.2 Barron, R. A. (1948): 'Consolidation of Fine Grained Soils by Drain Wells', *Transactions ASCE*, Vol. 113.

7.3 Bjerrum, L. (1967): 'Engineering Geology of Norwegian Normally-Consolidated Marine Clays as Related to Settlement of Buildings', *Geotechnique*, Vol. 17, No. 2.

7.4 British Standard 1377 (1975): *Methods of Test for Soils for Civil Engineering Purposes*, British Standards Institution, London.

7.5 Christie, I. F. (1959): 'Design and Construction of Vertical Drains to Accelerate the Consolidation of Soils', *Civil Engineering and Public Works Review*, Nos. 2, 3, 4.

7.6 Cour, F. R. (1971): 'Inflection Point Method for Computing c_v', Technical Note, *Journal ASCE*, Vol. 97, No. SM5.

7.7 Gibson, R. E. (1963): 'An Analysis of System Flexibility and its Effects on Time Lag in Pore Water Pressure Measurements', *Geotechnique*, Vol. 13, No. 1.

7.8 Gibson, R. E. (1966): 'A Note on the Constant Head Test to Measure Soil Permeability In-situ', *Geotechnique*, Vol. 16, No. 3.

7.9 Gibson, R. E. (1970): 'An Extension to the Theory of the Constant Head In-situ Permeability Test', *Geotechnique*, Vol. 20, No. 2.

7.10 Gibson, R. E. and Lumb, P. (1953): 'Numerical Solution of Some Problems in the Consolidation of Clay', *Proceedings ICE*, Part I.

7.11 Lambe, T. W. (1964): 'Methods of Estimating Settlement', *Journal ASCE*, Vol. 90, No. SM5.

7.12 Lambe, T. W. (1967): 'Stress Path Method', *Journal ASCE*, Vol. 93, No. SM6.

7.13 McGown, A. and Hughes, F. H. (1981): 'Practical Aspects of the Design and Installation of Deep Vertical Drains', *Geotechnique*, Vol. 31, No. 1.

7.14 Naylor, A. H. and Doran, I. G. (1948): 'Precise Determination of Primary Consolidation', *Proceedings 2nd International Conference SMFE, Rotterdam*, Vol. 1.

7.15 Rowe, P. W. (1968): 'The Influence of Geological Features of Clay Deposits on the Design and Performance of Sand Drains', *Proceedings ICE*.

7.16 Rowe, P. W. and Barden, L. (1966): 'A New Consolidation Cell', *Geotechnique*, Vol. 16, No. 2.

7.17 Schmertmann, J. H. (1953): 'Estimating the True Consolidation Behaviour of Clay from Laboratory Test Results', *Proceedings ASCE*, Vol. 79.

7.18 Scott, R. F. (1961): 'New Method of Consolidation Coefficient Evaluation', *Journal ASCE*, Vol. 87, No. SM1.

7.19 Scott, R. F. (1963): *Principles of Soil Mechanics*, Addison-Wesley, Reading, Massachusetts.

7.20 Simons, N. E. and Som, N. N. (1969): 'The Influence of Lateral Stresses on the Stress Deformation Characteristics of London Clay', *Proceedings 7th International Conference SMFE, Mexico City*, Vol. 1.

7.21 Skempton, A. W. and Bjerrum, L. (1957): 'A Contribution to the Settlement Analysis of Foundations on Clay', *Geotechnique*, Vol. 7, No. 4.

7.22 Taylor, D. W. (1948): *Fundamentals of Soil Mechanics*, John Wiley and Sons, New York.

7.23 Terzaghi, K. (1943): *Theoretical Soil Mechanics*, John Wiley and Sons, New York.

7.24 Wilkinson, W. B. (1968): 'Constant Head In-situ Permeability Tests in Clay Strata', *Geotechnique*, Vol. 18, No. 2.

CHAPTER 8

Bearing Capacity

8.1 Introduction

This chapter is concerned with the bearing capacity of soils on which foundations are supported, a foundation being that part of a structure which transmits loads directly to the underlying soil. If the soil near the surface is capable of adequately supporting the structural loads it is possible to use either *footings* or a *raft*. A footing is a relatively small slab giving separate support to part of the structure. A footing supporting a single column is referred to as an individual footing (or pad), one supporting a group of columns as a combined footing and one supporting a load-bearing wall as a strip footing. A raft is a relatively large single slab, usually stiffened, supporting the structure as a whole. If the soil near the surface is incapable of adequately supporting the structural loads, piles or piers are used to transmit the loads to suitable soil (or rock) at greater depth. Foundation level should be below the depth which is subjected to frost action (around $0 \cdot 5$ m in the United Kingdom) and, where appropriate, the depth to which seasonal swelling and shrinkage of the soil takes place.

A foundation must satisfy two fundamental requirements: (1) the factor of safety against shear failure of the supporting soil must be adequate, a value between 2.5 and 3 normally being specified; (2) the settlement of the foundation should be tolerable and, in particular, differential settlement should not cause any unacceptable damage nor interfere with the function of the structure. The *allowable* bearing capacity (q_a) is defined as the maximum pressure which may be applied to the soil such that the two fundamental requirements are satisfied. An indirect requirement is that the foundation, and the operations involved in its construction, should have no adverse effect on adjacent structures and services. For preliminary design purposes BS 8004 [8.7] gives presumed bearing values (Table 8.1) being the pressure which would normally result in an adequate factor of safety against shear failure for particular soil types, but without consideration of settlement.

Damage due to settlement may be classified as architectural, functional or structural. In the case of framed structures, settlement damage is usually confined to the cladding and finishes (i.e. architectural damage): such damage is due only to the settlement occurring subsequent to the application of the cladding and finishes. In some cases, structures can be

Table 8.1   Presumed Bearing Values (BS 8004:1986)

Soil type	Bearing value (kN/m^2)	Remarks
Dense gravel or dense sand and gravel	> 600	Width of foundation (B) not less than 1 m. Water table at least B below base of foundation
Medium dense gravel or medium dense sand and gravel	200–600	
Loose gravel or loose sand and gravel	< 200	
Compact sand	> 300	
Medium dense sand	100–300	
Loose sand	< 100	
Very stiff boulder clays and hard clays	300–600	Susceptible to long-term consolidation settlement
Stiff clays	150–300	
Firm clays	75–150	
Soft clays and silts	< 75	
Very soft clays and silts	—	

designed and constructed in such a way that a certain degree of movement can be accommodated without damage. In other cases a certain degree of cracking may be inevitable if the structure is to be economic. It may be that damage to services, and not to the structure, will be the limiting criterion. Based on observations of damage in buildings, Skempton and MacDonald [8.34] proposed limits for maximum settlement at which damage could be expected and related maximum settlement to angular distortion. The angular distortion between two points under a structure is equal to the differential settlement between the points divided by the distance between them. No damage was observed where the angular distortion was less than 1/300: for individual footings this figure corresponds roughly to a maximum settlement of 50 mm on sands and of 75 mm on clays. Angular distortion limits were subsequently proposed by Bjerrum [8.2] as a general guide for a number of structural situations (Table 8.2). It is recommended that the safe limit to avoid cracking in the panel walls of framed structures should be 1/500. In the case of load-bearing brickwork the criteria recommended by Polshin and Tokar [8.27] are generally used. These criteria are given in terms of the ratio of deflection to the length of the deflected part and depend on the length-to-height ratio of the building: recommended deflection ratios are within the range 0.3×10^{-3} to 0.7×10^{-3}. In the case of buildings subjected to hogging the criteria of Polshin and Tokar should be halved.

Table 8.2   Angular Distortion Limits

1/150	Structural damage of general buildings expected
1/250	Tilting of high rigid buildings may be visible
1/300	Cracking in panel walls expected
	Difficulties with overhead cranes
1/500	Limit for buildings in which cracking is not permissible
1/600	Overstressing of structural frames with diagonals
1/750	Difficulties with machinery sensitive to settlement

The above approach to settlement limits is empirical and is intended to be only a general guide for simple structures. A more fundamental damage criterion is the limiting tensile strain at which visible cracking occurs in a given material. Ideally the concept of limiting tensile strain should be used in conjunction with an elastic strain analysis using a simple idealisation of the structure, including foundations, partitions and finishes. A comprehensive discussion of settlement damage in buildings has been presented by Burland and Wroth [8.12].

Results from elastic theory (Fig. 5.8) indicate that the increase in vertical stress in the soil below the centre of a strip footing of width B is approximately 20% of the foundation pressure at a depth of $3B$. In the case of a square footing the corresponding depth is $1 \cdot 5B$. For practical purposes these depths can normally be accepted as the limits of the zones of influence of the respective foundations and are called the *significant depths*. It is essential that the soil conditions are known within the significant depth of any foundation.

8.2   Ultimate Bearing Capacity

The ultimate bearing capacity (q_f) is defined as the least pressure which would cause shear failure of the supporting soil immediately below and adjacent to a foundation.

Three distinct modes of failure have been identified and these are illustrated in Fig. 8.1: they will be described with reference to a strip footing. In the case of *general shear failure*, continuous failure surfaces develop between the edges of the footing and the ground surface as shown in Fig. 8.1. As the pressure is increased towards the value q_f the state of plastic equilibrium is reached initially in the soil around the edges of the footing then gradually spreads downwards and outwards. Ultimately the state of plastic equilibrium is fully developed throughout the soil above the failure surfaces. Heaving of the ground surface occurs on both sides of the footing although the final slip movement would occur only on one side, accompanied by tilting of the footing. This mode of failure is typical of soils of low

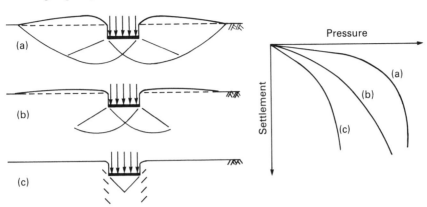

Fig. 8.1   Modes of failure: (a) general shear, (b) local shear, (c) punching shear.

compressibility (i.e. dense or stiff soils) and the pressure-settlement curve is of the general form shown in Fig. 8.1, the ultimate bearing capacity being well defined. In the mode of *local shear failure* there is significant compression of the soil under the footing and only partial development of the state of plastic equilibrium. The failure surfaces, therefore, do not reach the ground surface and only slight heaving occurs. Tilting of the foundation would not be expected. Local shear failure is associated with soils of high compressibility and, as indicated in Fig. 8.1, is characterized by the occurrence of relatively large settlements (which would be unacceptable in practice) and the fact that the ultimate bearing capacity is not clearly defined. *Punching shear failure* occurs when there is compression of the soil under the footing, accompanied by shearing in the vertical direction around the edges of the footing. There is no heaving of the ground surface away from the edges and no tilting of the footing. Relatively large settlements are also a characteristic of this mode and again the ultimate bearing capacity is not well defined. Punching shear failure will also occur in a soil of low compressibility if the foundation is located at considerable depth. In general the mode of failure depends on the compressibility of the soil and the depth of the foundation relative to its breadth.

The bearing capacity problem can be considered in terms of plasticity theory. The lower and upper bound theorems (Section 6.1) can be applied to give solutions for the ultimate bearing capacity of a soil. In certain cases, exact solutions can be obtained corresponding to the equality of the lower and upper bound solutions. However, such solutions are based on the assumption that the soil can be represented by a perfectly plastic stress-strain relationship, as shown in Fig. 6.1. This approximation is only realistic for soils of low compressibility, i.e. soils corresponding to the general shear mode of failure. However, for the other modes, settlement and not shear failure is normally the limiting criterion.

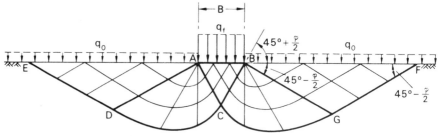

Fig. 8.2   Failure under a strip footing.

A suitable failure mechanism for a strip footing is shown in Fig. 8.2. The footing, of width B and infinite length, carries a uniform pressure q on the surface of a mass of homogeneous, isotropic soil. The shear strength parameters for the soil are c and ϕ but the unit weight is assumed to be zero. When the pressure becomes equal to the ultimate bearing capacity q_f the footing will have been pushed downwards into the soil mass producing a state of plastic equilibrium, in the form of an active Rankine zone, below the footing, the angles ABC and BAC being $(45° + \phi/2)$. The downward movement of the wedge ABC forces the adjoining soil sideways, producing outward lateral forces on both sides of the wedge. Passive Rankine zones ADE and BGF therefore develop on both sides of the wedge ABC, the angles DEA and GFB being $(45° - \phi/2)$. The transition between the downward movement of the wedge ABC and the lateral movement of the wedges ADE and BGF takes place through zones of radial shear (also known as slip fans) ACD and BCG, the surfaces CD and CG being logarithmic spirals (or circular arcs if $\phi = 0$) to which BC and ED, or AC and FG, are tangential. A state of plastic equilibrium thus exists above the surface EDCGF, the remainder of the soil mass being in a state of elastic equilibrium.

The following exact solution can be obtained, using plasticity theory, for the ultimate bearing capacity of a strip footing on the surface of a weightless soil, based on the mechanism described above. For the undrained condition ($\phi_u = 0$) in which the shear strength is given by c_u:

$$q_f = (2 + \pi)c_u = 5.14\,c_u \tag{8.1}$$

For the general case in which the shear strength parameters are c and ϕ it is necessary to consider a surcharge pressure q_o acting on the soil surface as shown in Fig. 8.2: otherwise if $c = 0$ the bearing capacity of a weightless soil would be zero. The solution for this case, attributed to Prandtl and Reissner, is:

$$q_f = c \cot \phi \left[\exp(\pi \tan \phi) \tan^2 (45° + \phi/2) - 1 \right]$$
$$+ q_o \left[\exp(\pi \tan \phi) \tan^2 (45° + \phi/2) \right] \tag{8.2}$$

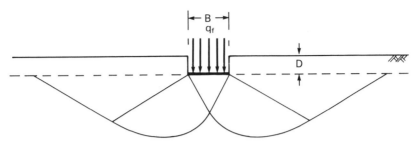

Fig. 8.3   Footing at depth D below the surface.

However, an additional term must be added to Equation 8.2 to take into account the component of bearing capacity due to the self-weight of the soil. This component can only be determined approximately, by numerical or graphical means, and is sensitive to the value assumed for the angles ABC and BAC in Fig. 8.2

Foundations are not normally located on the surface of a soil mass, as assumed in the above solutions, but at a depth D below the surface as shown in Fig. 8.3. In applying these solutions in practice it is assumed that the shear strength of the soil between the surface and depth D is neglected, this soil being considered only as a surcharge imposing a uniform pressure $q_o = \gamma D$ on the horizontal plane at foundation level. This is a reasonable assumption for a *shallow* foundation (interpreted as a foundation for which the depth D is not greater than the breadth B). The soil above foundation level is normally weaker, especially if backfilled, than the soil at greater depth.

The ultimate bearing capacity of the soil under a shallow strip footing can be expressed by the following general equation (due to Terzaghi):

$$q_f = \tfrac{1}{2}\gamma B N_\gamma + c N_c + \gamma D N_q \qquad (8.3)$$

where N_γ, N_c and N_q are bearing capacity factors depending only on the value of ϕ. The first term in Equation 8.3 represents the contribution to bearing capacity resulting from the self-weight of the soil, the second term is the contribution due to the constant component of shear strength and the third term is the contribution due to the surcharge pressure. It should be realized, however, that the superposition of the components of bearing capacity is theoretically incorrect for a plastic material: however, any resulting error is considered to be on the safe side.

For many years Terzaghi's bearing capacity factors were widely used. Terzaghi [8.36] assumed that the angles ABC and BAC in Fig. 8.2 were equal to ϕ (i.e. ABC was not considered to be an active Rankine zone). Values of N_γ were obtained by determining the total passive resistance and adhesion force on the planes AC and BC. Terzaghi's values of N_q and N_c

were obtained by modifying the Prandtl-Reissner solution. However, Terzaghi's values have now been largely superseded.

It is now considered that the values of N_q and N_c implicit in Equation 8.2 should be used in bearing capacity calculations, i.e.

$$N_q = \exp(\pi \tan \phi) \tan^2 (45° + \phi/2)$$
$$N_c = (N_q - 1) \cot \phi$$

Currently, the most widely used values for the factor N_y are those obtained by Brinch Hansen [8.18] and Meyerhof [8.23]. These values can be represented by the following approximations:

$$N_y = 1.80 (N_q - 1) \tan \phi \qquad \text{(Brinch Hansen)}$$
$$N_y = (N_q - 1) \tan (1.4 \phi) \qquad \text{(Meyerhof)}$$

Values of N_y, N_c and N_q are plotted in terms of ϕ in Fig. 8.4. Brinch Hansen's values of N_y are used in the examples in this chapter.

The problems involved in extending the two-dimensional solution for a strip footing to three dimensions would be considerable. Accordingly, the ultimate bearing capacities of square, rectangular and circular footings are determined by means of semi-empirical *shape factors* applied to the solution for a strip footing. The bearing capacity factors N_y, N_c and N_q should be multiplied by the respective shape factors s_y, s_c and s_q. The shape factors proposed by Terzaghi and Peck [8.37] are still widely used in practice although they are considered to give conservative values of ultimate bearing capacity for high values of ϕ. The factors are $s_y = 0.8$ for a square footing or 0.6 for a circular footing, $s_c = 1.2$ and $s_q = 1$. Thus Equation 8.3 becomes, for a square footing:

$$q_f = 0.4 \gamma B N_y + 1.2 c N_c + \gamma D N_q \qquad (8.4)$$

and for a circular footing:

$$q_f = 0.3 \gamma B N_y + 1.2 c N_c + \gamma D N_q \qquad (8.5)$$

For a rectangular footing of breadth B and length L, the shape factors are obtained by linear interpolation between the values for a strip footing ($B/L = 0$) and a square footing ($B/L = 1$), e.g. $s_y = 1 - 0.2B/L$.

Alternative proposals for shape factors have been made by DeBeer [8.15] and Brinch Hansen [8.18; 8.41].

It should be recognized that the results of bearing capacity calculations are very sensitive to the values assumed for the shear strength parameters, especially the higher values of ϕ. Due consideration must therefore be given to the probable degree of accuracy of the shear strength parameters employed.

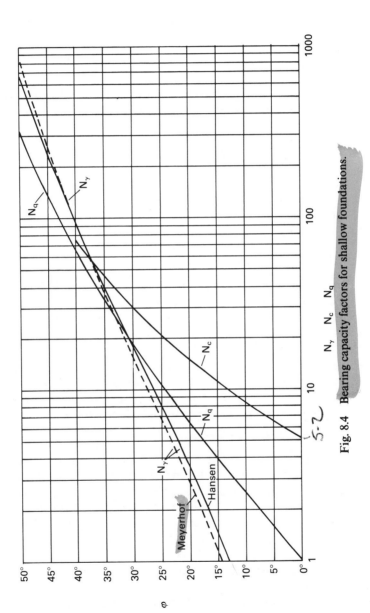

Fig. 8.4 Bearing capacity factors for shallow foundations.

Factor of Safety

The actual pressure on the soil due to the weight of the structure is called the *total* foundation pressure (q). The *net* foundation pressure (q_n) is the increase in pressure at foundation level, being the total foundation pressure less the effective weight of soil per unit area permanently removed, i.e.

$$q_n = q - \gamma D \tag{8.6}$$

The factor of safety (F) with respect to shear failure is defined in terms of the net ultimate bearing capacity (q_{nf}), i.e.

$$F = \frac{q_{nf}}{q_n} = \frac{q_f - \gamma D}{q - \gamma D} \tag{8.7}$$

However, in the case of shallow footings, when the value of ϕ is relatively high, there is no significant difference between the values of F defined in terms of net and gross pressures.

Skempton's Values of N_c

In a review of bearing capacity theory, Skempton [8.31] concluded that in the case of saturated clays under undrained conditions ($\phi_u = 0$) the ultimate bearing capacity of a footing could be expressed by the equation:

$$q_f = c_u N_c + \gamma D \tag{8.8}$$

the factor N_c being a function of the shape of the footing and the depth/breadth ratio. Skempton's values of N_c are given in Fig. 8.5. The factor for a rectangular footing of dimensions $B \times L$ (where $B < L$) is the value for a square footing multiplied by ($0.84 + 0.16 B/L$).

Eccentric and Inclined Loading

Footings may be subjected to eccentric and inclined loading and such conditions lead to a reduction in bearing capacity. If e is the eccentricity of the resultant load on the base of a footing of width B, it was suggested by Meyerhof that an effective foundation width B' be used in Equation 8.3, where

$$B' = B - 2e \tag{8.9}$$

The resultant load is assumed to be uniformly distributed over the effective width B'.

The effect of inclined loading on bearing capacity can be taken into account by means of inclination factors proposed by Meyerhof. If the angle of inclination of the resultant load is α to the vertical then the bearing capacity factors N_γ, N_c and N_q should be multiplied respectively by the following factors:

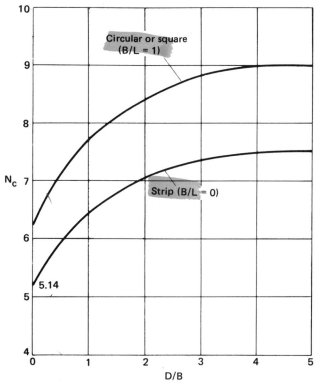

Fig. 8.5 Skempton's values of N_c for $\phi_u = 0$. (Reproduced from A. W. Skempton (1951) *Proceedings of the Building Research Congress*, Division 1, p. 181, by permission, Building Research Establishment: Crown Copyright).

$$i_\gamma = (1 - \alpha/\phi)^2 \tag{8.10a}$$

$$i_c = i_q = (1 - \alpha/90°)^2 \tag{8.10b}$$

Alternative inclination factors have been proposed by Brinch Hansen [8.18; 8.41].

An alternative approach in the case of inclined loading is to use the following empirical rule, given in BS 8004 [8.7]:

$$\frac{V}{P_v} + \frac{H}{P_h} < 1 \tag{8.11}$$

where V = vertical component of the inclined load, H = horizontal component of the inclined load, P_v = allowable vertical load, and P_h = allowable horizontal load (a fraction of the available passive resistance).

Example 8.1

A footing 2·25 m square is located at a depth of 1·5 m in a sand, the shear strength parameters being $c' = 0$ and $\phi' = 38°$. Determine the ultimate bearing capacity (a) if the water table is well below foundation level, (b) if the water table is at the surface. The unit weight of the sand above the water table is $18 \, \text{kN/m}^3$: the saturated unit weight is $20 \, \text{kN/m}^3$.

For a square footing the ultimate bearing capacity (with $c = 0$) is given by

$$q_f = 0.4\gamma BN_y + \gamma DN_q$$

For $\phi' = 38°$ the bearing capacity factors (Fig. 8.4) are $N_y = 67$ and $N_q = 49$. Therefore

$$q_f = (0.4 \times 18 \times 2.25 \times 67) + (18 \times 1.5 \times 49)$$
$$= 1085 + 1323$$
$$= 2408 \, \text{kN/m}^2$$

When the water table is at the surface, the ultimate bearing capacity is given by

$$q_f = 0.4\gamma' BN_y + \gamma' DN_q$$
$$= (0.4 \times 10.2 \times 2.25 \times 67) + (10.2 \times 1.5 \times 49)$$
$$= 615 + 750$$
$$= 1365 \, \text{kN/m}^2$$

Example 8.2

A strip footing is to be designed to carry a load of $800 \, \text{kN/m}$ at a depth of 0.7 m in a gravelly sand. The appropriate shear strength parameters are $c' = 0$ and $\phi' = 40°$. Determine the width of the footing if a factor of safety of 3 against shear failure is specified and assuming that the water table may rise to foundation level. Above the water table the unit weight of the sand is $17 \, \text{kN/m}^3$ and below the water table the saturated unit weight is $20 \, \text{kN/m}^3$.

For $\phi' = 40°$ the bearing capacity factors (Fig. 8.4) are $N_y = 95$ and $N_q = 64$. The ultimate bearing capacity (units kN/m^2) is given by

$$q_f = \tfrac{1}{2}\gamma' BN_y + \gamma DN_q$$
$$= (\tfrac{1}{2} \times 10.2 \times B \times 95) + (17 \times 0.7 \times 64)$$
$$= 485B + 762$$
$$\therefore \quad q_{nf} = q_f - \gamma D = 485B + 750$$

The net foundation pressure is

$$q_n = \frac{800}{B} - (17 \times 0.7)$$

Then, for a factor of safety of 3,

$$\frac{1}{3}(485B + 750) = \frac{800}{B} - 12$$

Hence,

$$B = 1.55\,\text{m}$$

Example 8.3

A footing 2 m square is located at a depth of 4 m in a stiff clay of saturated unit weight 21 kN/m³. The undrained strength of the clay at a depth of 4 m is given by the parameters $c_u = 120\,\text{kN/m}^2$ and $\phi_u = 0$. For a factor of safety of 3 with respect to shear failure, what load could be carried by the footing?

In this case $D/B = 2$ and from Fig. 8.5 the value of N_c for a square footing is 8·4. The ultimate bearing capacity is given by:

$$q_f = c_u N_c + \gamma D$$
$$\therefore \quad q_{nf} = c_u N_c = 120 \times 8.4 = 1008\,\text{kN/m}^2$$

For $F = 3$, $q_n = 1008/3 = 336\,\text{kN/m}^2$

$$\therefore \quad q = q_n + \gamma D = 336 + (21 \times 4) = 420\,\text{kN/m}^2$$

Allowable load $= 420 \times 2^2 = 1680\,\text{kN}$.

Example 8.4

The base of a long retaining wall is 3 m wide and is 1 m below the ground surface in front of the wall: the water table is well below base level. The vertical and horizontal components of the base reaction are 282 kN/m and 102 kN/m respectively: the eccentricity of the base reaction is 0·36 m. If the appropriate shear strength parameters for the foundation soil are $c' = 0$ and $\phi' = 35°$, and the unit weight of the soil is 18 kN/m³, determine the factor of safety against shear failure.

The effective width of the base is given by

$$B' = B - 2e = 2.28\,\text{m}$$

For $\phi' = 35°$ the bearing capacity factors (Fig. 8.4) are $N_\gamma = 41$ and $N_q = 33$.

The angle of inclination (to the vertical) of the resultant load is given by $\alpha = \tan^{-1}(102/282) = 20°$. Hence the inclination factors, according to Meyerhof, are:

$$i_\gamma = (1 - 20/35)^2 = 0.18$$
$$i_q = (1 - 20/90)^2 = 0.61$$

The ultimate bearing capacity is given by

$$q_f = \tfrac{1}{2}\gamma B' N_\gamma i_\gamma + \gamma D N_q i_q$$
$$= (\tfrac{1}{2} \times 18 \times 2\!\cdot\!28 \times 41 \times 0\!\cdot\!18) + (18 \times 1 \times 33 \times 0\!\cdot\!61)$$
$$= 151 + 362 = 513\,\text{kN/m}^2$$

$$\therefore \quad q_{nf} = q_f - \gamma D = 495\,\text{kN/m}^2$$

The net base pressure is

$$q_n = \frac{282}{2\!\cdot\!28} - 18 = 106\,\text{kN/m}^2$$

Then the factor of safety is

$$F = \frac{q_{nf}}{q_n} = \frac{495}{106} = 4\!\cdot\!7$$

Base Failure in Excavations

It was pointed out by Bjerrum and Eide [8.4] that Skempton's values of N_c could also be used in the analysis of base stability in temporary excavations in saturated clay (Fig. 8.6): base stability was referred to in Section 6.8. The analysis is confined to cases in which the sides of the excavation are strutted or anchored such that significant horizontal displacements are prevented. As excavation proceeds, shear stresses develop in the clay below the base due to the weight of adjacent soil outside the excavation. If these stresses exceed the undrained shear strength of the clay, local failure will occur by virtue of the base of the excavation being forced upwards: a corresponding subsidence of the clay will take place outside the excavation, adjacent to one or more of the sides. The problem is analogous to that of bearing capacity. In the base failure problem the soil is unloaded as excavation proceeds, as opposed to loading in the bearing capacity problem: corresponding shear stresses act in opposite directions in the two cases.

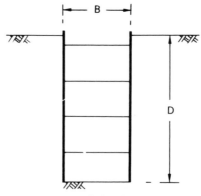

Fig. 8.6 Base failure in excavation.

Base failure in an excavation will occur at a critical depth D_c corresponding to $q_f = 0$ in Equation 8.8. Thus,

$$D_c = \frac{c_u N_c}{\gamma}$$

The appropriate value of c_u is that immediately below and adjacent to the base of the excavation. In general the factor of safety against base failure in an excavation of depth D is given by

$$F = \frac{c_u N_c}{\gamma D} \tag{8.12}$$

If a surcharge pressure acts on the surface adjacent to the excavation, its value is added to γD in the denominator in Equation 8.12.

8.3 Allowable Bearing Capacity of Clays

The allowable bearing capacity of clays, silty clays and plastic silts may be limited either by the requirement of an adequate factor of safety against shear failure or by settlement considerations. Shear strength, and hence the factor of safety, will increase whenever consolidation takes place. For homogeneous clays with low mass permeability, the factor of safety, therefore, should be checked for the condition immediately after construction, using the undrained shear strength parameters. However, in the case of clays exhibiting significant macro-fabric features the mass permeability may be relatively high and the undrained condition may be over-conservative at the end of construction. The methods of estimating the immediate settlement under undrained conditions and the long-term consolidation settlement are detailed in Chapters 5 and 7 respectively. For most cases in practice, simple settlement calculations are adequate provided the appropriate in-situ soil parameters have been determined. The precision of settlement predictions is much more influenced by inaccuracies in the values of soil parameters than by shortcomings in the methods of analysis. Sampling disturbance can have a serious effect on the values of parameters determined in the laboratory. In settlement analysis the same degree of precision should not be expected as, for example, in structural calculations.

The factor of safety and immediate settlement should be estimated on the basis of dead load plus initial (short-term) live load. Estimates of consolidation settlement should be based on dead load plus the average live load expected over a long period of time.

Settlement on overconsolidated clays depends on whether the preconsolidation pressure is exceeded, and if so to what extent, for a given foundation. The bearing pressure should normally be limited so that the preconsolidation pressure is not exceeded. In the case of a series of footings, differential settlement may be reduced by increasing the size of the largest footings above that required by the allowable bearing capacity. Foundations are not usually supported on normally consolidated clays because the resulting consolidation settlement would almost certainly be excessive.

If a stratum of soft clay lies below a firm stratum in which footings are located, there is a possibility that the footings may break through into the soft stratum. Such a possibility can be avoided if the vertical stress increments at the top level of the clay are less than the allowable bearing capacity of the clay by an adequate factor.

8.4 Allowable Bearing Capacity of Sands

In this section the term sand includes gravelly sands, silty sands and non-plastic silts. Most sand deposits are non-homogeneous and the allowable bearing capacity of shallow foundations is limited by settlement considerations except in the case of narrow footings. In most situations the allowable settlement is reached at a pressure for which the factor of safety against shear failure is greater than 3. In the case of narrow footings, however, shear failure may be the limiting consideration. Other factors being equal, the pressure that will produce the allowable settlement in a dense sand will be greater than that to produce the allowable settlement in a loose sand. Settlement in sand is rapid and occurs almost entirely during construction and initial loading. Settlement, therefore, should be estimated using the dead load plus the maximum live load.

Differential settlement between a number of footings is governed mainly by variations in the homogeneity of the sand within the significant depth and to a lesser extent by variations in foundation pressure. According to Terzaghi and Peck [8.37], settlement records indicate that the differential settlement between footings of approximately equal size carrying the same pressure is unlikely to exceed 50% of the maximum settlement. If the footings are of different size the differential settlement will be greater. The maximum settlement of footings carrying the same pressure increases with increasing footing size. There is no appreciable difference between the settlement of square and strip footings of the same width. For a given pressure and footing size the settlement decreases slightly with increasing footing depth below ground level due to the fact that the lateral confining pressure will be greater. In most cases, even under extreme variations of footing size and depth, it is unlikely that differential settlement will be greater than 75% of the maximum settlement. A few cases have been reported, however, in which the differential settlement was almost equal to the maximum settlement.

A reasonable design criterion for footings on sands is an allowable maximum settlement of 25 mm. The differential settlement between any two footings is then likely to be less than 20 mm. Differential settlement may be decreased by reducing the size of the smallest footings, provided the factor of safety with respect to shear failure remains above the specified value.

The settlement distribution under a raft is different from that for a series of footings. The settlement of footings is governed by the soil character-istics relatively near the surface and any one footing may be influenced by a weak pocket of soil. The settlement of a raft, on the other hand, is governed by the soil characteristics over a much greater depth. Weak pockets of soil may occur at random within this depth but they tend to be bridged over. The differential settlement of a raft as a percentage of the maximum settlement is roughly half the corresponding percentage for a series of footings. Thus for a differential settlement of 20 mm or less, the same as for a series of footings, the criterion for the maximum settlement of a raft is 50 mm.

The allowable bearing capacity of a sand depends primarily on the relative density, stress history, the position of the water table relative to foundation level and the size of the foundation. Of secondary importance are particle shape and grading. Both the magnitude of settlement and the value of the shear strength parameter ϕ' are strongly dependent on relative density. However the magnitude of settlement is also influenced by the stress history of the deposit, i.e. whether the sand is normally consolidated or overconsolidated and the previous stress path. If two sands were to exist at the same relative density but one were normally consolidated and the other overconsolidated, the settlement would be greater in the normally consolidated sand for identical loading conditions. The water table position affects both the settlement and the ultimate bearing capacity. If the sand within the significant depth is fully saturated the effective unit weight is roughly halved, resulting in a reduction in lateral confining pressure and a corresponding increase in settlement: the reduced effective unit weight will also result in a lower value of ultimate bearing capacity. The size of the foundation governs the depth to which the soil characteristics are relevant. It should be realized that unpredictable settlement can be caused by a reduction in relative density due to disturbance of the sand during construc-tion. Settlement can also be caused as a result of a reduction in lateral confining pressure, for example due to adjacent excavation. If a sand de-posit is loose, vibration may result in volume decrease, causing appreciable settlement. Loose sands should be compacted prior to construction, for example by using the technique of *vibro-compaction* (Section 8.6), or else piles should be used.

Due to the extreme difficulty of obtaining undisturbed sand samples for laboratory testing and to the inherent heterogeneity of sand deposits, the allowable bearing capacity is normally estimated by means of correlations based on the results of in-situ tests. The tests in question are plate bearing tests and dynamic or static penetration tests.

The Plate Bearing Test

In this test the sand is loaded through a steel plate at least 300 mm square, readings of load and settlement being observed up to failure or to at least 1·5 times the estimated allowable bearing capacity. The load increments should be approximately one-fifth of the estimated allowable bearing capacity. The test plate is generally located at foundation level in a pit at least 1·5 m square. The test is reliable only if the sand is reasonably uniform over the significant depth of the full-scale foundation. Minor local weaknesses near the surface will influence the results of the test while having no appreciable effect on the full-scale foundation. On the other hand, a weak stratum below the significant depth of the test plate but within the significant depth of the foundation, as shown in Fig. 8.7 would have no influence on the test results: the weak stratum, however, would have an appreciable effect on the performance of the foundation.

Settlement in a sand increases as the size of the loaded area increases and the main problem with the use of plate bearing tests is the extrapolation of the settlement of a test plate to that of a full-scale foundation. The required correlation appears to depend on the relative density, particle size distribution and stress history of the sand, and at present there is no reliable method of extrapolation. Bjerrum and Eggestad [8.3], for example, from a study of case records, showed that there is a considerable scatter in the relationship between settlement and the size of the loaded area for a given pressure. Ideally, plate bearing tests should be carried out at different depths and using plates of different sizes in order that extrapolations may be made, but this is generally ruled out on economic grounds: further problems would be introduced if the tests had to be carried out below water table level.

The screw-plate test is a form of bearing test in which no excavation is required. The plate penetrates the sand by rotation and can therefore be positioned, in turn, at a series of depths above or below the water table. Loading is carried out through the shaft of the screw plate.

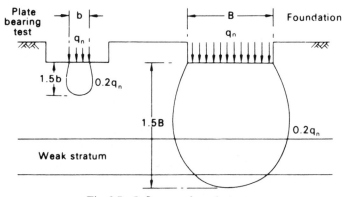

Fig. 8.7   Influence of weak stratum.

The Standard Penetration Test

This dynamic penetration test, specified in BS 1377, is used to assess the in-situ relative density of a sand deposit. The test is performed using a split barrel sampler (Fig. 10.5c), 50 mm external diameter, 35 mm internal diameter and about 650 mm in length, connected to the end of boring rods. The sampler is driven into the sand at the bottom of a cased borehole by means of a 65 kg hammer falling freely through a height of 760 mm onto the top of the boring rods. In the UK a trip-release mechanism and guiding assembly are normally used to control the fall of the hammer, and an anvil at the lower end of the assembly is used to transmit the blow to the boring rods. However, different methods of releasing the hammer are used in different countries. The borehole must be cleaned out to the required depth, care being taken to ensure that the material to be tested is not disturbed: jetting as part of the boring operation is undesirable. The casing must not be driven below the level at which the test is to begin.

Initially the sampler is driven 150 mm into the sand to seat the sampler and to by-pass any disturbed sand at the bottom of the borehole. The number of blows required to drive the sampler a further 300 mm is then recorded: this number is called the *standard penetration resistance* (N). The number of blows required for each 75 mm of penetration (including the initial drive) should be recorded separately. If 50 blows are reached before a penetration of 300 mm, no further blows should be applied but the actual penetration should be recorded. At the conclusion of a test the sampler is withdrawn and the sand extracted. Tests are normally carried out at intervals of between 0·75 m and 1·50 m to a depth at least equal to the width (B) of the foundation. If the test is to be carried out in gravelly soils the driving shoe is replaced by a solid 60° cone. There is evidence that slightly higher results are obtained in the same material when the normal driving shoe is replaced by the 60° cone.

When testing below the water table, care must be taken to avoid entry of water through the bottom of the borehole as this would tend to loosen the sand due to upward seepage pressure. Water should be added as necessary to maintain the water table level in the borehole (or at the level required to balance any excess pore water pressure). When the test is carried out in very fine sand or silty sand below the water table the measured N value, if greater than 15, should be corrected for the increased resistance due to negative excess pore water pressure set up during driving and unable to dissipate immediately: the corrected value is given by

$$N' = 15 + \tfrac{1}{2}(N - 15) \tag{8.13}$$

Skempton [8.33] summarized the evidence regarding the influence of test procedure on the value of standard penetration resistance. Measured N values should be corrected to allow for the different methods of releasing the hammer, the type of anvil and the total length of boring rods. Only the energy delivered to the sampler is applied in penetrating the sand, the ratio of the delivered energy to the free-fall energy of the hammer being referred to as the rod energy ratio. Rod energy ratios for the operating procedures used in several countries vary between 45% and 78%. For the trip-release mechanism, guiding assembly and anvil generally used in the UK the energy ratio for rod lengths exceeding 10 m is 60%. It has been recommended that a standard rod energy ratio of 60% should be adopted and that all measured N values should be normalized, by simple proportion of energy ratios, to this standard: the normalized values are denoted N_{60}. If a short length of boring rods (<10 m) is used in a test a reflection of energy occurs and a further loss in delivered energy results. A further correction should there- fore be applied to the measured N values if the total length of rods is less than 10 m: for example if a 3–4 m length is used a correction factor of 0·75 has been proposed. An additional effect relates to the borehole diameter, there being evidence that lower N values are obtained in 150 mm and 200 mm diameter boreholes than in those less than 115 mm in diameter. Tentative correction factors for 150 mm and 200 mm boreholes are 1·05 and 1·15, respectively.

The relative density of a sand was described by Terzaghi and Peck [8.37], in general terms, on the basis of standard penetration resistance, as shown in columns (1) and (2) of Table 8.3. Numerical values of relative density, as shown in column (3), were subsequently added by Gibbs and Holtz [8.17]. However standard penetration resistance depends not only on relative density but also on the effective stresses at the depth of measurement: effective stresses can be represented to a first approximation by effective overburden pressure. This dependence was first demonstrated in the laboratory by Gibbs and Holtz and was later confirmed in the field. Sand at the same relative density would thus give different values of standard penetration resistance at different depths. Several proposals have been made for the correction of measured N values following the work of Gibbs and Holtz. The corrected value (N_1) is related to the measured value (N) by the factor C_N, where:

$$N_1 = C_N N \qquad (8.14)$$

The relationship between C_N and effective overburden pressure shown in Fig. 8.8 represents a consensus of published proposals.

The following relationship between standard penetration resistance (N), relative density (D_r) and effective overburden pressure $(\sigma_0'$ kN/m$^2)$ was proposed by Meyerhof:

$$\frac{N}{D_r^2} = a + b\frac{\sigma_0'}{100} \qquad (8.15)$$

Values of the parameters a and b for a number of sands have been given by Skempton [8.33]. The characteristics of a sand can be represented by $(N_1)_{60}$ and $(N_1)_{60}/D_r^2$ where $(N_1)_{60}$ is the standard penetration resistance nor-

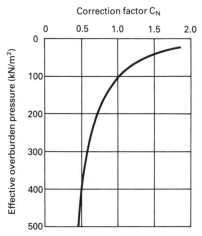

Fig. 8.8 Correction of measured values of standard penetration resistance.

malized to a rod energy ratio of 60% and an effective overburden pressure of 100 kN/m^2. Appropriate values of $(N_1)_{60}$ were added to the Terzaghi and Peck classification of relative density by Skempton, as shown in column (4) of Table 8.3. Table 8.3 should be considered to apply to normally consolidated sands.

There is evidence that standard penetration resistance is also influenced by the grading and shape of the particles, the degree of overconsolidation and the time during which the sand has been undergoing consolidation (referred to as the ageing effect). All other factors being equal, the evidence indicates that standard penetration resistance increases with increasing particle size, increasing overconsolidation ratio and ageing.

A correlation between the shear strength parameter ϕ', standard penetration resistance and effective overburden pressure, published by Schmertmann [8.30], but based on previous work by DeMello, is shown in Fig. 8.9. It must be appreciated that this chart provides only a rough estimate of the value of ϕ' and should not be used for very shallow depths.

Table 8.3 Relative Density of Sands

(1) N value	(2) Classification	(3) D_r (%)	(4) $(N_1)_{60}$
0–4	Very loose	0–15	0–3
4–10	Loose	15–35	3–8
10–30	Medium dense	35–65	8–25
30–50	Dense	65–85	25–42
>50	Very dense	85–100	42–58

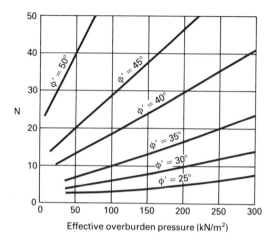

Effective overburden pressure (kN/m²)

Fig. 8.9   Correlation between shear strength parameter ϕ', standard penetration resistance and effective overburden pressure. (Reproduced from J. H. Schmertmann (1975), *Proceedings of Conference on In-Situ Measurement of Soil Properties*, by permission of the American Society of Civil Engineers).

Associated Design Methods

In 1948, Terzaghi and Peck [8.37] presented empirical correlations between standard penetration resistance, width of footing and the bearing pressure limiting maximum settlement to 25 mm (and differential settlement to 75% of maximum settlement). According to Terzaghi and Peck, the correlations, represented in Fig. 8.10, are applicable to situations in which the water table is not less than $2B$ below the footing, where B is the width of the footing. If the sand at foundation level is saturated the pressures obtained from Fig. 8.10 should be reduced by one-half if the depth/breadth ratio of the footing is zero, and reduced by one-third if the depth/breadth ratio is unity. For intermediate positions of the water table and intermediate values of depth/breadth ratio the appropriate value of bearing pressure may be obtained by linear interpolation. However these recommendations are now considered to produce too severe a reduction in allowable pressure, and a correction should be made only if the water table is within a depth B below the foundation. Peck, Hanson and Thornburn [8.26] proposed that linear interpolation should be used between a reduction of 50% if the water table is at ground level and zero reduction if the water table is at depth B below the foundation. Thus the provisional value of allowable bearing pressure obtained from Fig. 8.10 should be multiplied by a factor C_w, given by:

$$C_w = 0\cdot5 + 0\cdot5\frac{D_w}{D + B} \qquad (8.16)$$

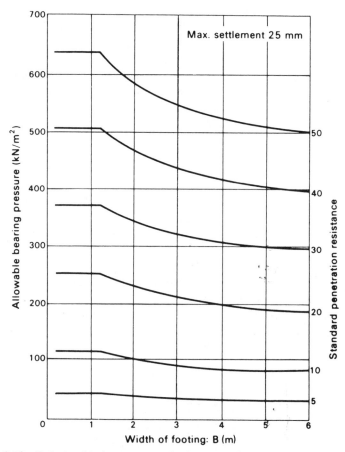

Fig. 8.10   Relationship between standard penetration resistance and allowable bearing pressure. (Reproduced from K. Terzaghi and R. B. Peck (1967) *Soil Mechanics in Engineering Practice*, by permission of John Wiley and Sons, Inc.)

where D_w is the depth of the water table below the surface and D is the depth of the foundation.

Terzaghi and Peck stated that their correlations were a conservative basis for the design of shallow footings. It was intended that the largest footing should not settle by more than 25 mm even if it were situated on the most compressible pocket of sand. It should be realized that the Terzaghi and Peck correlations were not intended to yield actual settlement values for particular footings but only to ensure that the maximum settlement *would not exceed* 25 mm.

In using the correlations the average N value is determined for each borehole on the site in question and the lowest average is then used in design. For a series of footings the bearing pressure is obtained for the largest footing: this value of pressure is then used to calculate the dimensions of all other footings, subject to a check on the factor of safety against shear failure. In the case of rafts the allowable bearing pressure obtained from the design chart should be doubled because a maximum settlement of 50 mm is considered acceptable.

Settlement measurements on actual structures have shown that the Terzaghi and Peck method is excessively conservative. Meyerhof [8.24] recommended that the allowable bearing pressure given by the Terzaghi and Peck method should be increased by 50% and that no correction should be applied for the position of the water table, arguing that its effect is reflected in the measured N values.

The influence of effective overburden pressure was not considered in the original Terzaghi and Peck correlations and it is now recognised that corrected values of standard penetration resistance (N_1), determined from Fig. 8.8, should be used in determining allowable bearing pressures. It should be noted that the stress history of the sand is not taken into account in the Terzaghi and Peck design procedure.

Burland, Broms and De Mello [8.9] collated settlement data from a number of sources and plotted settlement per unit pressure (s/q) against foundation breadth (B): lines representing upper limits were then drawn for dense and medium-dense sands. This graph, shown in Fig. 8.11, may be adequate for routine work. The 'probable' settlement could perhaps be taken as 50% of the upper limit value. In most cases the maximum

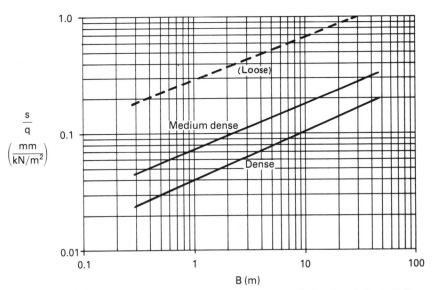

Fig. 8.11 Envelopes of settlement per unit pressure (Reproduced from J. B. Burland, B. B. Broms and V. F. B. De Mello (1977) Proceedings 9th International Conference SMFE, Tokyo, Vol. 2, by permission of the Japanese Society of Soil Mechanics and Foundation Engineering.)

settlement would be unlikely to exceed 75%, of the upper limit value. Factors such as foundation depth and water table position are not considered. Use of the graph implies that the settlement/pressure relationship remains approximately linear.

Burland and Burbidge [8.10] carried out a statistical analysis of over 200 settlement records of foundations on sands and gravels. A relationship was established between the compressibility of the soil (a_f), the width of the foundation (B) and the average value of standard penetration resistance ($\bar{N}$) over the depth of influence of the foundation. The compressibility is given by the slope of the pressure/settlement plot, units mm/(kN/m²), over the working range of pressure. Evidence was presented which indicated that if N tends to increase with depth or is approximately constant with depth then the ratio of the depth of influence to foundation width (z_I/B) decreases with increasing foundation width: values of z_I obtained from Fig. 8.12 can be used as a guide in design. If, however, N tends to decrease with depth the value of z_I should be taken as $2B$, provided the stratum thickness exceeds this value. The compressibility is related to foundation width by a compressibility index (I_c), where:

$$I_c = \frac{a_f}{B^{0.7}} \tag{8.17}$$

The compressibility index, in turn, is related to the average value of standard penetration resistance by the expression:

$$I_c = \frac{1.71}{\bar{N}^{1.4}} \tag{8.18}$$

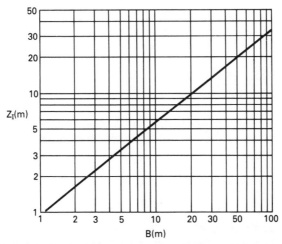

Fig. 8.12 Relationship between depth of influence and foundation width. (Reproduced from J. B. Burland and M. C. Burbidge (1985), *Proceedings Institution of Civil Engineers*, Part 1, Vol. 78, by permission of Thomas Telford Ltd.).

The N values should not be corrected for effective overburden pressure as this has a major influence on both standard penetration resistance and compressibility: this influence should not therefore be eliminated from the correlation. The results of the analysis tend to confirm Meyerhof's conclusion that the influence of water table level is reflected in the measured N values. However the position of the water table does influence *settlement* and if the level were to fall subsequent to the determination of the N values then a larger settlement would be expected. Equation 8.13 should be applied in the case of very fine sands and silty sands below the water table. It was further proposed that in the case of gravels or sandy gravels the measured N values should be increased by 25%.

In a normally consolidated sand the average settlement s_i(mm) at the end of construction for a foundation of width B(m) carrying a net foundation pressure q_n(kN/m²) is given by:

$$s_i = q_n B^{0.7} I_c \tag{8.19a}$$

If it can be established that the sand is overconsolidated and an estimate of preconsolidation pressure (σ_c') can be made, the settlement is given by one or other of the following expressions:

$$s_i = (q_n - \frac{2}{3}\sigma_c')B^{0.7} I_c \qquad \text{(if } q_n > \sigma_c') \tag{8.19b}$$

$$s_i = q_n B^{0.7}\frac{I_c}{3} \qquad \text{(if } q_n < \sigma_c') \tag{8.19c}$$

The analysis indicated that foundation depth had no significant influence on settlement for depth/breadth ratios less than 3. However a significant correlation was found between settlement and the length/breadth ratio (L/B) of the foundation: accordingly the settlement given by Equations 8.19 should be multiplied by a shape factor f_s, where:

$$f_s = \left(\frac{1.25 \; L/B}{L/B + 0.25}\right)^2 \tag{8.20}$$

It was tentatively proposed that if the thickness (H) of the sand stratum below foundation level is less than the depth of influence (z_I) the settlement should be multiplied by a factor f_l, where:

$$f_l = \frac{H}{z_I}\left(2 - \frac{H}{z_I}\right) \tag{8.21}$$

Although it is normally assumed that settlement in sands is virtually complete by the end of construction and initial loading the records indicated that continuing settlement can occur and it was proposed that the settlement should be multiplied by a factor f_t for time in excess of 3 years after the end of construction, where:

Table 8.4   Compressibility Classification for Normally Consolidated Sands and Gravels (Burland and Burbidge [8.10])

N value	Compressibility grade
0–4	VII
4–8	VI
9–15	V
16–25	IV
26–40	III
41–60	II
>60	I

$$f_t = \left(1 + R_3 + R_t \log \frac{t}{3}\right) \tag{8.22}$$

where R_3 is the time-dependent settlement, as a proportion of s_i, occurring during the first three years after construction and R_t is the settlement occurring during each log cycle of time in excess of three years. A conservative interpretation of the data indicates that after 30 years f_t can reach 1·5 for static loads and 2·5 for fluctuating loads.

It should be noted that unlike the Terzaghi and Peck method, the Burland and Burbidge method predicts a specific value of settlement for a given foundation pressure.

Burland and Burbidge also introduced the concept of compressibility grades, based on uncorrected N values, as detailed in Table 8.4, these grades being a function of both relative density and overburden pressure. Charts for the assessment of compressibility grades from the results of plate bearing tests were also presented.

Example 8.5

A footing 3 m square is to be located at a depth of 1·5 m in a sand deposit, the water table being 3·5 m below the surface. Values of standard penetration resistance were determined as detailed in Table 8.5. Determine the allowable bearing capacity using the various design methods.

Terzaghi and Peck recommended that N values should be determined between foundation level and a depth of approximately B below the foundation, in this case between depths of 1·5 m and 4·5 m: the values at depths of 0·75 m and 5·20 m are therefore superfluous. The measured N values are corrected using equation 8.14. Values of effective overburden pressure are calculated (using $\gamma = 17$ kN/m^3 above the water table and $\gamma' = 10$ kN/m^3 below the water table) and the corresponding values of C_N determined from Fig. 8.8. The average of the corrected values (N_1) is 16.

Table 8.5

Depth (m)	N	σ_v' (kN/m²)	C_N	N_1
0·75	8	—	—	—
1·55	7	26	2·0	14
2·30	9	39	1·6	14
3·00	13	51	1·4	18
3·70	12	65	1·25	15
4·45	16	70	1·2	19
5·20	20	—	—	—
				(av. 16)

Then referring to Fig. 8.10, for $B = 3$ m and $N = 16$, the provisional value of allowable bearing capacity is 165 kN/m². For the given water table level the provisional value should be multiplied by the factor C_w (Equation 8.16), where:

$$C_w = 0.5 + \frac{0.5 \times 3.5}{4.5} = 0.89$$

Hence the allowable bearing capacity is given by:

$$q_a = 0.89 \times 165 = 150 \text{ kN/m}^2$$

Using Meyerhof's method the average of the measured N values between depths of 1·5 m and 4·5 m is 11. For $B = 3$ m and $N = 11$ the provisional value of allowable bearing capacity is 100 kN/m². This value is increased by 50% with no correction being made for the position of the water table. Thus:

$$q_a = 1.5 \times 100 = 150 \text{ kN/m}^2$$

Using the Burland and Burbidge method, and assuming that the sand is normally consolidated, the depth of influence (Fig. 8.12) for $B = 3$ m is 2·2 m, i.e. 3·7 m below the surface. The average of the measured N values between depths of 1·5 m and 3·7 m is 10, hence the compressibility index (equation 8.17) is given by:

$$I_c = 1.71/10^{1.4} = 0.068$$

Then the allowable bearing capacity for a settlement of 25 mm at the end of construction is given by:

$$q_a = \frac{s_i}{B^{0.7}I_c} = \frac{25}{3^{0.7} \times 0.068} = 170 \text{ kN/m}^2$$

In design the usual requirement is to determine the foundation dimensions for the support of a given (gross) load and an iterative technique is necessary.

The Dutch Cone Test

The Dutch cone (Fig. 8.13) is a static penetrometer having an apex angle of 60° and an end area of 1000 mm² (35·7 mm diameter). The penetrometer is attached to a string of solid rods running inside hollow outer rods: the external diameter of the outer rods is equal to the cone diameter. The outer rods are attached to a union sleeve the lower end of which has a reduced diameter and runs inside the body of the cone. No boring is required for this test. The cone is pushed 80 mm into the sand at a uniform rate of 15-20 mm/s by means of the inner rods, the sleeve remaining stationary. The force is provided hydraulically, the required pressure being recorded on a Bourdon gauge. Reaction is usually provided by means of screw anchors in the ground. The end resistance of the cone at any depth is called the *cone penetration resistance* (q_c)—actually defined as the force required to advance the cone divided by the end area. After the penetration resistance has been determined the outer rods are pushed downwards. The cone and sleeve are thus advanced together after the travel inside the body of the cone has been taken up. The test is then repeated, cone penetration resistance usually being determined at depth intervals of 200 mm. It is vital that the verticality of the cone is maintained during penetration.

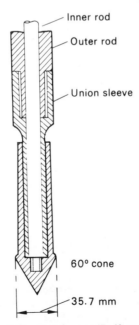

Fig. 8.13 Dutch cone (Delft model).

A more elaborate instrument is the friction cone penetrometer in which a friction sleeve is located above the cone. The cone and the friction sleeve can each be advanced separately by means of the inner rods. Initially the inner rods are pushed downwards a distance of 40 mm, causing the cone only to penetrate the sand, and the cone penetration resistance is recorded. The cone then engages the friction sleeve and when the inner rods are again pushed downwards (a further distance of 40 mm) the cone and sleeve move together, the combined cone and sleeve resistance being recorded. The outer rods are then pushed downwards (over a distance of 200 mm), first bringing the sleeve into contact with the top of the cone then advancing the cone and sleeve together. The test is then repeated. The side resistance on the sleeve, which can be used in estimating the allowable load on a pile, is obtained by subtracting the cone penetration resistance from the combined cone and sleeve resistance.

A further development is the 'electric' penetrometer in which the conical point is mounted on a cylindrical body of the same diameter. Cone penetration resistance is measured by means of a load cell inside the body of the instrument and can thus be recorded continuously as the penetrometer is pushed into the sand. The results are normally plotted automatically, against depth, by means of a chart recorder. Electric penetrometers are also available with a friction sleeve, mechanically separate from the conical point; side resistance is measured by means of a

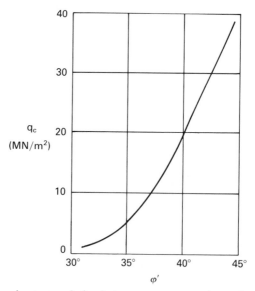

Fig. 8.14 Approximate correlation between cone penetration resistance and shear strength parameter ϕ'. (Reproduced from G. G. Meyerhof (1976) *Proceedings ASCE*, Vol. 102, No. GT3, by permission of the American Society of Civil Engineers.)

second load cell. Cone resistance and side resistance can thus be measured independently.

An approximate correlation between cone penetration resistance and the shear strength parameter ϕ', proposed by Meyerhof [8.25], is given in Fig. 8.14.

Associated Design Methods

The Buisman-DeBeer Method. This method, in which the settlement under a given pressure is estimated, depends on correlations between cone penetration resistance and the compressibility of the sand. Buisman proposed the following empirical equation to obtain a constant of compressibility (C) from a value of cone penetration resistance (q_c):

$$C = 1 \cdot 5 \frac{q_c}{\sigma'_0} \tag{8.23}$$

where σ'_0 is the effective overburden pressure at the depth of measurement. Settlement can be estimated by means of the equation:

$$s = \frac{H}{C} \ln \frac{\sigma'_0 + \Delta\sigma}{\sigma'_0} \tag{8.24}$$

where s is the settlement of a layer of thickness H and $\Delta\sigma$ is the increment of vertical stress at the centre of the layer (cf. Equation 7.8). If C represents the compressibility of the sand over an elemental layer of thickness Δz, the settlement can be expressed as:

$$s = \int_0^H \frac{1}{C} \left\{ \ln \left(\frac{\sigma'_0 + \Delta\sigma}{\sigma'_0} \right) \right\} dz$$

or approximately:

$$s = \sum_0^H \frac{2 \cdot 3 \sigma'_0}{1 \cdot 5 q_c} \Delta z \log \frac{\sigma'_0 + \Delta\sigma}{\sigma'_0}$$

$$= \sum_0^H 1 \cdot 53 \frac{\sigma'_0}{q_c} \Delta z \log \frac{\sigma'_0 + \Delta\sigma}{\sigma'_0} \tag{8.25}$$

In practice the q_c/depth profile is divided into suitable layers (thickness Δz) within each of which the value of q_c is assumed constant. In deep deposits the summation may be terminated at the depth at which the stress increment $(\Delta\sigma)$ becomes less than 10% of the effective overburden pressure (σ'_0). The Buisman-DeBeer method is strictly applicable only to normally loaded sands. In the case of pre-loaded sands the method will give settlements which are too high.

Based on a study of case records, Meyerhof [8.24] recommended that the foundation pressure producing the allowable settlement by the

Buisman-DeBeer method should be increased by 50%. This is approximately equivalent to using the following equation for the constant of compressibility:

$$C = 1 \cdot 9 \frac{q_c}{\sigma_0'} \tag{8.26}$$

Schmertmann's Method. This method of settlement estimation is based on a simplified distribution of vertical strain under the centre of a shallow footing, expressed in the form of a strain influence factor I_z. The vertical strain ε_z at a point under the centre of the footing, carrying a net pressure q_n, is written as

$$\varepsilon_z = \frac{q_n}{E} I_z \tag{8.27}$$

where E is the appropriate value of Young's modulus. The assumed distribution of strain influence factor with depth is shown in Fig. 8.15: depth is expressed in terms of the width (B) of the footing. This is a simplified distribution, based on both theoretical and experimental results, in which it is assumed that strains become insignificant at a depth of $2B$ below the footing. It should be noted that the maximum vertical strain does not occur immediately below the footing as is the case with vertical stress. Corrections can be applied to the strain distribution for the depth of the footing below the surface and for creep. Although it is usually assumed that settlement in sands is virtually complete by the end of construction, some case records indicate continued settlement with time thus suggesting a creep effect: however the creep correction is often omitted. The correction factor for footing depth is given by

$$C_1 = 1 - 0 \cdot 5 \frac{\sigma_0'}{q_n} \tag{8.28}$$

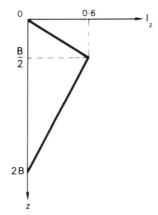

Fig. 8.15   Assumed distribution of strain influence factor with depth.

where σ'_0 = effective overburden pressure at foundation level, and q_n = net foundation pressure. The correction factor for creep is given by

$$C_2 = 1 + 0{\cdot}2\log\frac{t}{0{\cdot}1} \tag{8.29}$$

where t is the time in years at which the settlement is required.

The settlement of a footing carrying a net pressure q_n is written as

$$s = \int_0^{2B} \varepsilon_z \, dz$$

or, approximately,

$$s = C_1 C_2 q_n \sum_0^{2B} \frac{I_z}{E} \Delta z \tag{8.30}$$

Schmertmann obtained the following correlation between Young's modulus, determined from the results of screw-plate tests, and cone penetration resistance:

$$E = 2q_c \tag{8.31}$$

The measured q_c/depth profile, to a depth of $2B$ below the footing, is divided into suitable layers (thickness Δz) within each of which the value of q_c is assumed constant. The value of I_z at the centre of each layer is obtained from Fig. 8.15, it being assumed that this distribution is independent of the sand heterogeneity.

Example 8.6

A footing 2·5 m square carries a net foundation pressure of 150 kN/m² at a depth of 1 m in a deep deposit of fine sand. The water table is at a depth of 4 m. Above the water table the unit weight of the sand is 17 kN/m³ and below the water table the saturated unit weight is 20 kN/m³. The variation of cone penetration resistance with depth is given in Fig. 8.16. Estimate the settlement of the footing using (a) the Buisman-DeBeer method, (b) Schmertmann's method.

(a) In the Buisman-DeBeer method the q_c/z plot below foundation level is divided into a number of layers (thickness Δz) for each of which the value of q_c can be assumed constant. The depth below ground level (z_c) of the centre of each layer is obtained from the plot. The values of effective overburden pressure (σ'_0) and stress increment ($\Delta\sigma$) at the centre of each layer are calculated: the values of $\Delta\sigma$ are calculated using Fig. 5.10. Equation 8.25 is then applied to obtain the settlement. (See Table 8.6).

(b) In Schmertmann's method the q_c/z plot is divided into layers in the same way. In this case the distribution of strain influence factor (I_z) is

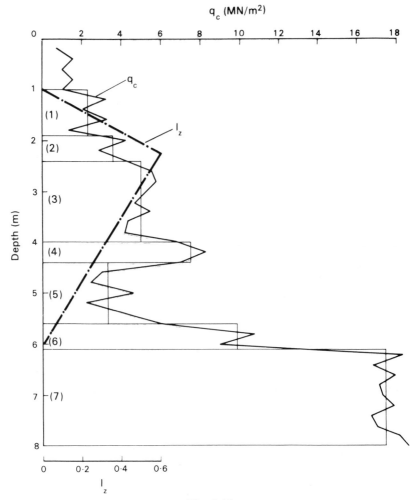

Fig. 8.16

superimposed on the plot and the value of I_z is determined at the centre of each layer. The settlement is calculated using Equation 8.30, in which E is given by Equation 8.31 (See Table 8.7.)

Schmertmann's correction factor for footing depth (Equation 8.28) is

$$C_1 = 1 - \frac{0 \cdot 5 \times 17}{150} = 0 \cdot 94$$

The correction factor for creep (C_2) will be taken as unity. Then

$$s = 0 \cdot 94 \times 150 \times 0 \cdot 1906 = 26 \cdot 9 \, \text{mm}$$

(from Table 8.7). The settlement of the footing is 27 mm.

Table 8.6 Buisman-DeBeer Method

	Δz (m)	q_c (MN/m²)	z_c (m)	σ'_0 (kN/m²)	$\Delta\sigma$ (kN/m²)	$\dfrac{\sigma'_0 + \Delta\sigma}{\sigma'_0}$	log	$1\cdot53\,\dfrac{\sigma'_0}{q_c}\,\Delta z$ (mm)	Δs (mm)
1.	0·90	2·3	1·45	25	144	6·83	0·834	14·8	12·4
2.	0·50	3·6	2·15	36	120	4·29	0·632	7·8	4·9
3.	1·60	5·0	3·20	54	60	2·11	0·323	26·6	8·6
4.	0·40	7·5	4·20	70	35	1·50	0·175	5·7	1·0
5.	1·20	3·3	5·00	78	23	1·30	0·113	43·5	4·9
6.	0·50	9·9	5·85	87	17	1·19	0·076	6·7	0·5
7.	1·90	17·5	7·05	99	12	1·12	0·050	16·5	0·8
									33·1

The settlement of the footing is 33 mm.

Table 8.7   Schmertmann's Method

	Δz (m)	q_c (MN/m^2)	E (MN/m^2)	I_z	$\dfrac{I_z}{E}\Delta z$ (m^3/MN)
1.	0·90	2·3	4·6	0·22	0·0431
2.	0·50	3·6	7·2	0·55	0·0382
3.	1·60	5·0	10·0	0·45	0·0720
4.	0·40	7·5	15·0	0·29	0·0077
5.	1·20	3·3	6·6	0·16	0·0291
6.	0·50	9·9	19·8	0·02	0·0005
					0·1906

8.5   Bearing Capacity of Piles

Piles may be divided into two main categories according to their method of installation. The first category consists of driven piles of steel or precast concrete and piles formed by driving tubes or shells which are fitted with a driving shoe: the tubes or shells are filled with concrete after driving. Also included in this category are piles formed by placing concrete as the driven tubes are withdrawn. The installation of any type of driven pile causes displacement and disturbance of the soil around the pile. However, in the case of steel H piles and tubes without a driving shoe, soil displacement is small. The second category consists of piles which are installed without soil displacement. Soil is removed by boring or drilling to form a shaft, concrete then being cast in the shaft to form the pile: the shaft may be cased or uncased depending on the type of soil. In clays the shaft may be enlarged at its base by a process known as under-reaming: the resultant pile then has a larger base area in contact with the soil.

The ultimate load which can be carried by a pile is equal to the sum of the base resistance and the shaft resistance. The base resistance is the product of the base area (A_b) and the ultimate bearing capacity (q_f) at base level. The shaft resistance is the product of the perimeter area of the shaft (A_s) and the average value of ultimate shearing resistance per unit area (f_s), generally referred to as the 'skin friction', between the pile and the soil. The weight of soil displaced or removed is generally assumed to be equal to the weight of the pile. Thus the ultimate load (Q_f) which can be applied to the top of the pile is given by the equation:

$$Q_f = A_b q_f + A_s f_s \qquad (8.32)$$

An appropriate load factor is applied to Q_f to obtain the allowable load on the pile. Different values of load factor may be applied to the base and shaft resistances.

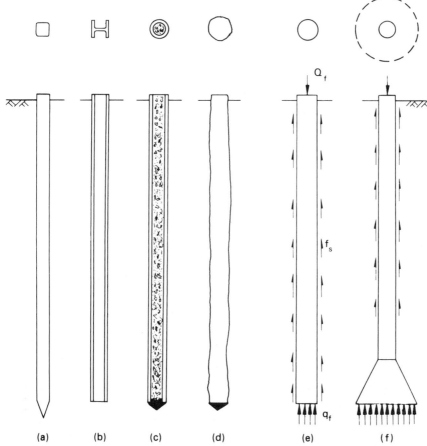

Fig. 8.17 Principal types of pile: (a) precast RC pile, (b) steel H pile, (c) shell pile, (d) concrete pile cast as driven tube withdrawn, (e) bored pile (cast in situ), (f) under-reamed bored pile (cast in situ).

Evidence from load tests on instrumented piles indicates that in the initial stages of loading, most of the load is supported by skin friction on the upper part of the pile. Subsequently, as the load is increased, further mobilization of skin friction takes place but gradually a greater proportion of the load is supported by base resistance. At failure the proportion of the load supported by skin friction may decrease slightly due to plastic flow of the soil near the base of the pile.

Piles in Sands

The ultimate bearing capacity and settlement of a pile depends mainly on the relative density of the sand. However, if a pile is driven into sand the

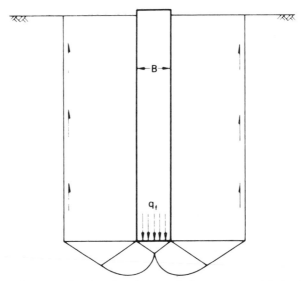

Fig. 8.18 Failure mechanism in theory of Berezantzev, Khristoforov, and Golubkov.

relative density adjoining the pile is increased by compaction due to soil displacement (except in dense sands, which may be loosened). The soil characteristics governing ultimate bearing capacity and settlement, therefore, are different from the original characteristics prior to driving. This fact, in addition to the heterogeneous nature of sand deposits, makes the prediction of pile behaviour by analytical methods extremely difficult.

The ultimate bearing capacity at base level can be expressed as

$$q_f = \sigma'_0 N_q \tag{8.33}$$

where σ'_0 is the effective overburden pressure at base level. (It should be noted that the N_y term for a pile is negligible because the width B is small compared with the length L.)

Berezantzev, Khristoforov and Golubkov [8.1] developed a theory for the ultimate bearing capacity of piles in which failure is assumed to have taken place when the failure surfaces reach the level of the base, as shown in Fig. 8.18. The surcharge at base level consists of the pressure due to the weight of an annulus of soil surrounding the pile, reduced by the frictional force on the outer surface of the annulus. The resulting factor N_q depends on the shear strength parameter ϕ' and the ratio L/B. For a given value of ϕ' the value of N_q decreases with increasing L/B ratio. Values of N_q for an L/B ratio of 25 are given in Table 8.8: extrapolated values for an L/B ratio of 50 are shown in brackets.

Meyerhof [8.25] presented a semi-empirical relationship between N_q for driven piles, the depth ratio D_b/B, and the shear strength parameter ϕ' of

Table 8.8 **Berezantzev, Khristoforov and Golubkov theory: Relationship between ϕ' and N_q**

ϕ'	28°	30°	32°	34°	36°	38°	40°
N_q	12	17	25	40	58	89	137
	(9)	(14)	(22)	(37)	(56)	(88)	(136)

the sand prior to driving (D_b being the length of the pile embedded in the sand).

The average value of skin friction over the length of pile embedded in sand can be expressed as

$$f_s = K_s \bar{\sigma}'_0 \tan \delta \tag{8.34}$$

where K_s = average coefficient of earth pressure along the embedded length, $\bar{\sigma}'_0$ = average effective overburden pressure along the embedded length, and δ = angle of friction between the pile and the sand. For concrete piles driven in sand, values of K_s of 1·0 and 2·0 for loose and dense sand, respectively, have been suggested for use in design. These values should be halved for steel H piles. Suggested values of δ are 0·75ϕ' for concrete piles and 20° for steel piles.

Equations 8.33 and 8.34 represent a linear increase with depth of both q_f and f_s. However, tests on full-scale and model piles have indicated that these equations are valid only above a critical depth of roughly $15B$ to $20B$. Below the critical depth both q_f and f_s remain approximately constant at limiting values in uniform soil conditions. This is thought to be due to arching of the soil around the lower part of the pile when the soil yields below the base.

Due to the critical depth limitation and to the difficulty of obtaining values of the required parameters, the above equations are difficult to apply in practice. It is preferable to use empirical correlations, based on the results of pile loading tests and dynamic or static penetration tests, to estimate values of q_f and f_s. The following correlations have been proposed by Meyerhof [8.25] for piles driven into a sand stratum:

$$q_f = 40N \frac{D_b}{B} \leqslant 400N \text{ (kN/m}^2) \tag{8.35}$$

where N is the value of standard penetration resistance in the vicinity of the pile base and D_b is the length of the pile embedded in the sand. For piles driven into non-plastic silts an upper limit of $300N$ is appropriate. Also,

$$f_s = 2\bar{N} \text{ (kN/m}^2) \tag{8.36}$$

where $\bar{N}$ is the average value of standard penetration resistance over the embedded length of the pile within the sand stratum. The value of f_s given by Equation 8.36 should be halved in the case of small-displacement piles such as steel H piles. For bored piles the values of q_f and f_s are approximately $\frac{1}{3}$ and $\frac{1}{2}$, respectively, of the corresponding values for driven piles.

The results of Dutch cone tests can also be used. The end bearing capacity of a pile is given approximately by the product of the cross-sectional area of the pile and the cone penetration resistance at base level. It has been proposed that the average value of cone penetration resistance between $3B$ above the base and B below the base should be used. The frictional resistance on the penetrometer sleeve will be lower than that on the pile shaft because different volumes of soil will be displaced in the two cases. The approximate value of skin friction on the pile can be obtained by scaling up the resistance on the penetrometer sleeve by the ratio of the pile to the sleeve circumferences (except in the case of steel H piles), with an upper limit of 3 on the ratio. When only cone resistance has been determined, it has been suggested that f_s is equal, approximately, to $\bar{q}_c/200$ in the case of sands or $\bar{q}_c/150$ in the case of non-plastic silts, where $\bar{q}_c$ is the average value of cone penetration resistance over the embedded length of the pile: these values should be halved for steel H piles. A pile should not be driven below the level at which a base resistance of 10 MN/m^2 is obtained. Driving against a resistance in excess of 10 MN/m^2 may result in damage to the pile.

The direct application of cone penetration resistance ignores the difference in scale between the penetrometer and the pile but DeBeer [8.14] proposed a design procedure that allows for this effect. The procedure is based on the concept that the failure surfaces adjacent to the cone and the pile must be fully developed before penetration can take place. Meyerhof [8.21] developed a bearing capacity theory for deep foundations when the failure surfaces have been fully developed. The shear zones assumed in Meyerhof's theory are shown in Fig. 8.19 and to achieve maximum resistance in a bearing stratum these zones must be entirely within that stratum. The q_c/depth profile for a loose sand stratum underlain by a dense stratum is also shown in Fig. 8.19. The penetration of the cone into the dense stratum is indicated by a significant increase in resistance, as shown by the line BC, followed by a much lower rate of increase: the depth between B and C is denoted by y_c. At C it can be assumed that the shear zones associated with the penetration of the cone are entirely within the dense stratum. However for a pile having a diameter (or equivalent diameter) n times that of the cone, a penetration of y_p, equal to ny_c, into the dense stratum would be required before the maximum base resistance were obtained and the line BD would then represent the corresponding increase in resistance during penetration of the pile. However it is recommended that an upper limit of 20 pile diameters should be placed on the value of y_p. If the resistance at D were greater than the generally accepted damage limit of 10 MN/m^2 the pile should be driven only to the depth corresponding to a resistance of

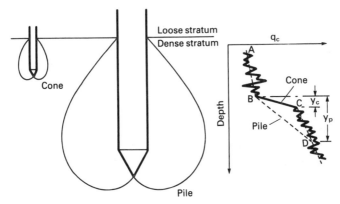

Fig. 8.19 DeBeer's method for scale effect between penetrometer and pile.

10 MN/m^2 on the line BD, subject to a minimum embedment of 5 pile diameters in the dense stratum.

Piles in Clays

In the case of driven piles, the clay adjacent to the pile is displaced both laterally and vertically. Upward displacement of the clay results in heaving of the ground surface around the pile and can cause a reduction in the bearing capacity of adjacent piles already installed. The clay in the disturbed zone around the pile is completely remoulded during driving. The excess pore water pressure set up by the driving stresses dissipates within a few months as the disturbed zone is relatively narrow (of the order of *B*): in general, dissipation is virtually complete before significant structural load is applied to the pile. Dissipation is accompanied by an increase in the shear strength of the remoulded clay and a corresponding increase in skin friction. Thus the skin friction at the end of dissipation is normally appropriate in design.

In the case of bored piles, a thin layer of clay (of the order of 25 mm) immediately adjoining the shaft will be remoulded during boring. In addition, a gradual softening of the clay will take place adjacent to the shaft due to stress release, pore water seeping from the surrounding clay towards the shaft. Water can also be absorbed from the wet concrete when it comes into contact with the clay. Softening is accompanied by a reduction in shear strength and a reduction in skin friction. Construction of a bored pile, therefore, should be completed as quickly as possible. Limited reconsolidation of the remoulded and softened clay takes place after installation of the pile.

The relevant shear strength for the determination of the base resistance of a pile in clay is the undrained strength at base level. The ultimate bearing capacity is expressed as

$$q_f = c_u N_c \tag{8.37}$$

Based on theoretical and experimental evidence, a value of N_c of 9 is appropriate (i.e. Skempton's value for $D/B > 4$). If the clay is fissured the shear strength of a small laboratory specimen (e.g. 38 mm diameter) will be greater than the in-situ strength because it will be relatively less fissured than the soil mass in the vicinity of the pile base: a reduction factor should thus be applied to the laboratory strength (e.g. 0·75 has been suggested for London clay).

The skin friction can be correlated empirically with the average undrained strength $(\bar{c}_u)$ of the undisturbed clay over the depth occupied by the pile, i.e.

$$f_s = \alpha \bar{c}_u \tag{8.38}$$

where α is a coefficient depending on the type of clay, the method of installation and the pile material. The appropriate value of α is obtained from the results of load tests. Values of α can range from around 0·3 to around 1·0. One difficulty with this approach is that there is usually a considerable scatter in the plot of undrained shear strength against depth and it may be difficult to define the value of $\bar{c}_u$.

An alternative approach is to express skin friction in terms of effective stress. The zone of soil disturbance around the pile is relatively thin, therefore dissipation of the positive or negative excess pore water pressure set up during installation should be virtually complete by the time the structural load is applied. In principle, therefore, an effective stress approach has more justification than one based on total stress. In terms of effective stress the skin friction can be expressed as

$$f_s = K_s \bar{\sigma}'_0 \tan \phi' \tag{8.39}$$

where K_s is the average coefficient of earth pressure and $\bar{\sigma}'_0$ is the average effective overburden pressure adjacent to the pile shaft. Failure is assumed to take place in the remoulded soil close to the pile shaft, therefore the angle of friction between the pile and the soil is represented by the angle of shearing resistance in terms of effective stress (ϕ') for the remoulded clay: the cohesion intercept for remoulded clay will be zero. The product $K_s \tan \phi'$ is written as a coefficient β, thus

$$f_s = \beta \bar{\sigma}'_0 \tag{8.40}$$

Approximate values of β can be deduced by making assumptions regarding the value of K_s, especially in the case of normally consolidated clays. However, the coefficient is generally obtained empirically from the results of load tests carried out a few months after installation. For normally consolidated clays the value of β is usually within the range 0·25 to 0·40 but for overconsolidated clays values are significantly higher and vary within relatively wide limits.

The base resistance requires a larger deformation for full mobilization than the shaft resistance, therefore different values of load factor may be appropriate for the two components, the higher factor being applied to the base resistance. In the case of large-diameter bored piles, including under-reamed piles, the shaft resistance may be fully mobilized at working load and it is advisable to ensure a load factor of 3 for base resistance, with a factor of 1 for shaft resistance, in addition to the specified overall load factor (generally 2) for the pile. In the case of under-reamed piles, as a result of settlement, there is a possibility that a small gap will develop between the top of the under-ream and the overlying soil, leading to a drag-down of soil on the pile shaft. Accordingly no skin friction should be taken into account below a level $2B$ above the top of the under-ream.

It should be noted that in the case of under-reamed piles the reduction in pressure on the soil at base level due to the removal of soil is greater than the subsequent increase in pressure due to the weight of the pile. The left hand side of Equation 8.32 must then be written as $(Q_f + W - \gamma D A_b)$, where W is the weight of the pile, A_b is the area of the enlarged base and D is the depth to base level.

Negative Skin Friction. Negative skin friction can occur on the perimeter of a pile driven through a layer of clay undergoing consolidation (e.g. due to a fill recently placed over the clay) into a firm bearing stratum (Fig. 8.20). The consolidating layer exerts a downward drag on the pile, therefore the direction of skin friction in this layer is reversed. The force due to this downward or negative skin friction is thus carried by the pile instead of helping to support the external load on the pile. Negative skin friction increases gradually as consolidation of the clay layer proceeds, the effective overburden pressure σ'_0 gradually increasing as the excess pore water pressure dissipates. Equation 8.40 can also be used to represent negative skin friction. In normally consolidated clays, present evidence indicates that a value of β of 0·25 represents a reasonable upper limit to negative skin friction for preliminary design purposes. It should be noted that there will be a reduction in effective overburden pressure adjacent to the pile in the bearing stratum due to the transfer of part of the overlying soil weight to the pile: If the bearing stratum is sand, this will result in a reduction in bearing capacity above the critical depth.

Load Tests

The loading of a test pile enables the ultimate load to be determined directly and provides a means of assessing the accuracy of predicted values. Tests may also be carried out in which loading is stopped when the proposed working load has been exceeded by a specified percentage. The results from a test on a particular pile will not necessarily reflect the performance of all other piles on the same site, therefore an adequate number of tests is required, depending on the extent of the ground investigation. Driven piles in clays should not be tested for at least a month after installation to allow

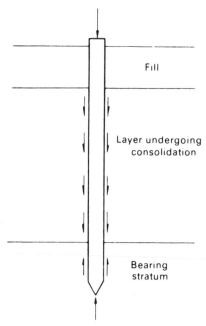

Fill

Layer undergoing
consolidation

Bearing
stratum

Fig. 8.20 Negative skin friction.

most of the increase in skin friction (a result of dissipation of the excess pore water pressure due to the driving stresses) to take place.

Two test procedures are detailed in BS 8004 [8.7]. In the *maintained load test* the load/settlement relationship for the test pile is obtained by loading in suitable increments, allowing sufficient time between increments for settlement to be substantially complete. The ultimate load is normally taken as that corresponding to a specified settlement, for example 10% of the pile diameter. Unloading stages are normally included in the test programme. In the *constant rate of penetration* (CRP) test the pile is jacked into the soil at a constant speed, the load applied in order to maintain the penetration being continuously measured. Suitable rates of penetration for tests in sands and clays are $1 \cdot 5$ mm/min and $0 \cdot 75$ mm/min respectively. The test is continued until either shear failure of the soil takes place or the penetration is equal to 10% of the diameter of the pile base, thus defining the ultimate load. Allowance should be made for the elastic deformation of the pile under test. The settlement of a pile under maintained load cannot be estimated from the results of a CRP test. Typical load/settlement plots are shown in Fig. 8.21.

Pile Groups

A piled foundation usually consists of a group of piles installed fairly close together (typically $2B–4B$ where B is the width or diameter of a single pile) and joined by a slab, known as the pile cap, cast on top of the piles. The cap is usually in contact with the soil in which case part of the structural load is

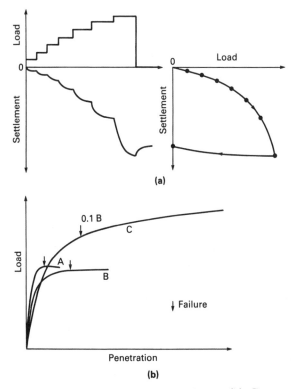

(a)

(b)

Fig. 8.21   Pile loading tests. (a) Maintained load test. (b) Constant rate of penetration test.

carried directly on the soil immediately below the surface. If the cap is clear of the ground surface, the piles in the group are referred to as free-standing. The principles described in this section also apply to piled rafts. In stiff clays, piles at spacings of $4B$ or greater may be installed under a raft for the prime purpose of reducing settlement. An excellent review of the design of piled rafts has been presented by Cooke [8.13].

In general the ultimate load which can be supported by a group of n piles is not equal to n times the ultimate load of a single isolated pile of the same dimensions in the same soil. The ratio of the average load per pile in a group at failure to the ultimate load for a single pile is defined as the efficiency of the group. It is generally assumed that the distribution of load between the piles in an axially loaded group is uniform. However experimental evidence indicates that for a group in sand the piles at the centre of the group carry greater loads than those on the perimeter: in clay, on the other hand, the piles on the perimeter of the group carry greater loads than those at the centre. It can generally be assumed that all piles in a group will settle by the same amount, due to the rigidity of the pile cap. The settlement of a pile group is always greater than the settlement of a corresponding single pile, as a result of the overlapping of the individual zones of influence of the piles

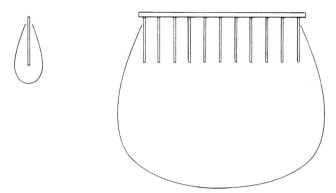

Fig. 8.22   Bulbs of pressure for a single pile and a pile group.

in the group. The bulbs of pressure of a single pile and a pile group (with piles of the same length as the single pile) are of the form illustrated in Fig. 8.22: significant stresses are thus developed over a much wider area and much greater depth in the case of a pile group than in the case of a corresponding single pile. The settlement ratio of a group is defined as the ratio of the settlement of the group to the settlement of a single pile when both are carrying the same proportion of their ultimate load.

The driving of a group of piles into loose or medium-dense sand causes compaction of the sand between the piles, provided that the spacing is less than about $8B$: consequently the efficiency of the group is greater than unity. A value of 1.2 is often used in design. However, for a group of bored piles the efficiency may be as low as $\frac{2}{3}$ because the sand between the piles is not compacted during installation but the zones of shear of adjacent piles will overlap. In the case of piles driven into dense sand, the group efficiency is less than unity due to loosening of the sand and the overlapping of zones of shear.

A closely spaced group of piles in clay may fail as a unit, with shear failure taking place around the perimeter of the group and below the area covered by the piles and the enclosed soil: this is referred to as block failure. Tests by Whitaker [8.43] on free-standing model piles showed that for groups comprising a given number of piles of a given length there was a critical spacing of the order of $2B$ at which the mode of failure changed. For spacings above the critical value, failure occurred below individual piles. For spacings below the critical value, the group failed as a block, like an equivalent pier comprising the piles and the enclosed soil. The group efficiency at the critical spacing was between 0·6 and 0·7. However, when the pile cap was in contact with the soil, no change in the mode of failure was indicated at pile spacings above $2B$ and the efficiency exceeded unity at spacings greater than around $4B$. However, it is now considered that the length of the piles and the size and shape of the group also influence the critical spacing. It is recommended that the minimum centre-line spacing of

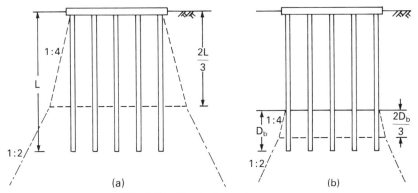

Fig. 8.23 Equivalent raft concept.

piles in clay should not be less than the pile perimeter. The ultimate load in the case of a pile group which fails as a block is given by

$$Q_f = A_b q_f + A_s c_s \qquad (8.41)$$

where A_b is equal to the base area of the group, A_s is equal to the perimeter area of the group and c_s is the average value of shearing resistance, per unit area, on the perimeter. The shearing resistance c_s should be taken as the undrained strength of the remoulded clay unless loading is to be delayed for at least six months, in which case the undrained strength of the undisturbed clay can be used. Dissipation of excess pore water pressure due to installation wil! take longer in the case of a pile group than in the case of a single pile and might not be complete before the early application of structural load. In design, provided the pile cap rests on the soil, the ultimate load should be taken as the lesser of the block failure value and the sum of the individual pile values. However if the piles are free-standing the ultimate load should be the lesser of the block failure value and $\frac{2}{3}$ of the sum of the individual pile values.

The settlement of a pile group in clay can be estimated by assuming that the total load is carried by an 'equivalent raft' located at a depth of $2L/3$ where L is the length of the piles. It may be assumed, as shown in Fig. 8.23a, that the load is spread from the perimeter of the group at a slope of 1 horizontal to 4 vertical to allow for that part of the load transferred to the soil by skin friction. The vertical stress increment at any depth below the equivalent raft may be estimated by assuming in turn that the total load is spread to the underlying soil at a slope of 1 horizontal to 2 vertical. The consolidation settlement is then calculated from Equation 7.10. The immediate settlement is determined by applying Equation 5.31 to the equivalent raft.

The settlement of a pile group underlain by a depth of sand can also be estimated by means of the equivalent raft concept. In this case it may be assumed, as shown in Fig. 8.23b, that the equivalent raft is located at a depth of $2D_b/3$ in the sand stratum with a 1:4 load spread from the perimeter of the group. Again a 1:2 load spread is assumed below the equivalent raft. The settlement is determined from values of standard penetration resistance or cone penetration resistance below the equivalent raft, using the methods detailed in Section 8.4.

It is also possible to estimate the settlement due to the consolidation of a clay layer situated below a sand stratum in which a pile group is supported. The possibility of a pile group in a sand stratum punching through into an underlying layer of soft clay should also be considered in relevant cases: the vertical stress increment at the top of the clay layer should not exceed the presumed bearing value of the clay.

An alternative proposal regarding the equivalent raft is that its area should be equal to that of the pile group. In clays the equivalent raft should, as above, be located at a depth of $2L/3$ but in sands it should be located at the base of the pile group. A 1:2 load spread should be assumed below the equivalent raft in each case. The alternative proposals should be used if shaft resistance is negligible compared with base resistance.

A method based on elastic theory for estimating the settlement of a pile group has been developed by Poulos [8.28].

It should be appreciated that settlement is normally the limiting design criterion for pile groups in both sands and clays.

Pile Driving Formulae

A number of formulae have been proposed in which the dynamics of the pile driving operation is considered in a very idealistic way and the dynamic resistance to driving is assumed to be equal to the static bearing capacity of the pile.

Upon striking the pile, the kinetic energy of the driving hammer is assumed to be

Wh − (energy losses)

where W is the weight of the hammer and h is the equivalent free fall. The energy losses may be due to friction, heat, hammer rebound, vibration and elastic compression of the pile, the packing assembly and the soil. The net kinetic energy is equated to the work done by the pile in penetrating the soil. The work done is Rs where R is the average resistance of the soil to penetration and s is the set or penetration of the pile per blow. The smaller the set the greater the resistance to penetration.

The *Engineering News* formula takes into account the energy loss due to temporary compression (c_p) resulting from elastic compression of the pile. Thus

$$R(s + c_p/2) = Wh \tag{8.42}$$

from which R can be determined. In practice, empirical values are given to the term $c_p/2$ (e.g. for drop hammers $c_p/2 = 25$ mm).

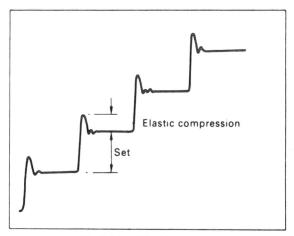

Fig. 8.24   Pile driving trace.

The Hiley formula takes into account the energy losses due to elastic compression of the pile, the soil and the packing assembly on top of the pile, all represented by a term c, and the energy losses due to impact, represented by an efficiency factor η. Thus

$$R(s + c/2) = \eta W h \qquad (8.43)$$

The elastic compression of the pile and the soil can be obtained from the driving trace of the pile (Fig. 8.24). The compression of the packing assembly must be estimated separately by assuming a value of the stress in the assembly during driving.

Driving formulae should be used only for piles in sands and gravels and if possible should be calibrated against the results of load tests.

The Wave Equation

The wave equation is a differential equation describing the transmission of compression waves, produced by the impact of a driving hammer, along the length of a pile. The pile is assumed to behave as a slender rod rather than as a rigid mass. A computer program can be written for the solution of the equation in finite difference form. Details of the method have been given by Smith [8.35]. The pile, packing assembly and driving ram are represented by a series of discrete weights and springs. The shaft and base resistances are represented by a series of springs and dashpots. Values of the parameters describing the behaviour of these elements and of the soil must be estimated.

The relationship can be obtained between the final set and the ultimate load which can be supported by the pile immediately after driving: no information can be obtained regarding the long-term behaviour of the pile.

The equation also enables the stresses in the pile during driving to be determined, for use in the structural design of the pile. An assessment can also be made of the adequacy of the driving equipment to produce the desired load capacity for a given pile.

Example 8.7

A precast concrete pile 450 mm square in section, to form part of a jetty, is to be driven into a river bed which consists of a depth of sand. The results of standard penetration tests in the sand are as follows:

Depth (m)	1.5	3.0	4.5	6.0	7.5	9.0	10.5	12.0
N	4	6	13	12	20	24	35	39

The pile is required to support a compressive load of 650 kN and to withstand an uplift load of 225 kN: the load factor in each case is to be at least 2.5. Determine the depth to which the pile must be driven.

$$\text{Ultimate compressive load} = A_b q_f + A_s f_s = 2.5 \times 650 = 1625 \text{ kN}$$
$$\text{Ultimate uplift load} = A_s f_s = 2\cdot5 \times 225 = 563 \text{ kN}$$

where $A_b = 0\cdot45^2$ (m²) and

$$A_s = 4 \times 0\cdot45 \times D_b \text{ (m}^2\text{)}$$

Using Meyerhof's correlations,

$$q_f = 40\, ND_b/B \leqslant 400N \quad \text{(kN/m}^2\text{)}$$
$$f_s = 2\bar{N} \quad \text{(kN/m}^2\text{)}$$

The calculations are set out in Table 8.9.

By inspection, uplift is the limiting consideration. By interpolation, the pile must be driven to at least 10·2 m. (The load factor under compressive load is then $3225/650 = 5\cdot0$.)

Example 8.8

The results of cone penetration tests at a particular site are shown in Fig. 8.25. On the basis of these results, to what minimum depth should a 600 mm diameter concrete pile be driven into the dense sand if the shear zones are to be fully developed? What load could the pile then support if a load factor of 2·5 were specified?

DeBeer's procedure will be used. By inspection of Fig. 8.25, the cone resistance is fully developed over a depth (y_c) of 0·3 m. Hence the base resistance of the pile will be fully developed over a depth y_p, where:

$$y_p = 0\cdot3 \times \frac{600}{36} = 5\cdot0 \text{ m} \quad \text{(check, } <20B)$$

Table 8.9

D_b	N	$\bar{N}$	$A_s f_s$	q_f (kN/m²)		$A_b q_f$	$A_b q_f + A_s f_s$
(m)			(kN)	$\dfrac{40}{0.45} N D_b$	$400N$	(kN)	(kN)
1·5	4	4	22	535		108	130
3·0	6	5	54	1600		324	378
4·5	13	8	130	5200	5200	1053	1183
6·0	12	9	194		4800	972	1166
7·5	20	11	297		8000	1620	1917
9·0	24	13	421		9600	1944	2365
10·5	35	16	605		14000	2835	3440
12·0	39	19	821		15600	3159	3980

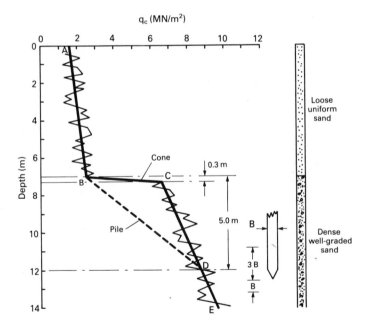

Fig. 8.25

Thus the pile should be driven 5 m into the dense sand, i.e. to a total depth of 12 m.

The increase in resistance during driving of the pile is represented by BD in Fig. 8.25. The ultimate value of base resistance to be used in design is

8·6 MN/m², i.e. the average value between $3B$ above and B below the base of the pile: this value is less than the damage limit of 10 MN/m².

The values of skin friction in the loose and dense sand are given by $\bar{q}_c/200$, where $\bar{q}_c$ is the average cone penetration resistance in each case. Thus:

$$f_s = (2\!\cdot\!0 \times 10^3)/200 = 10 \text{ kN/m}^2 \qquad \text{(loose sand)}$$

$$f_s = (7\!\cdot\!6 \times 10^3)/200 = 38 \text{ kN/m}^2 \qquad \text{(dense sand)}$$

The ultimate load on the pile is then given by:

$$Q_f = A_b q_f + \Sigma A_s f_s$$

$$= \left(\frac{\pi}{4} \times 0\!\cdot\!6^2 \times 8600\right) + (\pi \times 0\!\cdot\!6 \times 7 \times 10) + (\pi \times 0\!\cdot\!6 \times 5 \times 38)$$

$$= 2921 \text{ kN}$$

Then the allowable load is:

$$Q = Q_f/2\!\cdot\!5 = 1168 \text{ kN}$$

(It would be acceptable to disregard skin friction in the loose sand.)

Example 8.9

An under-reamed bored pile is to be installed in a stiff clay. The diameters of the pile shaft and under-reamed base are 1·05 m and 3·00 m respectively. The pile is to extend from a depth of 4 m to a depth of 22 m in the clay, the top of the under-ream being at a depth of 20 m. The relationship between undrained shear strength and depth is shown in Fig. 8.26 and the adhesion coefficient α is 0·4. Determine the allowable load on the pile to ensure (a) an

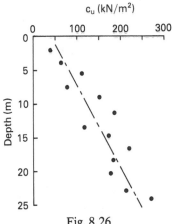

Fig. 8.26

overall load factor of 2, (b) a load factor of 3 under the base when shaft resistance is fully mobilized.

At base level (22 m) the undrained strength is 220 kN/m². Therefore

$$q_f = c_u N_c = 220 \times 9 = 1980 \text{ kN/m}^2$$

It is advisable to disregard skin friction over a length of $2B$ above the top of the under-ream, i.e. below a depth of 17·9 m. The average value of undrained strength between depths of 4 m and 17·9 m is 130 kN/m². Therefore

$$f_s = \alpha \bar{c}_u = 0·4 \times 130 = 52 \text{ kN/m}^2$$

The ultimate load is given by

$$Q_f = A_b q_f + A_s f_s$$

$$= \left(\frac{\pi}{4} \times 3^2 \times 1980 \right) + (\pi \times 1·05 \times 13·9 \times 52)$$

$$= 13\,996 + 2384$$

$$= 16\,380 \text{ kN}$$

The allowable load is the lesser of:

(a) $\dfrac{Q_f}{2} = \dfrac{16\,380}{2} = 8190$ kN

(b) $\dfrac{A_b q_f}{3} + A_s f_s = \dfrac{13\,996}{3} + 2384$

$$= 7049 \text{ kN}$$

However an allowance should be made for the difference between the pressure removed at the base of the under-ream due to boring of the shaft and the pressure subsequently applied due to the weight of the pile. Thus the allowable load may be increased by $(\gamma D A_b - W)/3$. Taking the unit weights of clay and concrete as 20 kN/m³ and 23·5 kN/m³, respectively, and neglecting the additional weight of the under-ream, the additional load is:

$$\left\{ \left(20 \times 18 \times \frac{\pi}{4} \times 3^2 \right) - \left(23·5 \times 18 \times \frac{\pi}{4} \times 1·05^2 \right) \right\}/3 = 726 \text{ kN}$$

Thus the allowable load on the pile is:

$$7049 + 726 = 7775 \text{ kN}$$

Example 8.10

A square group of 25 piles extends between depths of 1 m and 13 m in a deposit of stiff clay 25 m thick overlying rock. The piles are 0·6 m in

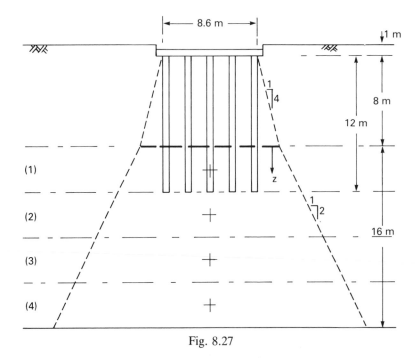

Fig. 8.27

diameter and are spaced at 2 m centres in the group (Fig. 8.27). The undrained shear strength of the clay at pile base level is 170 kN/m² and the average value of undrained strength over the depth of the piles is 105 kN/m². The adhesion coefficient α is 0·45, E_u is 65 MN/m² and m_v is 0·07 m²/MN. The pore pressure coefficient A is 0·24. If the total load on the pile group is 12 000 kN, determine the load factor and the total settlement.

At base level, $c_u = 170$ kN/m². Therefore

$$q_f = 9c_u = 9 \times 170 = 1530 \text{ kN/m}^2$$

Over the depth of the piles, $\bar{c}_u = 105$ kN/m². Therefore

$$f_s = \alpha \bar{c}_u = 0.45 \times 105 = 47 \text{ kN/m}^2$$

For a single pile the ultimate load is given by

$$Q_f = A_b q_f + A_s f_s$$

$$= \left(\frac{\pi}{4} \times 0.6^2 \times 1530 \right) + (\pi \times 0.6 \times 12 \times 47)$$

$$= 432 + 1063$$

$$= 1495 \text{ kN}$$

The ultimate load on the group assuming single pile failure and a group efficiency of 1

$$= 25 \times 1495 = 37\ 375\ \text{kN}$$

The width of the group is $8\cdot6$ m, therefore the ultimate load on the group assuming block failure and taking the full undrained strength on the perimeter

$$= (8\cdot6^2 \times 1530) + (4 \times 8\cdot6 \times 12 \times 105)$$
$$= 113\ 159 + 43\ 344 = 156\ 500\ \text{kN}$$

Hence the load factor is $37\ 375/12\ 000 = 3\cdot1$. Even if the remoulded strength were taken on the perimeter there would be no liklihood of block failure.

However, settlement is likely to be the limiting criterion. Referring to Fig. 8.23a, the equivalent raft is located 8 m ($2/3 \times 12$ m) below the top of the piles. The width of the equivalent raft is $12\cdot6$ m. The load on the equivalent raft ($12\ 000$ kN) is spread at a slope of $1:2$ to the underlying clay.

The pressure on the equivalent raft is

$$q = \frac{12\ 000}{12\cdot6^2} = 76\ \text{kN/m}^2$$

The immediate settlement is determined using Fig. 5.15. Now,

$$H/B = 16/12\cdot6 = 1\cdot3$$
$$D/B = 9/12\cdot6 = 0\cdot7$$
$$L/B = 1$$

Therefore $\mu_1 = 0\cdot50$ and $\mu_0 = 0\cdot80$; thus

$$s_i = \mu_0 \mu_1 \frac{qB}{E_u}$$

$$= 0\cdot80 \times 0\cdot50 \times \frac{76 \times 12\cdot6}{65}$$

$$= 6\ \text{mm}$$

To calculate the consolidation settlement the clay below the equivalent raft will be divided into four sublayers each of thickness (H) 4 m. The pressure increment ($\Delta\sigma$) at the centre of each sublayer is equal to the load of $12\ 000$ kN divided by the spread area (Table 8.10). The settlement coefficient is obtained from Fig. 7.12. The diameter of a circle having the same area as the equivalent raft is $14\cdot2$ m: thus $H/B = 16/14\cdot2 = 1\cdot1$. Then from Fig. 7.12, for $A = 0\cdot24$ and $H/B = 1\cdot1$, the value of μ is $0\cdot52$ and the consolidation settlement is:

$$s_c = \mu s_{\text{oed}} = 0\cdot52 \times 36\cdot9 = 19\ \text{mm}$$

Table 8.10

Layer	z(m)	Area (m^2)	$\Delta\sigma$(kN/m^2)	$m_v \Delta\sigma H$ (mm)
1	2	14.6^2	56.3	15.8
2	6	18.6^2	34.7	9.7
3	10	22.6^2	23.5	6.6
4	14	26.6^2	17.0	4.8
				$S_{oed} = 36.9$

The total settlement is

$$s = s_i + s_c = 6 + 19 = 25\,\text{mm}$$

8.6 Ground Improvement Techniques

An alternative to the use of deep foundations is the improvement of the soil properties near the surface: shallow foundations are then a possibility. The improvement techniques described below all require the services of a specialist contractor.

Vibro-compaction

The relative density of loose sand deposits can be increased by means of vibro-compaction. The process employs a depth vibrator suspended from the jib of a crane. The vibrator section is located at the lower end and is isolated from the main body of the unit: in most units the vibrator is operated hydraulically. The vibrator operates with a gyratory motion in a horizontal plane, produced by the rotation of eccentric masses. The unit penetrates the soil under its own weight, sometimes assisted by jets of water emitted from the conical point of the vibrator. Significant compaction of the sand can usually be achieved up to a distance of 2·5 m from the axis of the vibrator. Compaction should take place at least to the significant depth of the foundations in question: depths of up to 12 m can be penetrated. In general, however the increase in relative density depends on the spacing of vibration centres, the smaller the spacing the greater the relative density (and hence the greater the bearing capacity). Vibro-compaction cannot be used in cohesive soils because the vibrations are damped within a relatively small radius.

Vibro-replacement

This technique involves the reinforcement of cohesive soil deposits with 'stone columns' to provide adequate support for relatively light foundation loads. The method is not usually suitable for supporting heavy loads because the columns do not transfer the imposed stresses to the soil at depth. The stone columns also fulfill the same functions as a vertical sand drain in accelerating the rate of consolidation of the soil.

A depth vibrator is used to displace the soil radially as it penetrates under its own weight. The vibrator is then withdrawn (compressed air being introduced to break the suction) and the resulting cylindrical hole is filled with successive layers of 50–75 mm angular aggregate, each layer being compacted by re-inserting the vibrator. Further radial displacement of the soil occurs as the aggregate is compacted. The compacted aggregate forms what is known as a stone column. As a result of the method of formation of the column, the passive resistance of the surrounding soil is fully mobilized at small radial strains when the column is loaded. In very soft clays, material is removed by means of water emitted under pressure through the jet holes in the point of the vibrator, i.e. the soil is not displaced.

The strength and stiffness of a stone column depends on its degree of lateral confinement within the surrounding soil. It is uncertain if adequate support can be relied upon in a soft clay if the rate of load application is slow. A soft clay may be gradually squeezed into the voids in the column, in which case there will be very little lateral resistance and the efficiency of the column as a drain will be reduced.

Dynamic Consolidation

This method involves increasing the density of the soil near the surface by tamping and can be used in most soil conditions. Density improvement up to a depth of 10 m is possible. The technique involves dropping a heavy mass of 8–40 tonnes, called the pounder, onto the surface from a height of 5–30 m. A crawler crane or tripod is used to raise the pounder then release it in free fall. The high energy impact of the falling mass creates a hole, called the print, in the ground surface and causes shock waves to be transmitted through the soil to a considerable depth. The shock waves induce partial liquefaction in the soil and create fissures, resulting in a temporary increase in mass permeability. This, in turn, results in compaction or rapid consolidation (depending on the type of soil) and an increase in shear strength (and hence in bearing capacity).

Trials are carried out initially to determine the optimum number of drops at each point, the optimum energy per drop and the optimum print spacing. The time necessary for adequate pore pressure dissipation (using piezometers) and the safe proximity to adjacent structures (vibration measurements) are also assessed.

Typically the pounder is dropped 5–10 times at each point and the prints are spaced at 5–15 m centres in a square grid. The prints are filled in after each pass. Ground settlement is measured after each pass. Several passes may be necessary to achieve the required result, with the print spacing being reduced with each successive pass.

8.7 Excavations

Foundation works may require a relatively deep excavation with vertical sides. The sides may be supported by soldier piles with timber sheeting, sheet pile walls or diaphragm walls: these structures can be braced by means of horizontal or inclined struts or by tie-backs. In addition to the design of the supporting structure, consideration must be given to the ground movements which will occur around the excavation, especially if the excavation is close to existing structures. The following movements (Fig. 8.28) should be considered:

1. settlement of the ground surface adjacent to the excavation,
2. lateral movement of the vertical supports,
3. heave of the base of the excavation.

To a large extent the above movements are interdependent because they are a result of strains in the soil mass due to stress relief when excavation takes place.

The magnitude and distribution of the ground movements depends on the type of soil, the dimensions of the excavation, details of the construction procedure and the standard of workmanship. Current knowledge regarding ground movements is based mainly on observational data rather than on theoretical analysis. Ground movements should be monitored during excavation so that advance warning of excessive movement or possible instability can be obtained.

Assuming comparable construction techniques and workmanship, the magnitude of settlement adjacent to an excavation is likely to be relatively small in dense cohesionless soils but can be excessive in soft plastic clays. Envelopes of the upper limits of observed settlements in various types of soil have been produced by Peck, settlement being given in relation to maximum depth of excavation and distance from the edge of the excavation. These envelopes, shown in Fig. 8.29, are applicable to excavations supported by sheet piling or soldier piles, with bracing or tie-backs, and relate to average workmanship. For excavations supported by diaphragm walls the settlements are likely to be significantly lower than indicated by Peck's envelopes.

Settlement can be reduced by adopting construction procedures which decrease lateral movement and base heave. For a given type of soil, therefore, settlement can be kept to a minimum by installing the struts or tie-backs as soon as possible and before excavation proceeds significantly below the point of support. Care should also be taken to ensure that no voids are left between the supporting structure and the soil. In cohesionless soils it is vital that groundwater flow is controlled otherwise erratic settlement may be caused by a loss of soil into the excavation. It should be realized that for a given method of construction and the best possible standard of workmanship, the settlement at a given point cannot be

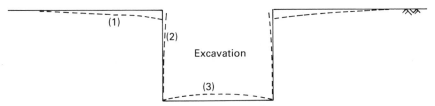

Fig. 8.28   Ground movements associated with deep excavation.

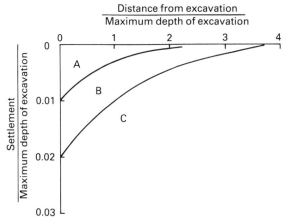

Fig. 8.29   Envelopes of settlement adjacent to excavation. (A) Sand and firm to stiff clay. (B) Very soft to soft clay of limited depth. (C) Very soft to soft clay of considerable depth. (Reproduced by permission of the Mexican Society SMFE.)

reduced below a minimum value which depends on the type of soil and the depth of excavation.

The magnitude and distribution of lateral movement depends to a large extent on the mode of deformation of the supporting structure (e.g. whether the structure is allowed to deflect as a cantilever or whether it is braced near the surface with the maximum deflection then taking place a greater depth). Lateral movement thus depends on the spacing and timing of installation of the struts or tie-backs. As in the case of settlement, excessive movements can occur if excavation is allowed to proceed too far before the first strut or tie-back is installed. The other main factor is the type of soil. Under comparable conditions, lateral movements in soft to medium clays are substantially greater than those in dense cohesionless soils.

Mana and Clough [8.20] developed a method for the prediction of lateral movement and surface settlement associated with braced excavations in soft to medium clays. The method is based on in-situ measurements and finite element analysis which show that correlations exist between the ground movements and the factor of safety against base failure of the excavation. The effects of wall stiffness, strut spacing, stiffness and preload, depth to a firm stratum, excavation width and soil modulus are all considered in the method. The method may also be applied to tied-back walls provided the anchors are unyielding.

In wide excavations it is advantageous to leave a sloping berm against the lower part of the supporting structure while foundation works proceed in the centre of the excavation.

Base heave is generally a problem only in cohesive soils. The soil outside the excavation acts as a surcharge with respect to that below the base of the excavation, therefore upward deformation, and in extreme cases shear failure (Section 8.2), will occur. Short-term heave will be mainly elastic, unless the factor of safety against base failure is low, but additional heave will occur due to swelling if the base remains unloaded for any length of time. In heavily overconsolidated clays, heave can be associated with the relief of the high lateral stresses existing in the clay prior to excavation.

8.8   Ground Anchors

A ground anchor normally consists of a high tensile steel cable or bar, called the tendon, one end of which is held securely in the soil by a mass of cement grout or grouted soil: the other end of the tendon is anchored against a bearing plate on the structural unit to be supported. The main application of ground anchors is in the construction of tie-backs for diaphragm or sheet pile walls. Other applications are in the anchoring of structures subjected to overturning, sliding or buoyancy, in the provision of reaction for in-situ load tests and in pre-loading to reduce settlement. Ground anchors can be constructed in sands (including gravelly sands and silty sands) and stiff clays, and they can be used in situations where either temporary or permanent support is required.

The grouted length of tendon, through which force is transmitted to the surrounding soil, is called the fixed anchor length. The length of tendon between the fixed anchor and the bearing plate is called the free anchor length: no force is transmitted to the soil over this length. For temporary anchors the tendon is normally greased and covered with plastic tape over the free anchor length. This allows for free movement of the tendon and gives protection against corrosion. For permanent anchors the tendon is normally greased and sheathed with polythene under factory conditions: on site the tendon is stripped and degreased over what will be the fixed anchor length.

The ultimate load which can be carried by an anchor depends on the soil resistance (principally skin friction) mobilized adjacent to the fixed anchor length. (This, of course, assumes that there will be no prior failure at the grout-tendon interface or of the tendon itself.) Anchors are usually prestressed in order to reduce the movement required to mobilize the soil resistance. Each anchor is subjected to a test loading after installation:

temporary anchors are usually tested to 1·2 times the working load and permanent anchors to 1·5 times the working load. Finally, prestressing of the anchor takes place. Creep displacements under constant load will occur in ground anchors. A creep coefficient, defined as the displacement per unit log time, can be determined by means of a load test. It has been suggested that this coefficient should not exceed 1 mm for 1·5 times the working load.

A comprehensive ground investigation is essential in any location where ground anchors are to be employed. The soil profile must be determined accurately, any variations in the level and thickness of strata being particularly important. In the case of sands the particle size distribution should be determined, in order that permeability and grout acceptability can be estimated. The relative density of sands is also required to allow an estimate of ϕ' to be made. In the case of stiff clays the undrained shear strength should be determined.

Anchors in Sands

In general the sequence of construction is as follows. A cased borehole (diameter usually within the range 75–125 mm) is advanced through the soil to the required depth. The tendon is then positioned in the hole and cement grout is injected under pressure over the fixed anchor length as the casing is withdrawn. The grout penetrates the soil around the borehole, to an extent depending on the permeability of the soil and on the injection pressure, forming a zone of grouted soil, the diameter of which can be up to four times that of the borehole (Fig. 8.30a). Care must be taken to ensure that the injection pressure does not exceed the overburden pressure of the soil above the anchor, otherwise heaving or fissuring may result. When the grout has achieved adequate strength the other end of the tendon is anchored against the bearing plate. The space between the sheathed tendon and the sides of the borehole, over the free anchor length, is normally filled with grout (under low pressure): this grout gives additional corrosion protection to the tendon.

The ultimate resistance of an anchor to pull-out is equal to the sum of the side resistance and the end resistance of the grouted mass. The following theoretical expression was proposed by Littlejohn [8.19]:

$$Q_f = A\gamma'\left(h + \frac{L}{2}\right)\pi DL \tan\phi' + B\gamma' h\frac{\pi}{4}(D^2 - d^2) \tag{8.44}$$

where Q_f = ultimate load capacity of anchor, A = ratio of normal pressure at interface to effective overburden pressure, B = bearing capacity factor, h = depth of overburden, L = fixed anchor length, D = diameter of fixed anchor, and d = diameter of borehole.

It was suggested that the value of A is normally within the range 1 to 2. The factor B is analogous to the bearing capacity factor N_q in the case of piles and it was suggested that the ratio N_q/B is within the range 1·3 to 1·4,

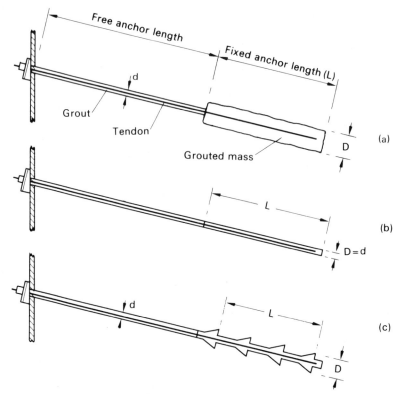

Fig. 8.30 Ground anchors: (a) grouted mass formed by pressure injection, (b) grout cylinder, (c) multiple under-reamed anchor.

using the N_q values of Berezantzev, Khristoforov and Golubkov. However, the above expression is unlikely to represent all the relevant factors in a complex problem. The ultimate resistance also depends on details of the installation technique and a number of empirical formulae have been proposed by specialist contractors, suitable for use with their particular technique.

Anchors in Stiff Clays

The simplest construction technique for anchors in stiff clays is to auger a hole to the required depth, position the tendon and grout the fixed anchor length using a tremie pipe (Fig. 8.30b) However, such a technique would produce an anchor of relatively low capacity because the skin friction at the grout-clay interface would be unlikely to exceed $0.3c_u$ (i.e. $\alpha = 0.3$).

Anchor capacity can be increased by the technique of gravel injection. The augered hole is filled with pea gravel over the fixed anchor length, then a casing, fitted with a pointed shoe, is driven into the gravel, forcing it into the surrounding clay. The tendon is then positioned and grout is injected into the gravel as the casing is withdrawn (leaving the shoe behind). This technique results in an increase in the effective diameter of the fixed anchor (of the order of 50%) and an increase in side resistance: a value of α of around 0·6 can be expected. In addition there will be some end resistance. The borehole is again filled with grout over the free anchor length.

Another technique employs an expanding cutter to form a series of enlargements (or under-reams) of the augered hole at close intervals over the fixed anchor length (Fig. 8.30c): the cuttings are generally removed by flushing with water. The cable is then positioned and grouting takes place. A value of α of around 0·8 can be assumed along the cylindrical surface through the extremities of the enlargements.

The following design formula can be used for anchors in stiff clays:

$$Q_f = \pi DL\alpha c_u + \frac{\pi}{4}(D^2 - d^2)c_u N_c \qquad (8.45)$$

where Q_f = ultimate load capacity of anchor, L = fixed anchor length, D = diameter of fixed anchor, d = diameter of borehole, α = skin friction coefficient, and N_c = bearing capacity factor (generally assumed to be 9).

Problems

8.1 A load of 425 kN/m is carried on a strip footing 2 m wide at a depth of 1 m in a stiff clay of saturated unit weight 21 kN/m³, the water table being at ground level. Determine the factor of safety with respect to shear failure (a) when $c_u = 105$ kN/m² and $\phi_u = 0$, (b) when $c' = 10$ kN/m² and $\phi' = 28°$.

8.2 A strip footing 1·5 m wide is located at a depth of 0·75 m in a sand of unit weight 18 kN/m³, the water table being well below foundation level. The shear strength parameters are $c' = 0$ and $\phi' = 38°$. The footing carries a load of 500 kN/m. Determine the factor of safety with respect to shear failure (a) if the load is vertical, (b) if the load is inclined at 10° to the vertical.

8.3 Determine the allowable load on a footing 4·50 m × 2·25 m at a depth of 3·50 m in a stiff clay if a factor of safety of 3 with respect to shear failure is specified. The saturated unit weight of the clay is 20 kN/m³ and the relevant shear strength parameters are $c_u = 135$ kN/m² and $\phi_u = 0$.

8.4 A footing 2·5 m square carries a pressure of 400 kN/m² at a depth of 1 m in a sand. The saturated unit weight of the sand is 20 kN/m³ and the unit weight above the water table is 17 kN/m³. The shear strength parameters are $c' = 0$ and $\phi' = 40°$. Determine the factor of safety with respect to shear failure for the following cases:

Table 8.11

Depth (m)	BH 1	BH 2	BH 3	BH 4
0·75	15	18	16	14
1·50	17	21	15	18
2·25	21	19	18	20
3·00	22	22	20	21
3·75	25	24	22	26
4·50	29	27	21	26

(a) the water table is 5 m below ground level,
(b) the water table is 1 m below ground level,
(c) the water table is at ground level and there is seepage vertically upwards under a hydraulic gradient of 0·2.

8.5 A footing 4 m square is located at a depth of 1 m in a layer of saturated clay 13 m thick, the water table being at ground level. The following parameters are known for the clay: $c_u = 100\,\text{kN/m}^2$, $\phi_u = 0$, $c' = 15\,\text{kN/m}^2$, $\phi' = 27°$, $m_v = 0·065\,\text{m}^2/\text{MN}$, $A = 0·42$, $\gamma_{\text{sat}} = 21\,\text{kN/m}^3$. Determine the allowable bearing capacity if the factor of safety with respect to shear failure is not to be less than 3 and if the maximum consolidation settlement is to be limited to 30 mm.

8.6 A long braced excavation in soft clay is 4 m wide and 8 m deep. The saturated unit weight of the clay is $20\,\text{kN/m}^3$ and the undrained shear strength adjacent to the bottom of the excavation is given by $c_u = 40\,\text{kN/m}^2$ ($\phi_u = 0$). Determine the factor of safety against base failure of the excavation.

8.7 The following column loads are to be carried on individual footings at a depth of 1 m in a sand of unit weight $18\,\text{kN/m}^3$.

500, 550, 850, 900, 1075, 1200, 880, 700 kN.

Standard penetration tests were carried out in four boreholes on the site, the recorded N values being as shown in Table 8.11.
The water table is at a depth of 2 m. Determine the allowable bearing capacity for the design of the footings.

8.8 A building, with a basement, is to be supported on a raft foundation 20 m × 50 m at a depth of 5 m in a deep sand deposit. However, a layer of fine silty sand exists between depths of 8·5 m and 13·0 m below ground level. The water table is at a depth of 3 m but is to be lowered temporarily to a depth of 7 m during construction of the foundation. Recorded values of standard penetration resistance (in the borehole having the least average value), at intervals of 1·5 m, between depths of 1·5 m and 25·5 m inclusive, are as follows:

7, 9, 16, 23, 18, 31, 27, 33, 21, 30, 23, 28, 36, 42, 38, 44, 50.
Determine the allowable bearing capacity.

8.9 A footing 3 m square carries a net foundation pressure of $130 \, kN/m^2$ at a depth of $1.2 \, m$ in a deep deposit of sand: the water table is at a depth of 3 m. Above the water table the unit weight of the sand is $16 \, kN/m^3$ and below the water table the saturated unit weight is $19 \, kN/m^3$. The variation of cone penetration resistance (q_c) with depth (z) is as follows:

z(m)	1·2	1·6	2·0	2·4	2·6	3·0	3·4	3·8	4·2
q_c(MN/m²)	3·2	2·1	2·8	2·3	6·1	5·0	6·6	4·5	5·5

z(m)	4·6	5·0	5·4	5·8	6·2	6·6	7·0	7·4	8·0
q_c(MN/m²)	4·5	5·4	10·4	8·9	9·9	9·0	15·1	12·9	14·8

Determine the settlement of the footing using (a) the Buisman-DeBeer method, (b) Schmertmann's method.

8.10 A bored pile with an enlarged base is to be installed in a stiff clay, the undrained shear strength at base level being $220 \, kN/m^2$. The saturated unit weight of the clay is $21 \, kN/m^3$. The diameters of the pile shaft and base are $1.05 \, m$ and $3.00 \, m$ respectively. The pile extends from a depth of 4 m to a depth of 22 m, the top of the under-ream being at a depth of 20 m. Past experience indicates that a skin friction coefficient β of 0.70 is appropriate for the clay. Determine the allowable load on the pile to ensure (a) an overall load factor of 2 and (b) a load factor of 3 under the base when shaft resistance is fully mobilized.

8.11 At a particular site the soil profile consists of a layer of soft clay underlain by a depth of sand. The values of standard penetration resistance at depths of 0.75 m, 1.50 m, 2.25 m, 3.00 m and 3.75 m in the sand are 18, 24, 26, 34 and 32 respectively. Nine precast concrete piles, in a square group, are driven through the clay and 2 m into the sand. The piles are 0.25 m square in section and are spaced at 0.75 m centres in the group. Neglecting skin friction in the clay, determine the allowable load on the group if a load factor of 2.5 is specified and if the settlement is not to exceed 25 mm.

8.12 A ground anchor in a stiff clay, formed by the gravel injection technique, has a fixed anchor length of 5 m and an effective fixed anchor diameter of 200 mm: the diameter of the borehole is 100 mm. The relevant shear strength parameters for the clay are $c_u = 110 \, kN/m^2$ and $\phi_u = 0$. What would be the expected ultimate load capacity of the anchor, assuming a skin friction coefficient of 0.6?

References

8.1 Berezantzev, V. G., Khristoforov, V. S. and Golubkov, V. N. (1961): 'Load Bearing Capacity and Deformation of Piled Foundations', *Proceedings 5th International Conference SMFE, Paris*, Vol. 2.

8.2   Bjerrum, L. (1963): Discussion, *Proceedings European Conference SMFE, Wiesbaden*, Vol. 3.

8.3   Bjerrum, L. and Eggestad, A. (1963): 'Interpretation of Loading Tests on Sand', *Proceedings European Conference SMFE, Wiesbaden*, Vol. 1.

8.4   Bjerrum, L. and Eide, O. (1956): 'Stability of Strutted Excavations in Clay', *Geotechnique*, Vol. 6, No. 1.

8.5   British Standard 1377 (1975): *Methods of Test for Soils for Civil Engineering Purposes*, British Standards Institution, London.

8.6   British Standard Code of Practice, CP 101 (1972): *Foundations and Substructures for Non-industrial Buildings of not more than Four Storeys*, British Standards Institution, London.

8.7   British Standard 8004 (1986): *Code of Practice for Foundations*, British Standards Institution, London.

8.8   Burland, J. B. (1973): 'Shaft Friction of Piles in Clay', *Ground Engineering*, Vol. 6, No. 3.

8.9   Burland, J. B., Broms, B. B. and De Mello, V. F. B. (1977): 'Behaviour of Foundations and Structures', *Proc. 9th International Conference SMFE, Tokyo*, Vol. 2.

8.10   Burland, J. B. and Burbidge, M. C. (1985): 'Settlement of Foundations on Sand and Gravel', *Proc. Institution of Civil Engineers*, Part 1, Vol. 78, December.

8.11   Burland, J. B. and Cooke, R. W. (1974): 'The Design of Bored Piles in Stiff Clays', *Ground Engineering*, Vol. 7, No. 4.

8.12   Burland, J. B. and Wroth, C. P. (1975): 'Settlement of Buildings and Associated Damage', *Proceedings of Conference on Settlement of Structures* (British Geotechnical Society), Pentech Press, London.

8.13   Cooke, R. W. (1986): 'Piled Raft Foundations on Stiff Clays – a Contribution to Design Philosophy', *Geotechnique*, Vol. 36, No. 2.

8.14   DeBeer, E. E. (1963): 'The Scale Effect in the Transposition of the Results of Deep Sounding Tests on the Ultimate Bearing Capacity of Piles and Caisson Foundations', *Geotechnique*, Vol. 13, No. 1.

8.15   DeBeer, E. E. (1970): 'Experimental Determination of the Shape Factors and the Bearing Capacity Factors of Sand', *Geotechnique*, Vol. 20, No. 4.

8.16   DeBeer, E. E. and Martens, A. (1957): 'Method of Computation of an Upper Limit for the Influence of Heterogeneity of Sand Layers on the Settlement of Bridges', *Proceedings 4th International Conference SMFE, London*, Vol. 1, Butterworths.

8.17   Gibbs, H. J. and Holtz, W. G. (1957): 'Research on Determining the Density of Sands by Spoon Penetration Testing', *Proceedings 4th International Conference SMFE, London*, Vol. 1, Butterworths.

8.18   Hansen, J. B. (1968): 'A Revised Extended Formula for Bearing Capacity', *Danish Geotechnical Institute Bulletin*, No. 28.

8.19   Littlejohn, G. S. (1970): 'Soil Anchors', *Proceedings of Conference on Ground Engineering*, ICE, London.

8.20 Mana, A. I. and Clough, G. W. (1981): 'Prediction of Movements for Braced Cuts in Clay', *Journal ASCE*, Vol. 107, No. GT6.

8.21 Meyerhof, G. G. (1955): 'Influence of Roughness of Base and Groundwater Conditions on the Ultimate Bearing Capacity of Foundations', *Geotechnique*, Vol. 5, No. 3.

8.22 Meyerhof, G. G. (1956): 'Penetration Tests and Bearing Capacity of Cohesionless Soils', *Proceedings ASCE*, Vol. 82, No. SM1.

8.23 Meyerhof, G. G. (1963): 'Some Recent Research on the Bearing Capacity of Foundations', *Canadian Geotechnical Journal*, Vol. 1, No. 1.

8.24 Meyerhof, G. G. (1965): 'Shallow Foundations', *Proceedings ASCE*, Vol. 91, No. SM2.

8.25 Meyerhof, G. G. (1976): 'Bearing Capacity and Settlement of Pile Foundations', *Proceedings ASCE*, Vol. 102, No. GT3.

8.26 Peck, R. B., Hanson, W. E. and Thornburn, T. H. (1974): *Foundation Engineering*, John Wiley and Sons, New York.

8.27 Polshin, D. E. and Tokar, R. A. (1957): 'Maximum Allowable Non-Uniform Settlement of Structures' *Proceedings 4th International Conference SMFE, London*, Vol. 1, Butterworths.

8.28 Poulos, H. G. (1977): 'Estimation of Pile Group Settlements', *Ground Engineering*, Vol. 10, No. 2.

8.29 Schmertmann, J. H. (1970): 'Static Cone to Compute Static Settlement over Sand', *Proceedings ASCE*, Vol. 96, No. SM3.

8.30 Schmertmann, J. H. (1975): 'Measurement of In-Situ Shear Strength', *Proceedings of Conference on In Situ Measurement of Soil Properties*, ASCE, New York.

8.31 Skempton, A. W. (1951): 'The Bearing Capacity of Clays', *Proceedings Building Research Congress*, Vol. 1.

8.32 Skempton, A. W. (1959): 'Cast-in-situ Bored Piles in London Clay', *Geotechnique*, Vol. 9, No. 4.

8.33 Skempton, A. W. (1986): 'Standard Penetration Test Procedures and the Effects in Sands of Overburden Pressure, Relative Density, Particle Size, Ageing and Overconsolidation', *Geotechnique*, Vol. 36, No. 3.

8.34 Skempton, A. W. and MacDonald, D. H. (1956): 'Allowable Settlement of Buildings', *Proceedings ICE*, Vol. 5, Part 3.

8.35 Smith, E. A. L. (1960): 'Pile Driving Analysis by the Wave Equation', *Proceedings ASCE*, Vol. 86, No. SM4.

8.36 Terzaghi, K. (1943): *Theoretical Soil Mechanics*, John Wiley and Sons, New York.

8.37 Terzaghi, K. and Peck, R. B. (1967): *Soil Mechanics in Engineering Practice* (2nd edition), John Wiley and Sons, New York.

8.38 Thorburn, S. (1963): 'Tentative Correction Chart for the Standard Penetration Test in Non-Cohesive Soils', *Civil Engineering and Public Works Review*, Vol. 58.

8.39 Thorburn, S. (1975): 'Building Structures Supported by Stabilised

Ground', *Geotechnique*, Vol. 25, No. 1.

8.40 Tomlinson, M. J. (1977): *Pile Design and Construction Practice*, Cement and Concrete Association, London.

8.41 Tomlinson, M. J. (1986): *Foundation Design and Construction* (5th edition), Pitman, London.

8.42 Vesic, A. S. (1973): 'Analysis of Ultimate Loads of Shallow Foundations', Journal ASCE, Vol. 99, No. SM1.

8.43 Whitaker, T. (1957): 'Experiments with Models Piles in Groups', *Geotechnique*, Vol. 7, No. 4.

8.44 Whitaker, T. (1976): *The Design of Piled Foundations*, Pergamon Press, Oxford.

CHAPTER 9

Stability of Slopes

9.1 Introduction

Gravitational and seepage forces tend to cause instability in natural slopes, in slopes formed by excavation and in the slopes of embankments and earth dams. The most important types of slope failure are illustrated in Fig. 9.1. In *rotational* slips the shape of the failure surface in section may be a circular arc or a non-circular curve. In general, circular slips are associated with homogeneous soil conditions and non-circular slips with non-homogeneous conditions. *Translational* and *compound* slips occur where the form of the failure surface is influenced by the presence of an adjacent stratum of significantly different strength. Translational slips tend to occur where the adjacent stratum is at a relatively shallow depth below the surface of the slope: the failure surface tends to be plane and roughly parallel to the slope. Compound slips usually occur where the adjacent stratum is at greater depth, the failure surface consisting of curved and plane sections.

In practice, limiting equilibrium methods are used in the analysis of slope stability. It is considered that failure is on the point of occurring along an assumed or a known failure surface. The shear strength required to maintain a condition of limiting equilibrium is compared with the available shear strength of the soil, giving the average factor of safety along the

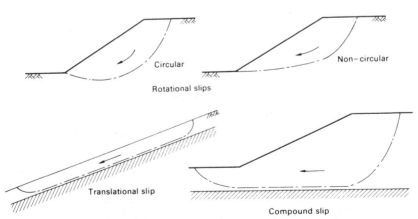

Fig. 9.1   Types of slope failure.

failure surface. The problem is considered in two dimensions, conditions of plane strain being assumed. It has been shown that a two-dimensional analysis gives a conservative result for a failure on a three-dimensional (dish-shaped) surface.

9.2 Analysis for the Case of $\phi_u = 0$

This analysis, in terms of total stress, covers the case of a fully saturated clay under undrained conditions, i.e. for the condition immediately after construction. Only moment equilibrium is considered in the analysis. In section, the potential failure surface is assumed to be a circular arc. A trial failure surface (centre O, radius r and length L_a) is shown in Fig. 9.2. Potential instability is due to the total weight of the soil mass (W per unit length) above the failure surface. For equilibrium the shear strength which must be mobilized along the failure surface is expressed as

$$\tau_m = \frac{\tau_f}{F} = \frac{c_u}{F}$$

where F is the factor of safety with respect to shear strength. Equating moments about O:

$$Wd = \frac{c_u}{F} L_a r$$

therefore

$$F = \frac{c_u L_a r}{Wd} \tag{9.1}$$

The moments of any additional forces must be taken into account. In the event of a tension crack developing, as shown in Fig. 9.2, the arc length L_a is shortened and a hydrostatic force will act normal to the crack if the crack fills with water. It is necessary to analyse the slope for a number of trial

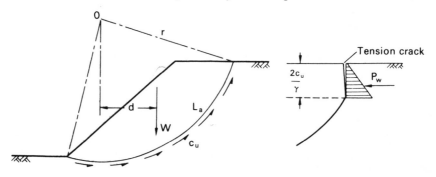

Fig. 9.2   The $\phi_u = 0$ analysis.

failure surfaces in order that the minimum factor of safety can be determined.

Based on the principle of geometric similarity, Taylor [9.16] published *stability coefficients* for the analysis of homogeneous slopes in terms of total stress. For a slope of height H the stability coefficient (N_s) for the failure surface along which the factor of safety is a minimum is

$$N_s = \frac{c_u}{F\gamma H} \qquad (9.2)$$

For the case of $\phi_u = 0$, values of N_s can be obtained from Fig. 9.3. The coefficient N_s depends on the slope angle β and the depth factor D, where DH is the depth to a firm stratum.

Gibson and Morgenstern [9.6] published stability coefficients for slopes in normally consolidated clays in which the undrained strength $c_u(\phi_u = 0)$ varies linearly with depth.

Example 9.1

A 45° slope is excavated to a depth of 8 m in a deep layer of saturated clay of unit weight 19 kN/m³: the relevant shear strength parameters are $c_u = 65\,kN/m^2$ and $\phi_u = 0$. Determine the factor of safety for the trial failure surface specified in Fig. 9.4.

In Fig. 9.4, the cross-sectional area ABCD is 70 m².

Weight of soil mass = 70 × 19 = 1330 kN/m

The centroid of ABCD is 4·5 m from O. The angle AOC is $89\frac{1}{2}°$ and radius OC is 12·1 m. The arc length ABC is calculated as 18·9 m. The factor of safety is given by:

$$F = \frac{c_u L_a r}{Wd}$$

$$= \frac{65 \times 18{\cdot}9 \times 12{\cdot}1}{1330 \times 4{\cdot}5} = 2{\cdot}48$$

This is the factor of safety for the trial failure surface selected and is not necessarily the minimum factor of safety.

The minimum factor of safety can be estimated by using Equation 9.2. From Fig. 9.3, $\beta = 45°$ and assuming that D is large, the value of N_s is 0.18. Then

$$F = \frac{c_u}{N_s \gamma H}$$

$$= \frac{65}{0{\cdot}18 \times 19 \times 8}$$

$$= 2{\cdot}37$$

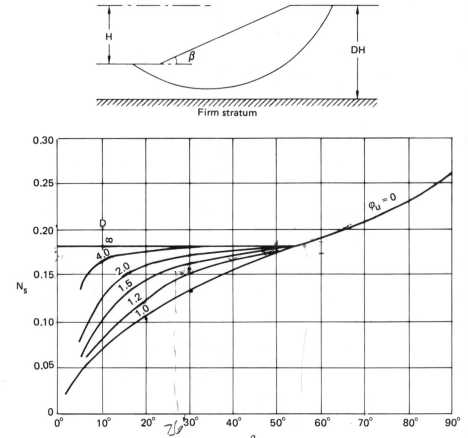

Fig. 9.3   Taylor's stability coefficients for $\phi_u = 0$. (Reproduced by permission of the Boston Society of Civil Engineers.)

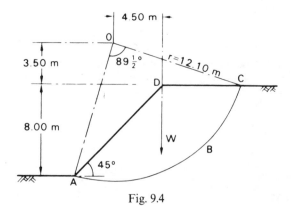

Fig. 9.4

9.3 The Method of Slices

In this method the potential failure surface, in section, is again assumed to be a circular arc with centre O and radius r. The soil mass (ABCD) above a trial failure surface (AC) is divided by vertical planes into a series of slices of width b, as shown in Fig. 9.5. The base of each slice is assumed to be a straight line. For any slice the inclination of the base to the horizontal is α and the height, measured on the centre-line, is h. The factor of safety is defined as the ratio of the available shear strength (τ_f) to the shear strength (τ_m) which must be mobilized to maintain a condition of limiting equilibrium, i.e.

$$F = \frac{\tau_f}{\tau_m}$$

The factor of safety is taken to be the same for each slice, implying that there must be mutual support between slices, i.e. forces must act between the slices.

The forces (per unit dimension normal to the section) acting on a slice are:

1. The total weight of the slice, $W = \gamma bh$ (γ_{sat} where appropriate).
2. The total normal force on the base, N (equal to σl). In general this force has two components, the effective normal force N' (equal to $\sigma' l$) and the boundary water force U (equal to ul), where u is the pore water pressure at the centre of the base and l is the length of the base.
3. The shear force on the base, $T = \tau_m l$.
4. The total normal forces on the sides, E_1 and E_2.
5. The shear forces on the sides, X_1 and X_2.

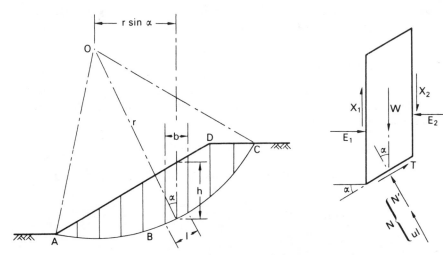

Fig. 9.5 The method of slices.

Any external forces must also be included in the analysis.

The problem is statically indeterminate and in order to obtain a solution assumptions must be made regarding the interslice forces E and X: the resulting solution for factor of safety is not exact.

Considering moments about O, the sum of the moments of the shear forces T on the failure arc AC must equal the moment of the weight of the soil mass ABCD. For any slice the lever arm of W is $r \sin \alpha$, therefore

$$\Sigma Tr = \Sigma Wr \sin \alpha$$

Now,

$$T = \tau_m l = \frac{\tau_f}{F} l$$

$$\therefore \quad \Sigma \frac{\tau_f}{F} l = \Sigma W \sin \alpha$$

$$\therefore \quad F = \frac{\Sigma \tau_f l}{\Sigma W \sin \alpha}$$

For an analysis in terms of effective stress,

$$F = \frac{\Sigma(c' + \sigma' \tan \phi')l}{\Sigma W \sin \alpha}$$

or

$$F = \frac{c'L_a + \tan \phi' \Sigma N'}{\Sigma W \sin \alpha} \tag{9.3}$$

where L_a is the arc length AC. Equation 9.3 is exact but approximations are introduced in determining the forces N'. For a given failure arc the value of F will depend on the way in which the forces N' are estimated.

The Fellenius Solution

In this solution it is assumed that for each slice the resultant of the interslice forces is zero. The solution involves resolving the forces on each slice normal to the base, i.e.

$$N' = W \cos \alpha - ul$$

Hence the factor of safety in terms of effective stress (Equation 9.3) is given by

$$F = \frac{c'L_a + \tan \phi' \Sigma(W \cos \alpha - ul)}{\Sigma W \sin \alpha} \tag{9.4}$$

The components $W \cos \alpha$ and $W \sin \alpha$ can be determined graphically for each slice. Alternatively, the value of α can be measured or calculated. Again, a series of trial failure surfaces must be chosen in order to obtain the minimum factor of safety. This solution underestimates the factor of safety: the error, compared with more accurate methods of analysis, is usually within the range 5–20%.

For an analysis in terms of total stress the parameters c_u and ϕ_u are used and the value of u in Equation 9.4 is zero. If $\phi_u = 0$ the factor of safety is given by

$$F = \frac{c_u L_a}{\Sigma W \sin \alpha} \tag{9.5}$$

As N' does not appear in Equation 9.5 an exact value of F is obtained.

The Bishop Simplified Solution

In this solution it is assumed that the resultant forces on the sides of the slices are horizontal, i.e.

$$X_1 - X_2 = 0$$

For equilibrium the shear force on the base of any slice is

$$T = \frac{1}{F}(c'l + N' \tan \phi')$$

Resolving forces in the vertical direction:

$$W = N' \cos \alpha + ul \cos \alpha + \frac{c'l}{F} \sin \alpha + \frac{N'}{F} \tan \phi' \sin \alpha$$

$$\therefore \quad N' = \left(W - \frac{c'l}{F} \sin \alpha - ul \cos \alpha \right) \Big/ \left(\cos \alpha + \frac{\tan \phi' \sin \alpha}{F} \right) \tag{9.6}$$

It is convenient to substitute

$$l = b \sec \alpha$$

From Equation 9.3, after some rearrangement,

$$F = \frac{1}{\Sigma W \sin \alpha} \Sigma \left[\{c'b + (W - ub) \tan \phi'\} \frac{\sec \alpha}{1 + \dfrac{\tan \alpha \tan \phi'}{F}} \right] \tag{9.7}$$

The pore water pressure can be related to the total 'fill pressure' at any point by means of the dimensionless *pore pressure ratio*, defined as

$$r_u = \frac{u}{\gamma h} \tag{9.8}$$

(γ_{sat} where appropriate). For any slice,

$$r_u = \frac{u}{W/b}$$

Hence Equation 9.7 can be written:

$$F = \frac{1}{\Sigma W \sin \alpha} \Sigma \left[\{c'b + W(1 - r_u)\tan \phi'\} \frac{\sec \alpha}{1 + \frac{\tan \alpha \tan \phi'}{F}} \right] \tag{9.9}$$

As the factor of safety occurs on both sides of Equation 9.9, a process of successive approximation must be used to obtain a solution but convergence is rapid.

Due to the repetitive nature of the calculations and the need to select an adequate number of trial failure surfaces, the method of slices is particularly suitable for solution by computer. More complex slope geometry and different soil strata can be introduced.

In most problems the value of the pore pressure ratio r_u is not constant over the whole failure surface but, unless there are isolated regions of high pore pressure, an average value (weighted on an area basis) is normally used in design. Again, the factor of safety determined by this method is an underestimate but the error is unlikely to exceed 7% and in most cases is less than 2%.

Spencer [9.15] proposed a method of analysis in which the resultant interslice forces are parallel and in which both force and moment equilibrium are satisfied. Spencer showed that the accuracy of the Bishop simplified method, in which only moment equilibrium is satisfied, is due to the insensitivity of the moment equation to the slope of the interslice forces.

Dimensionless stability coefficients for homogeneous slopes, based on Equation 9.9, have been published by Bishop and Morgenstern [9.4]. It can be shown that for a given slope angle and given soil properties the factor of safety varies linearly with r_u and can thus be expressed as

$$F = m - nr_u \tag{9.10}$$

where m and n are the stability coefficients. The coefficients m and n are functions of β, ϕ', the dimensionless number $c'/\gamma H$ and the depth factor D.

Example 9.2

Using the Fellenius method of slices, determine the factor of safety, in terms of effective stress, of the slope shown in Fig. 9.6 for the given failure surface. The unit weight of the soil, both above and below the water table, is $20\,kN/m^3$ and the relevant shear strength parameters are $c' = 10\,kN/m^2$ and $\phi' = 29°$.

The factor of safety is given by Equation 9.4. The soil mass is divided into slices 1·5 m wide. The weight (W) of each slice is given by

$$W = \gamma b h = 20 \times 1{\cdot}5 \times h = 30h \text{ kN/m}$$

The height h for each slice is set off below the centre of the base and the normal and tangential components $h\cos\alpha$ and $h\sin\alpha$ respectively are determined graphically, as shown in Fig. 9.6. Then

$$W\cos\alpha = 30h\cos\alpha$$

$$W\sin\alpha = 30h\sin\alpha$$

The pore water pressure at the centre of the base of each slice is taken to be $\gamma_w z_w$, where z_w is the vertical distance of the centre point below the water table (as shown in the figure). This procedure slightly overestimates the pore water pressure which strictly should be $\gamma_w z_e$, where z_e is the vertical distance below the point of intersection of the water table and the equipotential through the centre of the slice base. The error involved is on the safe side.

The arc length (L_a) is calculated as 14·35 mm. The results are given in Table 9.1.

$$\Sigma W \cos\alpha = 30 \times 17{\cdot}50 = 525 \text{ kN/m}$$

$$\Sigma W \sin\alpha = 30 \times 8{\cdot}45 = 254 \text{ kN/m}$$

$$\Sigma(W\cos\alpha - ul) = 525 - 132 = 393 \text{ kN/m}$$

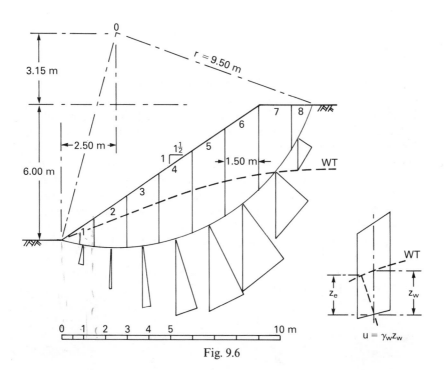

Fig. 9.6

Table 9.1

Slice no.	$h\cos\alpha$ (m)	$h\sin\alpha$ (m)	u (kN/m^2)	l (m)	ul (kN/m)
1	0·75	−0·15	5·9	1·55	9·1
2	1.80	−0·10	11·8	1·50	17·7
3	2·70	0·40	16·2	1·55	25·1
4	3·25	1·00	18·1	1·60	29·0
5	3·45	1·75	17·1	1·70	29·1
6	3·10	2·35	11·3	1·95	22·0
7	1·90	2·25	0	2·35	0
8	0·55	0·95	0	2·15	0
	17·50	8·45		14·35	132·0

$$F = \frac{c'L_a + \tan\phi'\,\Sigma(W\cos\alpha - ul)}{\Sigma W\sin\alpha}$$

$$= \frac{(10 \times 14\cdot35) + (0\cdot554 \times 393)}{254}$$

$$= \frac{143\cdot5 + 218}{254} = 1\cdot42$$

9.4 Analysis of a Plane Translational Slip

It is assumed that the potential failure surface is parallel to the surface of the slope and is at a depth that is small compared with the length of the slope. The slope can then be considered as being of infinite length, with end effects being ignored. The slope is inclined at angle β to the horizontal and the depth of the failure plane is z, as shown in section in Fig. 9.7. The water table is taken to be parallel to the slope at a height of mz ($0 < m < 1$) above the failure plane. Steady seepage is assumed to be taking place in a direction parallel to the slope. The forces on the sides of any vertical slice are equal and opposite and the stress conditions are the same at every point on the failure plane.

In terms of effective stress, the shear strength of the soil along the failure plane is

$$\tau_f = c' + (\sigma - u)\tan\phi'$$

and the factor of safety is

$$F = \frac{\tau_f}{\tau}$$

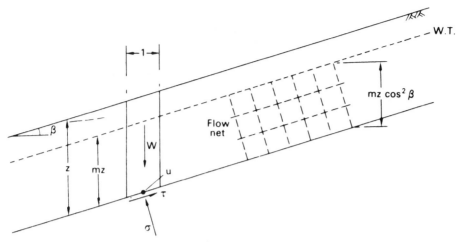

Fig. 9.7 Plane translational slip.

The expressions for σ, τ and u are:

$$\sigma = \{(1 - m)\gamma + m\gamma_{sat}\} z \cos^2 \beta$$
$$\tau = \{(1 - m)\gamma + m\gamma_{sat}\} z \sin \beta \cos \beta$$
$$u = mz\gamma_w \cos^2 \beta$$

The following special cases are of interest. If $c' = 0$ and $m = 0$ (i.e. the soil between the surface and the failure plane is not fully saturated), then

$$F = \frac{\tan \phi'}{\tan \beta} \tag{9.11}$$

If $c' = 0$ and $m = 1$ (i.e. the water table coincides with the surface of the slope), then:

$$F = \frac{\gamma'}{\gamma_{sat}} \frac{\tan \phi'}{\tan \beta} \tag{9.12}$$

It should be noted that when $c' = 0$ the factor of safety is independent of the depth z. If c' is greater than zero, the factor of safety is a function of z, and β may exceed ϕ' provided z is less than a critical value.

For a total stress analysis the shear strength parameters c_u and ϕ_u are used with a zero value of u.

Example 9.3

A long natural slope in a fissured overconsolidated clay is inclined at 12° to the horizontal. The water table is at the surface and seepage is roughly parallel to the slope. A slip has developed on a plane parallel to the surface at a depth of 5 m. The saturated unit weight of the clay is $20\,\text{kN/m}^3$. The peak strength parameters are $c' = 10\ \text{kN/m}^2$ and $\phi'_{max} = 26°$; the residual strength parameters are $c'_r = 0$ and $\phi'_r = 18°$. Determine the factor of safety along the slip plane (a) in terms of the peak strength parameters, (b) in terms of the residual strength parameters.

With the water table at the surface $(m = 1)$, at any point on the slip plane,

$$\sigma = \gamma_{sat}\, z \cos^2 \beta$$
$$= 20 \times 5 \times \cos^2 12° = 95\text{·}5\,\text{kN/m}^2$$
$$\tau = \gamma_{sat}\, z \sin \beta \cos \beta$$
$$= 20 \times 5 \times \sin 12° \times \cos 12° = 20\text{·}3\,\text{kN/m}^2$$
$$u = \gamma_w z \cos^2 \beta$$
$$= 9\text{·}8 \times 5 \times \cos^2 12° = 46\text{·}8\,\text{kN/m}^2$$

Using the peak strength parameters,

$$\tau_f = c' + (\sigma - u) \tan \phi'_{max}$$
$$= 10 + (48\text{·}7 \times \tan 26°) = 33\text{·}8\,\text{kN/m}^2$$

Then the factor of safety is given by

$$F = \frac{\tau_f}{\tau} = \frac{33\text{·}8}{20\text{·}3} = 1\text{·}66$$

Using the residual strength parameters, the factor of safety can be obtained from Equation 9.12:

$$F = \frac{\gamma' \tan \phi'_r}{\gamma_{sat} \tan \beta}$$
$$= \frac{10\text{·}2}{20} \times \frac{\tan 18°}{\tan 12°}$$
$$= 0\text{·}78$$

9.5   General Methods of Analysis

Morgenstern and Price [9.10] developed a general analysis in which all boundary and equilibrium conditions are satisfied and in which the failure surface may be any shape, circular, non-circular or compound. The ground surface is represented by a function $y = z(x)$ and the trial failure surface by $y = y(x)$ as shown in Fig. 9.8. The forces acting on an infinitesimal slice of width dx are also shown in the figure. The forces are denoted as follows:

E' = effective normal force on a side of the slice,

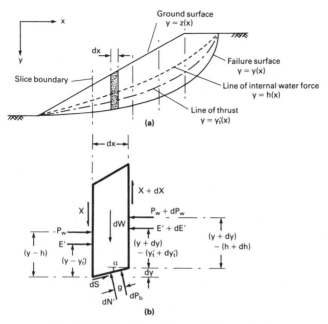

Fig. 9.8　The Morgenstern–Price method.

X = shear force on a side,
P_w = boundary water force on a side,
dN' = effective normal force on the base of the slice,
dS = shear force on the base,
dP_b = boundary water force on the base,
dW = total weight of the slice.

The line of thrust of the effective normal forces (E') is represented by a function $y = y'_t(x)$ and that of the internal water forces (P_w) by $y = h(x)$. Two governing differential equations are obtained by equating moments about the mid point of the base, and forces perpendicular and parallel to the base, to zero. The equations are simplified by working in terms of the total normal force (E), where:

$$E = E' + P_w$$

The position of force E on a side of the slice is obtained from the expression:

$$Ey_t = E'y'_t + P_w h$$

The problem is rendered statically determinate by assuming a relationship between the forces E and X of the form:

$$X = \lambda f(x)E \tag{9.13}$$

where $f(x)$ is a function chosen to represent the pattern of variation of the ratio X/E across the failure mass and λ is a scale factor. The value of λ is obtained as part of the solution along with the factor of safety F.

To obtain a solution the soil mass above a trial failure surface is divided into a series of slices of finite width, such that the failure surface within each slice can be assumed to be linear. The boundary conditions at each end of the failure surface are in terms of the force E and a moment M which is given by the integral of an expression containing both E and X: normally both E and M are zero at each end of the failure surface. The method of solution involves choosing trial values of λ and F, setting the force E to zero at the beginning of the failure surface and integrating across each slice in turn, obtaining values of E, X and y_t: the resulting values of E and M at the end of the failure surface will in general not be zero. A systematic iteration technique, based on the Newton–Raphson method and described by Morgenstern and Price [9.11], is used to modify the values of λ and F until the resulting values of E and M at the end of the failure surface are both zero. The factor of safety is not significantly affected by the choice of the function $f(x)$ and as a consequence $f(x) = 1$ is a common assumption.

For any assumed failure surface it is necessary to examine the solution to ensure that it is valid in respect of the implied state of stress within the soil mass above that surface. Accordingly, a check is made to ensure that neither shear failure nor a state of tension is implied within the mass. The first condition is satisfied if the available shearing resistance on each vertical interface is greater than the corresponding value of the force X: the ratio of these two forces represents the local factor of safety against shear failure along the interface. The requirement that no tension should be developed is satisfied if the line of thrust of the E forces, as given by the computed values of y_t, lies wholly above the failure surface.

Computer programs for the Morgenstern–Price analysis are readily available. The method can be fully exploited if an interactive approach, using computer graphics, is adopted.

Bell [9.1] proposed a method of analysis in which all the conditions of equilibrium are satisfied and the assumed failure surface may be of any shape. The soil mass is divided into a number of vertical slices and statical determinacy is obtained by means of an assumed distribution of normal stress along the failure surface.

Sarma [9.12] developed a method, based on the method of slices, in which the critical earthquake acceleration required to produce a condition of limiting equilibrium is determined. An assumed distribution of vertical interslice forces is used in the analysis. Again, all the conditions of equilibrium are satisfied and the assumed failure surface may be of any shape. The static factor of safety is the factor by which the shear strength of the soil must be reduced such that the critical acceleration is zero.

The use of a computer is also essential for the Bell and Sarma methods and all solutions must be checked to ensure that they are physically acceptable.

9.6   End-of-Construction and Long-Term Stability

When a slope is formed, either by excavation or by the construction of an embankment, the changes in total stress result in changes in pore water pressure in the vicinity of the slope and, in particular, along a potential failure surface. Prior to construction the initial pore water pressure (u_0) at any point is governed either by a static water table level or by a flow net for conditions of steady seepage. The change in pore water pressure at any point is given theoretically by Equation 4.27 or 4.28. The final pore water pressure, after dissipation of the excess pore water pressure is complete, is governed by the static water table level or the steady seepage flow net for the final conditions after construction.

If the permeability of the soil is low, a considerable time will elapse before any significant dissipation of excess pore water pressure will have taken place. At the end of construction the soil will be virtually in the undrained condition and a total stress analysis will be relevant. In principle an effective stress analysis is also possible for the end of construction condition using the pore water pressure (u) for this condition, where

$$u = u_0 + \Delta u$$

However, because of its greater simplicity, a total stress analysis is generally used. It should be realized that the same factor of safety will not generally be obtained from a total stress and an effective stress analysis of the end-of-construction condition. In a total stress analysis it is implied that the pore water pressures are those for a failure condition: in an effective stress analysis the pore water pressures used are those predicted for a non-failure condition. In the long term, the fully drained condition will be reached and only an effective stress analysis will be appropriate.

If, on the other hand, the permeability of the soil is high, dissipation of excess pore water pressure will be largely complete by the end of construction. An effective stress analysis is relevant for all conditions with values of pore water pressure being obtained from the static water table level or the appropriate flow net.

Pore water pressure may thus be an independent variable, determined from the static water table level or from the flow net for conditions of steady seepage, or may be dependent on the total stress changes tending to cause failure.

It is important to identify the most dangerous condition in any practical problem in order that the appropriate shear strength parameters are used in design.

Excavated and Natural Slopes in Saturated Clays

Equation 4.27, with $B = 1$ for a fully saturated clay, can be rearranged as follows:

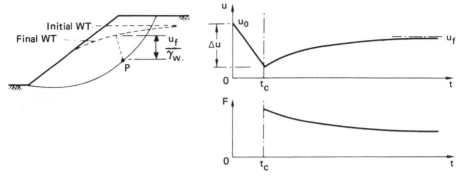

Fig. 9.9   Pore pressure dissipation and factor of safety after excavation. (Reproduced from A. W. Bishop and L. Bjerrum (1960) *Proceedings ASCE Research Conference on Shear Strength of Cohesive Soils, Boulder, Colorado*, by permission of the American Society of Civil Engineers.)

$$\Delta u = \tfrac{1}{2}(\Delta\sigma_1 + \Delta\sigma_3) + (A - \tfrac{1}{2})(\Delta\sigma_1 - \Delta\sigma_3) \qquad (9.14)$$

For a typical point P on a potential failure surface (Fig. 9.9) the first term in Equation 9.14 is negative and the second term will also be negative if the value of A is less than 0·5. Overall, the pore water pressure change Δu is negative. The effect of the rotation of the principal stress directions is neglected. As dissipation proceeds the pore water pressure increases to the final value as shown in Fig. 9.9. The factor of safety will therefore have a lower value in the long term, when dissipation is complete, than at the end of construction.

Slopes in overconsolidated fissured clays require special consideration. A number of cases are on record in which failures in this type of clay have occurred long after dissipation of excess pore water pressure had been completed. Analysis of these failures showed that the average shear strength at failure was well below the peak value. It is probable that large strains occur locally due to the presence of fissures, resulting in the peak strength being reached, followed by a gradual decrease towards the critical state value. The development of large local strains can lead eventually to a progressive slope failure. However fissures may not be the only cause of progressive failure; there is considerable non-uniformity of shear stress along a potential failure surface and local overstressing may initiate progressive failure. It is also possible that there could be a pre-existing slip surface in this type of clay and that it could be reactivated by excavation. In such cases a considerable slip movement could have taken place previously, sufficiently large for the shear strength to fall below the critical state value and towards the residual value.

Thus for an initial failure (i.e. a 'first time' slip) in overconsolidated fissured clay the relevant strength for the analysis of long-term stability is the critical state value. However for failure along a pre-existing slip surface the relevant strength is the residual value. Clearly it is vital that the presence of a pre-existing slip surface in the vicinity of a projected excavation should be detected during the ground investigation.

The strength of an overconsolidated clay at the critical state, for use in the analysis of a potential first-time slip, is difficult to determine accurately. Skempton [9.14] has suggested that the peak strength of the remoulded clay in the normally consolidated condition can be taken as a practical approximation to the strength of the overconsolidated clay at the critical state, i.e. when it has fully softened adjacent to the slip plane as the result of expansion during shear.

Embankments

The construction of an embankment results in increases in total stress, both within the embankment itself as successive layers of soil are placed and in the foundation soil. The construction period of a typical embankment is relatively short and, if the permeability of the compacted fill is low, no significant dissipation is likely during construction. Dissipation proceeds after the end of construction with the pore water pressure decreasing to the final value in the long-term, as shown in Fig. 9.10. The factor of safety of an embankment at the end of construction is therefore lower than in the long term. Shear strength parameters for the fill material should be determined from tests on specimens compacted to the values of dry density and water content to be specified for the embankment.

The stability of an embankment may also depend on the shear strength of the foundation soil. The possibility of failure along a surface such as that illustrated in Fig. 9.11 should be considered in appropriate cases.

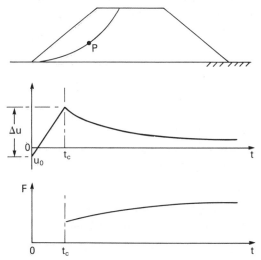

Fig. 9.10 Pore pressure dissipation and factor of safety in an embankment.

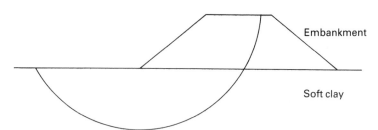

Fig. 9.11   Failure below an embankment.

9.7   Stability of Earth Dams

In the design of earth dams the factor of safety of both slopes must be
determined as accurately as possible for the most critical conditions. For
economic reasons an unduly conservative design must be avoided. In the
case of the upstream slope the most critical stages are at the end of
construction and during rapid drawdown of the reservoir level. The critical
stages for the downstream slope are at the end of construction and during
steady seepage when the reservoir is full. The pore water pressure distribution at any stage has a dominant influence on the factor of safety
and in large earth dams it is common practice to install a piezometer system
so that the actual pore water pressures can be measured at any stage and
compared with the predicted values used in design (provided an effective
stress analysis has been used). Remedial action can then be taken if the
factor of safety, based on the measured values, is considered to be too low.
Instrumentation to measure deformations within a dam, both during and
subsequent to construction, may also be installed. Non-uniform deformation could result in the development of cracks, in which case prompt
remedial action would be essential.

End of Construction

The construction period of an earth dam is likely to be long enough to allow
partial dissipation of excess pore water pressure before the end of
construction, especially in a dam with internal drainage. A total stress
analysis, therefore, would result in too conservative a design. An effective
stress analysis is preferable, using predicted values of r_u.

The pore pressure (u) at any point can be written as

$$u = u_0 + \Delta u$$

where u_0 is the initial value and Δu is the change in pore water pressure
under undrained conditions. In terms of the change in total major principal
stress,

$$u = u_0 + \bar{B}\Delta\sigma_1$$

Then

$$r_u = \frac{u_0}{\gamma h} + \bar{B}\frac{\Delta\sigma_1}{\gamma h}$$

If it is assumed that the increase in total major principal stress is approximately equal to the fill pressure along a potential failure surface,

$$r_u = \frac{u_0}{\gamma h} + \bar{B} \qquad (9.15)$$

The soil is partially saturated when compacted, therefore the initial pore water pressure (u_0) is negative. The actual value of u_0 depends on the placement water content, the higher the water content the closer the value of u_0 to zero. The value of $\bar{B}$ also depends on the placement water content, the higher the water content the higher the value of $\bar{B}$. Thus, for an upper bound,

$$r_u = \bar{B} \qquad (9.16)$$

The value of $\bar{B}$ must correspond to the stress conditions in the dam. Equations 9.15 and 9.16 assume no dissipation during construction. A factor of safety as low as 1·3 may be acceptable at the end of construction provided there is reasonable confidence in the design data.

If high values of r_u are anticipated, dissipation of excess pore water pressure can be accelerated by means of horizontal drainage layers incorporated in the dam, drainage taking place vertically towards the layers: a typical dam section is shown in Fig. 9.12. The efficiency of drainage layers has been examined theoretically by Gibson and Shefford [9.7] and it was shown that in a typical case the layers, in order to be fully effective, should have a permeability at least 10^6 times that of the embankment soil: an acceptable efficiency would be obtained with a permeability ratio of about 10^5.

Steady Seepage

After the reservoir has been full for some time, conditions of steady seepage become established through the dam with the soil below the top flow line in the fully saturated state. This condition must be analysed in terms of

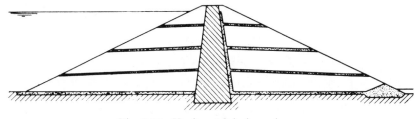

Fig. 9.12 Horizontal drainage layers.

effective stress with values of pore pressure being determined from the flow net. Values of r_u up to 0·45 are possible in homogeneous dams but much lower values can be achieved in dams having internal drainage. The factor of safety for this condition should be at least 1·5.

Rapid Drawdown

After a condition of steady seepage has become established, a drawdown of the reservoir level will result in a change in the pore water pressure distribution. If the permeability of the soil is low, a drawdown period measured in weeks may be 'rapid' in relation to the dissipation time and the change in pore water pressure can be assumed to take place under undrained conditions. Referring to Fig. 9.13, the pore water pressure before drawdown at a typical point P on a potential failure surface is given by

$$u_0 = \gamma_w(h + h_w - h')$$

where h' is the loss in total head due to seepage between the upstream slope surface and the point P. It is again assumed that the total major principal stress at P is equal to the fill pressure. The change in total major principal stress is due to the total or partial removal of water above the slope on the vertical through P. For a drawdown depth exceeding h_w:

$$\Delta\sigma_1 = -\gamma_w h_w$$

and the change in pore water pressure is then given by

$$\Delta u = \bar{B}\Delta\sigma_1$$
$$= -\bar{B}\gamma_w h_w$$

Therefore the pore water pressure at P immediately after drawdown is

$$u = u_0 + \Delta u$$
$$= \gamma_w\{h + h_w(1 - \bar{B}) - h'\}$$

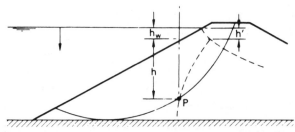

Fig. 9.13 Rapid drawdown conditions. (Reproduced from A. W. Bishop and L. Bjerrum (1960) *Proceedings ASCE Research Conference on Shear Strength of Cohesive Soils, Boulder, Colorado*, by permission of the American Society of Civil Engineers.)

Hence:

$$r_u = \frac{u}{\gamma_{sat} h}$$

$$= \frac{\gamma_w}{\gamma_{sat}} \left\{ 1 + \frac{h_w}{h}(1 - \bar{B}) - \frac{h'}{h} \right\} \tag{9.17}$$

For total stress decreases the value of $\bar{B}$ is slightly greater than 1. A conservative value of r_u could be obtained by assuming $\bar{B} = 1$ and neglecting h'. Typical values of r_u immediately after drawdown are within the range 0·3 to 0·4. A minimum factor of safety of 1·2 may be acceptable after rapid drawdown.

Morgenstern [9.9] published stability coefficients for the approximate analysis of homogeneous slopes after rapid drawdown.

The pore water pressure distribution after drawdown in soils of high permeability varies as pore water drains out of the soil above the drawdown level. The saturation line moves downwards at a rate depending on the permeability of the soil. A series of flow nets can be drawn for different positions of the saturation line and values of pore water pressure obtained. The factor of safety can thus be determined, using an effective stress analysis, for any position of the saturation line.

Problems

9.1   For the given failure surface, determine the factor of safety in terms of total stress for the slope detailed in Fig. 9.14. The unit weight for both soils is 19 kN/m³. For soil 1 the relevant shear strength parameters are

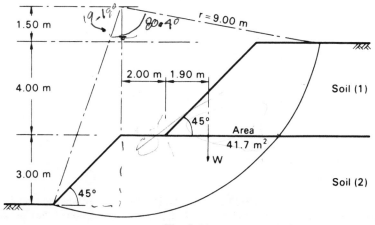

Fig. 9.14

$c_u = 20\,\mathrm{kN/m^2}$ and $\phi_u = 0$: for soil 2, $c_u = 35\,\mathrm{kN/m^2}$ and $\phi_u = 0$.
What is the factor of safety if allowance is made for the development of a tension crack which fills with water?

9.2 A cutting 9 m deep is to be excavated in a saturated clay of unit weight
$19\,\mathrm{kN/m^3}$. The relevant shear strength parameters are $c_u = 30\,\mathrm{kN/m^2}$
and $\phi_u = 0$. A hard stratum underlies the clay at a depth of 11 m below
ground level. Using Taylor's stability coefficients, determine the slope
angle at which failure would occur. What is the allowable slope angle
if a factor of safety of 1·2 is specified?

9.3 For the given failure surface, determine the factor of safety in terms of
effective stress for the slope detailed in Fig. 9.15, using the Fellenius
method of slices. The unit weight of the soil is $21\,\mathrm{kN/m^3}$ and the

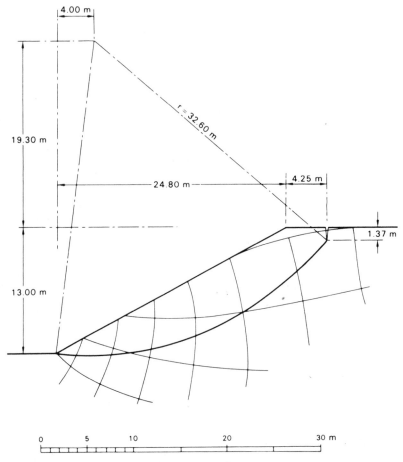

Fig. 9.15   (Reproduced from Skempton and Brown (1961) A landslide in boulder
clay at Selset, Yorkshire, *Geotechnique* Vol.11, p. 280, by permission of the Council of
the Institution of Civil Engineers.)

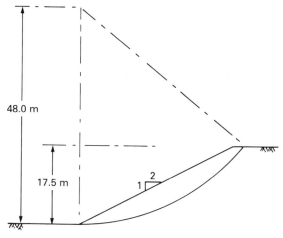

Fig. 9.16

relevant shear strength parameters are $c' = 8\,\text{kN/m}^2$ and $\phi' = 32°$.

9.4 Repeat the analysis of the slope detailed in Problem 9.3 using the Bishop simplified method of slices.

9.5 Using the Bishop simplified method of slices, determine the factor of safety in terms of effective stress for the slope detailed in Fig. 9.16 for the specified failure surface. The value of r_u is 0·45 and the unit weight of the soil is $20\,\text{kN/m}^3$. The relevant shear strength parameters are $c' = 16\,\text{kN/m}^2$ and $\phi' = 32°$.

9.6 A slope is to be constructed in a soil for which $c' = 0$ and $\phi' = 36°$. It is to be assumed that the water table may occasionally reach the surface of the slope, with seepage taking place parallel to the slope. Determine the maximum slope angle for a factor of safety of 1·5, assuming a potential failure surface parallel to the slope. What would be the factor of safety of the slope, constructed at this angle, if the water table should be well below the surface? The saturated unit weight of the soil is $19\,\text{kN/m}^3$.

References

9.1 Bell, J. M. (1968): 'General Slope Stability Analysis', *Journal ASCE*, Vol. 94, No. SM6.

9.2 Bishop. A. W. (1955): 'The Use of the Slip Circle in the Stability Analysis of Slopes', *Geotechnique*, Vol. 5, No. 1.

9.3 Bishop, A. W. and Bjerrum, L. (1960): 'The Relevance of the Triaxial Test to the Solution of Stability Problems', *Proceedings ASCE Research Conference on Shear Strength of Cohesive Soils*, Boulder, Colorado, p. 437.

9.4 Bishop, A. W. and Morgenstern, N. R. (1960): 'Stability Coefficients for Earth Slopes', *Geotechnique*, Vol. 10, No. 4.

9.5 British Standard 6031 (1981): *Code of Practice for Earthworks*, British Standards Institution, London.

9.6 Gibson, R. E. and Morgenstern, N. R. (1962): 'A Note on the Stability of Cuttings in Normally Consolidated Clays', *Geotechnique*, Vol. 12, No. 3.

9.7 Gibson, R. E. and Shefford, G. C. (1968): 'The Efficiency of Horizontal Drainage Layers for Accelerating Consolidation of Clay Embankments', *Geotechnique*, Vol. 18, No. 3.

9.8 Lo, K. Y. (1965): 'Stability of Slopes in Anisotropic Soils', *Journal ASCE*, Vol. 91, No. SM4.

9.9 Morgenstern, N. R. (1963): 'Stability Charts for Earth Slopes During Rapid Drawdown', *Geotechnique*, Vol. 13, No. 2.

9.10 Morgenstern, N. R. and Price, V. E. (1965): 'The Analysis of the Stability of General Slip Surfaces', *Geotechnique*, Vol. 15, No. 1.

9.11 Morgenstern, N. R. and Price, V. E. (1967): 'A Numerical Method for Solving the Equations of Stability of General Slip Surfaces', *Computer Journal*, Vol. 9, p. 388.

9.12 Sarma, S. K. (1973): 'Stability Analysis of Embankments and Slopes', *Geotechnique*, Vol. 23, No. 2.

9.13 Skempton, A. W. (1964): 'Long-Term Stability of Clay Slopes', *Geotechnique*, Vol. 14, No. 2.

9.14 Skempton, A. W. (1970): 'First-Time Slides in Overconsolidated Clays' (Technical Note), *Geotechnique*, Vol. 20, No. 3.

9.15 Spencer, E. (1967): 'A Method of Analysis of the Stability of Embankments Assuming Parallel Inter-Slice Forces', *Geotechnique*, Vol. 17, No. 1.

9.16 Taylor, D. W. (1937): 'Stability of Earth Slopes', *Journal of the Boston Society of Civil Engineers*, Vol. 24, No. 3.

9.17 Whitman, R. V. and Bailey, W. A. (1967): 'Use of Computers for Slope Stability Analysis', *Journal ASCE*, Vol. 93, No. SM4.

CHAPTER 10

Ground Investigation

10.1 Introduction

An adequate ground investigation is an essential preliminary to the execution of a civil engineering project. Sufficient information must be obtained to enable a safe and economic design to be made and to avoid any difficulties during construction. The principal objects of the investigation are: (1) to determine the sequence, thicknesses and lateral extent of the soil strata and, where appropriate, the level of bedrock; (2) to obtain representative samples of the soils (and rock) for identification and classification and, if necessary, for use in laboratory tests to determine relevant soil parameters; (3) to identify the groundwater conditions. The investigation may also include the performance of in-situ tests to assess appropriate soil characteristics. The results of a ground investigation should provide adequate information, for example, to enable the most suitable type of foundation for a proposed structure to be selected and to indicate if special problems are likely to arise during excavation.

A study of geological maps and memoirs, if available, should give an indication of the probable soil conditions of the site in question. If the site is extensive and if no existing information is available, the use of aerial photographs can be useful in identifying features of geological significance. Before the start of field work an inspection of the site and the surrounding area should be made on foot. River banks, existing excavations, quarries and road or railway cuttings, for example, can yield valuable information regarding the nature of the strata and groundwater conditions: existing structures should be examined for signs of settlement damage. Previous experience of conditions in the area may have been obtained by adjacent owners or local authorities. All information obtained in advance enables the most suitable type of investigation to be decided.

The actual investigation procedure depends on the nature of the strata and the type of project but will normally involve the excavation of boreholes or trial pits. The number and location of boreholes or pits should enable the basic geological structure of the site to be determined and significant irregularities in the sub-surface conditions to be detected. The greater the degree of variability of the ground conditions the greater the number of boreholes or pits required. The locations should be off-set from areas on which it is known that foundations are to be sited. A preliminary investigation on a modest scale may be carried out to obtain the general characteristics of the strata, followed by a more extensive and carefully planned investigation including sampling and in-situ testing.

It is essential that the investigation is taken to an adequate depth. This depth depends on the type and size of the project but must include all strata liable to be significantly affected by the structure and its construction. The investigation must extend below all strata which might have inadequate shear strength for the support of foundations or which would give rise to significant settlement. If the use of piles is anticipated the investigation will thus have to extend to a considerable depth below the surface. A general rule often applied in the case of foundations is that the ground conditions must be known within the significant depth (Section 8.1) provided there is no weak stratum below this depth which would cause unacceptable settlement. If rock is encountered it should be penetrated at least 3 m to confirm that bedrock, and not a large boulder, has been reached, unless geological knowledge indicates otherwise. The investigation may have to be taken to depths greater than normal in areas of old mine workings or other underground cavities.

Boreholes and trial pits should be backfilled after use. Backfilling with compacted soil may be adequate in many cases but if the groundwater conditions are altered by a borehole and the resultant flow could produce adverse effects then it is necessary to use a cement-based grout to seal the hole.

The cost of an investigation depends on the location and extent of the site, the nature of the strata and the type of project under consideration. In general, the larger the project, and the less critical the ground conditions are to the design and construction of the project, the lower the cost of the ground investigation as a percentage of the total cost. The cost is generally within the range of 0·1% to 2% of the project cost: therefore to reduce the scope of an investigation for financial reasons alone is seldom justified.

10.2   Methods of Investigation

Trials Pits

The excavation of trial pits is a simple and reliable method of investigation but is limited to a maximum depth of 4–5 m. The soil is generally removed by means of the back-shovel of a mechanical excavator. Before any person enters the pit the sides must always be supported unless they are sloped at a safe angle or are stepped: the excavated soil should be placed at least 1 m from the edge of the pit. If the pit is to extend below the water table, some form of dewatering is necessary in the more permeable soils, resulting in increased costs. The use of trial pits enables the in-situ soil conditions to be examined visually, thus the boundaries between strata and the nature of any macro-fabric can be accurately determined. It is relatively easy to obtain disturbed or undisturbed soil samples: in cohesive soils block samples can be cut by hand from the sides or bottom of the pit and tube samples can be obtained below the bottom of the pit. Trial pits are suitable for investigations in all types of soil, including those containing cobbles or boulders.

Shafts and Headings

Deep pits or shafts are usually advanced by hand excavation, the sides being supported by timbering. Headings or adits are excavated laterally from the bottoms of shafts or from the surface into hillsides, both the sides and roof being supported. It is unlikely that shafts or headings would be excavated below the water table. Shafts and headings are very costly and their use would be justified only in investigations for very large structures, such as dams, if the ground conditions could not be ascertained adequately by other means.

Percussion Boring

The boring rig (Fig. 10.1) consists of a derrick, a power unit and a winch carrying a light steel cable which passes through a pulley on top of the derrick. Most rigs are fitted with road wheels and when folded down can be towed behind a vehicle. In hard or dense soil the borehole is excavated by means of a heavy *chisel* (or *chopping bit*) attached to a string of solid boring rods of square section, the rods providing the necessary weight to penetrate the soil. Occasionally, however, a heavy element called a sinker bar is fitted immediately above the boring tool. The tool and rods are carried by the steel cable and are alternately raised and dropped, by means of the winch unit, to break up the soil. Cobbles and boulders can also be broken up by the chisel but this can be a very slow process.

Below the water table the loosened soil forms a slurry with the groundwater. Above the water table, water is introduced into the borehole to form a slurry. Periodically the chisel and rods are removed from the borehole and the slurry is extracted by means of a *shell* or *baler*: it is convenient if the rig is fitted with a second cable to carry the shell. The shell is a heavy steel tube fitted with a cutting shoe and a non-return flap or dart valve at the lower end. The shell is moved upwards and downwards to collect the slurry and when full is raised to the surface to be emptied. In loose sands and gravels below the water table the shell can be used directly as a boring tool, with a sinker bar if necessary, prior use of a chisel not being required.

The borehole must be cased if the sides are liable to collapse. The casing consists of lengths of steel pipe, screwed together, which are driven or jacked into the hole. At shallow depths the casing may slide into the hole

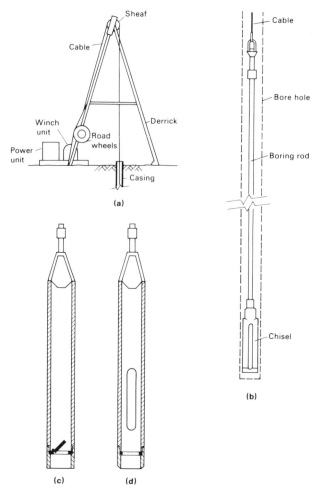

Fig. 10.1   (a) Percussion boring rig, (b) boring rods and chisel, (c) shell, (d) clay cutter.

under its own weight. On completion of the investigation the casing is recovered by means of the winch or by the use of jacks: excessive driving may make recovery of the casing difficult.

Other tools which can be used directly with the percussion rig are the *clay cutter* and the *auger*. The cutter, an open steel tube with a cutting shoe and retaining ring at the lower end, is used for boring in clays: it is used in a dry borehole. The cutter is alternately raised and dropped by means of the winch unit, a sinker bar being fitted above the tool if necessary. The clay gradually fills the cutter which is then raised to the surface to be emptied. The auger is also used in clays and is operated by rotating the boring rods by hand at the surface by means of a tiller bar. The auger is also used for cleaning out the hole prior to sampling.

Borehole diameters can range from 150 mm to 300 mm. The maximum borehole depth is generally between 50 m and 60 m. Percussion boring can be employed in most types of soil, including those containing cobbles and boulders. However, there is generally some disturbance of the soil below the bottom of the borehole, from which samples are taken, and it is extremely difficult to detect thin soil layers and minor geological features with this method. The rig is extremely versatile and can normally be fitted with a hydraulic power unit and attachments for mechanical augering, rotary core drilling and cone penetration testing.

Mechanical Augers

Power-operated augers are generally mounted on vehicles or are in the form of attachments to the derrick used for percussion boring. The power required to rotate the auger depends on the type and size of the auger itself and the type of soil to be penetrated. Downward pressure on the auger can be applied hydraulically, mechanically or by dead weight. The types of tool generally used are the *flight auger* and the *bucket auger*. The diameter of a flight auger is usually between 75 mm and 300 mm, although diameters as large as 1 m are available: the diameter of a bucket auger can range from 300 mm to 2 m. However, the larger sizes are used principally for excavating shafts for bored piles. Augers are used mainly in soils in which the borehole requires no support and remains dry, i.e. mainly in clays. The use of casing is inconvenient because of the necessity of removing the auger before driving the casing: however, it is possible to use bentonite slurry (Section 6.9) to support the sides of unstable holes. The presence of cobbles or boulders creates difficulties with the smaller-sized augers.

Short-flight augers (Fig. 10.2a) consist of a helix of limited length, with cutters below the helix. The auger is attached to a steel shaft, known as a Kelly bar, which passes through the rotary head of the rig. The auger is advanced until it is full of soil, then it is raised to the surface where the soil is ejected by rotating the auger in the reverse direction. Clearly, the shorter the helix the more often the auger must be raised and lowered for a given borehole depth. The depth of the hole is limited by the length of the Kelly bar.

Continuous-flight augers (Fig. 10.2b) consist of rods with a helix covering the entire length. The soil rises to the surface along the helix, obviating the necessity for withdrawal: additional lengths of auger are added as the hole is advanced. Borehole depths up to 50 m are possible with continuous-flight augers, but there is a possibility that different soil types may become mixed as they rise to the surface and it may be difficult to determine the depths at which changes of strata occur.

Continuous-fight augers with hollow stems are also used. When boring is in progress the hollow stem is closed at the lower end by a plug fitted to a rod running inside the stem. Additional lengths of auger (and internal rod)

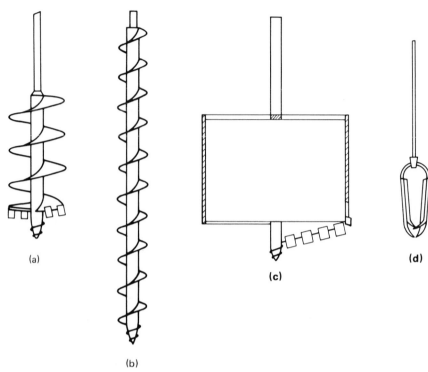

Fig. 10.2 (a) Short-flight auger, (b) continuous-flight auger, (c) bucket auger, (d) Iwan (hand) auger.

are again added as the hole is advanced. At any depth the rod and plug may be withdrawn from the hollow stem to allow undisturbed samples to be taken, a sample tube mounted on rods being lowered down the stem and driven into the soil below the auger. If bedrock is reached, drilling can also take place through the hollow stem. The internal diameter of the stem can range from 75 mm to 150 mm. As the auger performs the function of a casing it can be used in sands below the water table, although difficulty may be experienced with sand being forced upwards into the stem by hydrostatic pressure: this can be avoided by filling the stem with water up to water-table level.

Bucket augers (Fig. 10.2c) consist of a steel cylinder, open at the top but fitted with a base plate on which cutters are mounted, adjacent to slots in the plate: the auger is attached to a Kelly bar. When the auger is rotated and pressed downwards the soil removed by the cutters passes through the slots into the bucket. When the bucket is full it is raised to the surface to be emptied by releasing the hinged base plate.

Augered holes of 1 m diameter and larger can be used for the examination of the soil strata in situ, the person carrying out the examination being lowered down the hole in a special cage. The hole must be cased when used for this purpose and adequate ventilation is essential.

Hand and Portable Augers

Hand augers can be used to excavate boreholes to depths of around 5 m using a set of extension rods. The auger is rotated and pressed down into the soil by means of a T-handle on the upper rod. The two common types are the Iwan or post-hole auger (Fig. 10.2d) with diameters up to 200 mm, and the small helical auger, with diameters of about 50 mm. Hand augers are generally used only if the sides of the hole require no support and if particles of coarse gravel size and above are absent. The auger must be withdrawn at frequent intervals for the removal of soil. Undisturbed samples can be obtained by driving small-diameter tubes below the bottom of the borehole.

Small portable power augers, generally transported and operated by two persons, are suitable for boring to depths of 10–15 m; the hole diameter may range from 75 mm to 300 mm. The borehole may be cased if necessary, therefore the auger can be used in most soil types provided the larger particle sizes are absent.

Wash Boring

In this method, water is pumped through a string of hollow boring rods and is released under pressure through narrow holes in a chisel attached to the lower end of the rods (Fig. 10.3). The soil is loosened and broken up by the water jets and the up-and-down movement of the chisel. There is also provision for the manual rotation of the chisel by means of a tiller attached to the boring rods above the surface. The soil particles are washed to the surface between the rods and the side of the borehole and are allowed to settle out in a sump. The rig consists of a derrick with a power unit, a winch and a water pump. The winch carries a light steel cable which passes through the sheaf of the derrick and is attached to the top of the boring rods. The string of rods is raised and dropped by means of the winch unit, producing the chopping action of the chisel. The borehole is generally cased but the method can be used in uncased holes. Drilling mud may be used as an alternative to water in the method, eliminating the need for casing.

Wash boring can be used in most types of soil but progress becomes slow if particles of coarse gravel size and larger are present. The accurate identification of soil types is difficult due to particles being broken up by the chisel and to mixing as the material is washed to the surface: in addition, segregation of particles takes place as they settle out in the sump. However, a change in the feel of the boring tool can sometimes be detected, and there may be a change in the colour of the water rising to the surface, when the boundaries between different strata are reached. The method is unacceptable as a means of obtaining soil samples. It is used only as a

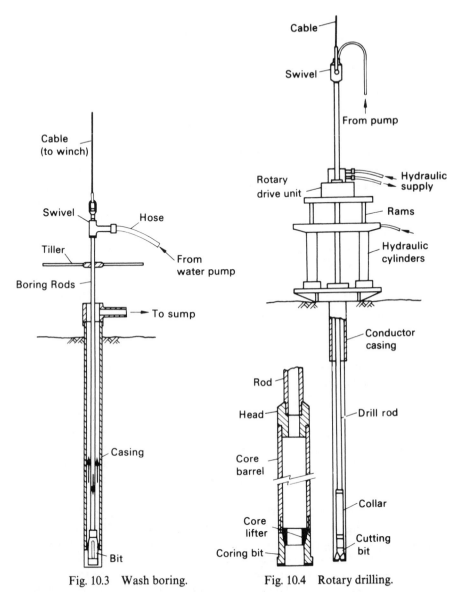

Fig. 10.3 Wash boring.   Fig. 10.4 Rotary drilling.

means of advancing a borehole to enable tube samples to be taken or in-situ tests to be carried out below the bottom of the hole. An advantage of the method is that the soil immediately below the hole remains relatively undisturbed.

Rotary Drilling

Although primarily intended for investigations in rock, the method is also used in soils. The drilling tool, which is attached to the lower end of a string of hollow drilling rods (Fig. 10.4) may be either a cutting bit or a coring bit: the coring bit is fixed to the lower end of a core barrel which in turn is carried by the drilling rods. Water or drilling mud is pumped down the hollow rods and passes under pressure through narrow holes in the bit or barrel: this is the same principle as used in wash boring. The drilling fluid cools and lubricates the drilling tool and carries the loose debris to the surface between the rods and the side of the hole. The fluid also provides some support to the sides of the hole if no casing is used.

The rig consists of a derrick, power unit, winch, pump and a drill head to apply high-speed rotary drive and downward thrust to the drilling rods. A rotary head attachment can be supplied as an accessory to a percussion boring rig.

There are two forms of rotary drilling, open-hole drilling and core drilling. Open-hole drilling, which is generally used in soils and weak rock, uses a cutting bit to break down all the material within the diameter of the hole. Open-hole drilling can thus be used only as a means of advancing the hole: the drilling rods can then be removed to allow tube samples to be taken or in-situ tests to be carried out. In core drilling, which is used in rocks and hard clays, the bit cuts an annular hole in the material and an intact core enters the barrel, to be removed as a sample. However, the natural water content of the material is liable to be increased due to contact with the drilling fluid. Typical core diameters are 41 mm, 54 mm and 76 mm, but can range up to 165 mm.

The advantage of rotary drilling in soils is that progress is much faster than with other investigation methods and disturbance of the soil below the borehole is slight. The method is not suitable if the soil contains a high percentage of gravel (or larger) particles as they tend to rotate beneath the bit and are not broken up.

Groundwater Observations

An important part of any ground investigation is the determination of water-table level and of any artesian pressure in particular strata. The variation of level or pressure over a given period of time may also require determination. Groundwater observations are of particular importance if deep excavations are to be carried out.

Water-table level can be determined by measuring the depth to the water surface in a borehole. Water levels in boreholes may take a considerable time to stabilize, this time, known as the response time, depending on the permeability of the soil. Measurements, therefore, should be taken at regular intervals until the water level becomes constant. It is preferable that the level should be determined as soon as it is considered that the borehole has reached water-table level. If the borehole is further advanced it may penetrate a stratum under artesian pressure, resulting in the water level in the hole being above water-table level. It is important that a stratum of low permeability below a perched water table should not be penetrated before the water level has been established. If a perched water table exists, the borehole must be cased in order that the main water-table level is correctly determined: if the perched aquifer is not sealed, the water level in the borehole will be above the main water-table level.

When it is desired to obtain the water pressure in a particular stratum a piezometer should be used. The simplest type is the Casagrande piezometer (Fig. 4.30) with the porous element sealed at the appropriate depth. However, this type of piezometer has a long response time in soils of low permeability and in such cases it is preferable to install a hydraulic piezometer (Fig. 4.31) having a relatively short response time.

Groundwater samples may be required for chemical analysis to determine if they contain sulphates (which may attack Portland cement concrete) or other corrosive constituents. It is important to ensure that samples are not contaminated or diluted. A sample should be taken immediately the water-bearing stratum is reached in boring. It is preferable to obtain samples from the standpipe piezometers if these have been installed.

10.3 Sampling

Soil samples are divided into two main categories, undisturbed and disturbed. Undisturbed samples, which are required mainly for shear strength and consolidation tests are obtained by techniques which aim at preserving the in-situ structure and water content of the soil. In boreholes, undisturbed samples can be obtained by withdrawing the boring tools (except when hollow-stem continuous-flight augers are used) and driving or pushing a sample tube into the soil at the bottom of the hole. When the tube is brought to the surface, some soil is removed from each end and molten wax is applied, in thin layers, to form a seal approximately 25 mm thick: the ends of the tube are then covered by protective caps. Undisturbed block samples can be cut by hand from the bottom or sides of a trial pit. During cutting, the samples must be protected from water, wind and sun to avoid any change in water content: the samples should be covered with molten wax immediately they have been brought to the surface. It is impossible to obtain a sample that is completely undisturbed, no matter how elaborate or careful the ground investigation and sampling technique might be. In the case of clays, for example, swelling will take place adjacent to the bottom of a borehole due to the reduction in total stresses when soil is removed and structural disturbance may be caused by the action of the boring tools: when a sample subsequently is removed from the ground the total stresses are reduced to zero.

Soft clays are extremely sensitive to sampling disturbance, the effects being more pronounced in clays of low plasticity than in those of high plasticity. The central core of a soft clay sample will be relatively less disturbed than the outer zone adjacent to the sampling tube. Immediately after sampling, the pore water pressure in the relatively undisturbed core will be negative due to the release of the in-situ total stresses. Swelling of the relatively undisturbed core will gradually take place due to water being drawn from the more disturbed outer zone and resulting in the dissipation of the negative excess pore water pressure: the outer zone of soil will consolidate due to the redistribution of water within the sample. The dissipation of the negative excess pore water pressure is accompanied by a corresponding reduction in effective stresses. The soil structure of the sample will thus offer less resistance to shear and will be less rigid than the in-situ soil.

A disturbed sample is one having the same particle size distribution as the in-situ soil but in which the soil structure has been significantly damaged or completely destroyed: in addition, the water content may be different from that of the in-situ soil. Disturbed samples, which are used mainly for soil classification tests, visual classification and compaction tests, can be excavated from trial pits or obtained from the tools used to advance boreholes (e.g. from augers and the clay cutter). The soil recovered from the shell in percussion boring will be deficient in fines and will be unsuitable for use as a disturbed sample. Samples in which the natural water content has been preserved should be placed in air-tight, non-corrosive containers: all containers should be completely filled so that there is negligible air space above the sample.

All samples should be clearly labelled to show the project name, date, location, borehole number, depth and method of sampling: in addition, each sample should be given a serial number. Special care is required in the handling, transportation and storage of samples (particularly undisturbed samples) prior to testing.

The principal types of tube samplers are described below.

Open Drive Sampler

An open drive sampler (Fig. 10.5a) consists of a length of steel tube with a screw thread at each end. A cutting shoe is attached to one end of the tube. The other end of the tube screws into a sampler head to which, in turn, the boring rods are connected. The sampler head also incorporates a non-return valve to allow air and water to escape as the soil fills the tube and to help to retain the sample as the tube is withdrawn. The inside of the tube should have a smooth surface and must be maintained in a clean condition.

The internal diameter of the cutting edge (d_c) should be approximately 1% smaller than that of the tube to reduce frictional resistance between the tube and the sample. This size difference also allows for slight elastic expansion of the sample on entering the tube and assists in sample retention. The external diameter of the cutting shoe (d_w) should be slightly greater than that of the tube to reduce the force required to withdraw the

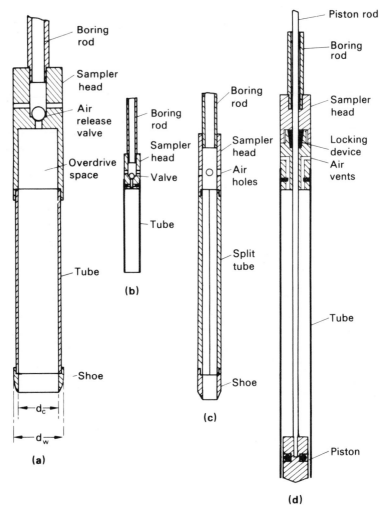

Fig. 10.5   (a) Open drive sampler, (b) thin-walled sampler, (c) split-barrel sampler, (d) stationary piston sampler.

tube. The volume of soil displaced by the sampler as a proportion of the sample volume is represented by the *area ratio* (A_r) of the sampler, where

$$A_r = \frac{d_w^2 - d_c^2}{d_c^2} \tag{10.1}$$

The area ratio is generally expressed as a percentage. Other factors being equal, the lower the area ratio the lower the degree of sample disturbance.

The sampler can be driven dynamically by means of a drop weight, or statically by hydraulic or mechanical jacking, usually working from the boring rig. Prior to sampling all loose soil should be removed from the bottom of the borehole. Care should be taken to ensure that the sampler is not driven beyond its capacity otherwise the sample will be compressed against the sampler head. Some types of sampler head have an overdrive space below the valve to reduce the risk of sample damage. After withdrawal, the cutting shoe and sampler head are detached and the ends of the sample are sealed.

The most widely used sample tube has an internal diameter of 100 mm and a length of 450 mm: the area ratio is approximately 30%. This sampler is suitable for all clay soils. When used to obtain samples of sand a core-catcher, a short length of tube with spring-loaded flaps, should be fitted between the tube and cutting shoe to prevent loss of soil.

Thin-walled samplers (Fig. 10.5b) are used in soils which are sensitive to disturbance such as soft to firm clays and plastic silts. The sampler does not employ a separate cutting shoe, the lower end of the tube itself being machined to form a cutting edge. The internal diameter may range from 35 mm to 100 mm. The area ratio is approximately 10% and samples of first-class quality can be obtained provided the soil has not been disturbed in advancing the borehole. In trial pits and shallow boreholes the tube can often be driven manually.

Split-barrel samplers (Fig. 10.5c) consist of a tube which is split longitudinally into two halves: a shoe and a sampler head incorporating air release holes are screwed onto the ends. The two halves of the tube can be separated when the shoe and head are detached to allow the sample to be removed. The internal and external diameters are 35 mm and 50 mm respectively, the area ratio being approximately 100%, with the result that there is considerable disturbance of the sample. This sampler is used mainly in sands, being the tool specified in the standard penetration test.

Stationary Piston Sampler

This type of sampler (Fig. 10.5d) consists of a thin-walled tube fitted with a piston. The piston is attached to lengths of rod which pass through the sampler head and run inside the hollow boring rods. The sampler is lowered into the borehole with the piston located at the lower end of the tube, the tube and piston being locked together by means of a clamping device at the top of the rods. The piston prevents water or loose soil from entering the tube. In soft soils the sampler can be pushed below the bottom of the borehole, by-passing any disturbed soil. The piston is held against the soil (generally by clamping the piston rod to the casing) and the tube is pushed past the piston (until the sampler head meets the top of the piston) to obtain the sample. The sampler is then withdrawn, a locking device in the sampler head holding the piston at the top of the tube as this takes place. The vacuum between the piston and the sample helps to retain the soil in the tube: the piston thus serves as a non-return valve.

Piston samplers should always be pushed down by hydraulic or mechanical jacking: they should never be driven. The diameter of the sampler is usually between 35 mm and 100 mm but can be as large as 250 mm. The samplers are generally used for soft clays and produce samples of first-class quality: they can also be used for silts and silty sands which have some cohesion.

Continuous Sampler

This is a highly specialized type of sampler which is capable of obtaining undisturbed samples up to 25 m in length: the sampler is used mainly in soft clays. Details of the soil fabric can be determined more easily if a continuous sample is available. An essential requirement of continuous samplers is the elimination of frictional resistance between the sample and the inside of the sampler tube. In one type of sampler, developed in Sweden [10.6], this is achieved by superimposing thin strips of metal foil between the sample and the tube. The lower end of the sampler (Fig. 10.6) has a sharp cutting edge above which the external diameter is enlarged to enable 16 rolls of foil to be housed in recesses within the wall of the sampler. The ends of the foil are attached to a piston which fits loosely inside the sampler: the piston is supported on a cable which is fixed at the surface. Lengths of sample tube (68 mm diameter) are attached as required to the upper end of the sampler.

As the sampler is pushed into the soil the foil unrolls and encases the sample, the piston being held at a constant level by means of the cable. As the sampler is withdrawn the lengths of tube are uncoupled and a cut is made, between adjacent tubes, through the foil and sample.

Compressed Air Sampler

This sampler (Fig. 10.7) is used to obtain undisturbed samples of sand below the water table. The sample tube, usually 60 mm in diameter, is attached to a sampler head having a relief valve which can be closed by a rubber diaphragm. Attached to the sampler head is a hollow guide rod surmounted by a guide head. An outer tube, or bell, surrounds the sample tube, the bell being attached to a weight which slides on the guide rod. The boring rods fit loosely into a plain socket in the top of the guide head, the weight of the bell and sampler being supported by means of a shackle which hooks over a peg in the lower length of boring rod: a light cable, leading to the surface, is fixed to the shackle. Compressed air, produced by a foot pump, is supplied through a tube leading to the guide head, the air passing down the hollow guide rod to the bell.

The sampler is lowered on the boring rods to the bottom of the borehole,

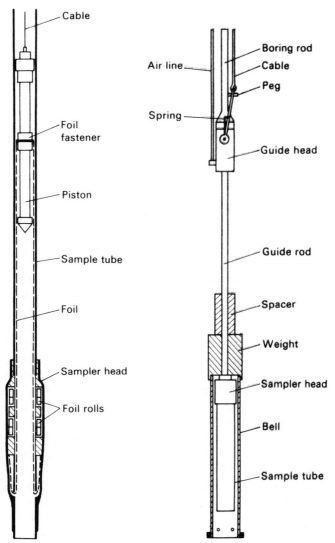

Fig. 10.6   Continuous sampler.   Fig. 10.7   Compressed air sampler.

which will contain water below the level of the water table. When the sampler comes to rest at the bottom of the borehole the shackle springs off the peg, removing the connection between the sampler and the boring rods. The tube is pushed into the soil by means of the boring rods, a stop on the guide rod preventing overdriving: the boring rods are then withdrawn. Compressed air is now introduced to expel the water from the bell and to close the valve in the sampler head by pressing the diaphragm downwards. The tube is withdrawn into the bell by means of the cable then the tube and bell together are raised to the surface. The sand sample remains in the tube by virtue of arching and the slight negative pore water pressure in the soil. A plug is placed in the bottom of the tube before the suction is released and the tube removed from the sampler head.

10.4   Borehole Logs

After an investigation has been completed and the results of any laboratory tests are available, the ground conditions discovered in each borehole (or trial pit) are summarized in the form of a borehole log. An example of such a log appears in Table 10.1, but details of the layout can vary. The last few columns are originally left without headings to allow for variations in the data presented. The method of investigation and details of the equipment used should be stated on each log. The location, ground level and diameter of the hole should be specified together with details of any casing used. The names of the client and the project should be stated.

The log should enable a rapid appraisal of the soil profile to be made. The log is prepared with reference to a vertical scale. A detailed description of each stratum is given and the levels of strata boundaries clearly shown: the level at which boring was terminated should be indicated. The different soil (and rock) types are represented by means of a legend using standard symbols. The depths, or ranges of depth, at which samples were taken or at which in-situ tests were performed are recorded: the type of sample is also specified. The results of certain laboratory or in-situ tests may be given in the log. The depths at which ground water was encountered and subsequent changes in levels, with times, should be detailed.

The soil description should be based on particle size distribution and plasticity, generally using the rapid procedure in which these characteristics are assessed by means of visual inspection and feel: disturbed samples are generally used for this purpose. The description should include details of soil colour, particle shape and composition: if possible the geological formation and type of deposit should be given. The structural characteristics of the soil mass should also be described but this requires an examination of undisturbed samples or of the soil in situ (e.g. in a trial pit). Details should be given of the presence and spacing of bedding features, fissures and other relevant characteristics. The relative density of sands (Table 8.3) and the consistency of clays (Table 4.2) should be indicated.

10.5   Geophysical Methods

Under certain conditions geophysical methods may be useful in ground investigation, especially at the reconnaissance stage. However, the methods

Table 10.1   BOREHOLE LOG

Location: Downfield   BOREHOLE NO. 1
Client: RFC Consultants   Sheet 1 of 1
Boring Method: Shell and Auger to 14·4 m   Ground level: 36.30
                 Rotary Core Drilling to 17·8 m   Date: 30:7:77
   Diameter: 150 mm   Scale: 1:100
                  NX
   Casing: 150 mm to 5 m

Description of Strata	Level	Legend	Depth	Samples	N	c_u kN/m^2	
TOPSOIL	35·6		0·7				
Loose, light brown SAND				D	6		
	33·7		2·6				
Medium dense, brown gravelly SAND ▽	32·5			D	15		
	31·9		4·4				
				U		80	
				U		86	
Firm to stiff, yellowish-brown, closely fissured CLAY of high plasticity				U		97	
				U		105	
	24·1		12·2				
Very dense, red, silty SAND with decomposed SANDSTONE				D	50 for 210 mm		
	21·9		14·4				
Red, medium grained, granular, fresh SANDSTONE, moderately weak, thickly bedded							
	18·5		17·8				

U: Undisturbed Sample   REMARKS: Water level (0930 hrs)
D: Disturbed Sample   29:7:77   32·2 m
B: Bulk Disturbed Sample   30:7:77   32·5 m
W: Water Sample   31:7:77   32·5 m
▽: Water Table

are not suitable for all ground conditions and there are limitations to the information that can be obtained: they must be considered mainly as supplementary methods. It is possible to locate strata boundaries only if the physical properties of the adjacent materials are significantly different. It is always necessary to check the results against data obtained by direct methods such as boring. Provided such correlations are established, geophysical methods can produce rapid and economic results, for example for the filling in of detail between widely spaced boreholes or to indicate where additional boreholes may be required. The methods can be useful in estimating the depth to bedrock or to the water table.

Seismic Refraction Method

The method depends on the fact that seismic waves have different velocities in different types of soil (or rock): in addition, the waves are refracted when they cross the boundary between different types of soil. The method enables the general soil types and the approximate depths to strata boundaries, or to bedrock, to be determined.

Waves are generated either by the detonation of explosives or by striking a metal plate with a large hammer. The equipment consists of one or more sensitive vibration transducers, called geophones, and an extremely accurate time-measuring device called a seismograph. A circuit between the detonator or hammer and the seismograph starts the timing mechanism at the instant of detonation or impact. The geophone is also connected electrically to the seismograph: when the first wave reaches the geophone the timing mechanism stops and the time interval is recorded in milliseconds.

When detonation or impact takes place, waves are emitted in every direction. One particular wave, called the direct wave, will travel parallel to the surface in the direction of the geophone. Other waves travel in a downward direction, at various angles to the horizontal, and will be refracted if they pass into a stratum of different seismic velocity. If the seismic velocity of the lower stratum is higher than that of the upper stratum, one particular wave will travel along the top of the lower stratum, parallel to the boundary, as shown in Fig. 10.8a: this wave continually 'leaks' energy back to the surface. Energy from this refracted wave can be detected by the geophone.

The procedure (Fig. 10.8a) consists of installing a geophone in turn at a number of points in a straight line, at increasing distances from the source of wave generation. The length of the line of points should be 3–5 times the required depth of investigation. A series of detonations or impacts is produced and the arrival time of the first wave at each geophone position is recorded in turn. When the distance between source and geophone is short, the arrival time will be that of the direct wave. When the distance between source and geophone exceeds a certain value (depending on the thickness of

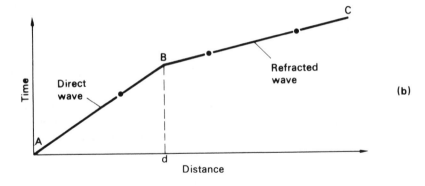

(b)

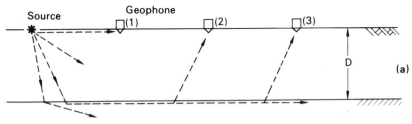

(a)

Fig. 10.8   Seismic refraction method.

the upper stratum) the refracted wave will be the first to be detected by the geophone. This is because the path of the refracted wave, although longer than that of the direct wave, is partly through a stratum of higher seismic velocity. The use of explosives is generally necessary if the source-geophone distance exceeds 30–50 m or if the upper soil stratum is loose.

An alternative procedure consists of using a single geophone position and producing a series of detonations or impacts at increasing distances from the geophone.

Arrival time is plotted against the distance between source and geophone, a typical plot being shown in Fig. 10.8b. If the source-geophone spacing is less than d the direct wave reaches the geophone in advance of the refracted wave and the time-distance relationship is represented by a straight line (AB) through the origin. If, on the other hand, the source-geophone distance is greater than d the refracted wave arrives in advance of the direct wave and the time-distance relationship is represented by a straight line (BC) at a different slope to that of AB. The slopes of the lines AB and BC are the seismic velocities (v_1 and v_2) of the upper and lower strata respectively. The general types of soil or rock can be determined from a knowledge of these velocities. The depth (D) of the boundary between the two strata (provided the thickness of the upper stratum is constant) can be estimated from the formula:

$$D = \frac{d}{2} \sqrt{\left(\frac{v_2 - v_1}{v_2 + v_1}\right)} \qquad\qquad (10.2)$$

The method can also be used where there are more than two strata and procedures exist for the identification of inclined strata boundaries and vertical discontinuities.

The formulae used to estimate the depths of strata boundaries are based on the assumptions that each stratum is homogeneous and isotropic, the boundaries are plane, each stratum is thick enough to produce a change in slope on the time-distance plot and the seismic velocity increases in each successive stratum from the surface downwards. A layer of clay lying below a layer of compact gravel, for example, would not be detected. Other difficulties arise if the velocity ranges of adjacent strata overlap, making it difficult to distinguish between them, and if the velocity increases with depth in a particular stratum. It is important that the results are correlated with data from borings.

Electrical Resistivity Method

The method depends on differences in the electrical resistance of different soil (and rock) types. The flow of current through a soil is mainly due to electrolytic action and therefore depends on the concentration of dissolved salts in the pore water: the mineral particles of a soil are poor conductors of current. The resistivity of a soil therefore decreases as both the water content and the concentration of salts increase. A dense, clean sand above the water table, for example, would exhibit a high resistivity due to its low degree of saturation and the virtual absence of dissolved salts. A saturated clay of high void ratio, on the other hand, would exhibit a low resistivity due to the relative abundance of pore water and the free ions in that water.

In its usual form (Fig. 10.9a) the method involves driving four electrodes into the ground at equal distances (L) apart in a straight line. Current (I), from a battery, flows through the soil between the two outer electrodes, producing an electrical field within the soil. The potential drop (E) is then measured between the two inner electrodes. The apparent resistivity (R) is given by the equation:

$$R = \frac{2\pi LE}{I} \qquad\qquad (10.3)$$

The apparent resistivity represents a weighted average of true resistivity in a large volume of soil, the soil close to the surface being more heavily weighted than the soil at depth. The presence of a stratum of soil of high resistivity lying below a stratum of low resistivity forces the current to flow closer to the surface resulting in a higher voltage drop and hence a higher value of apparent resistivity. The opposite is true if a stratum of low resistivity lies below a stratum of high resistivity.

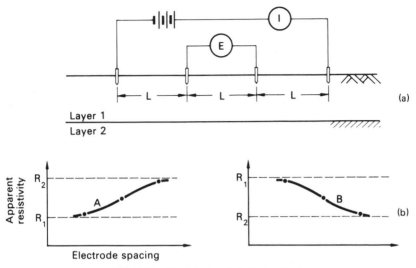

Fig. 10.9 Electrical resistivity method.

The procedure known as *sounding* is used when the variation of resistivity with depth is required: this enables rough estimates to be made of the types and depths of strata. A series of readings is taken, the (equal) spacing of the electrodes being increased for each successive reading: however the centre of the four electrodes remains at a fixed point. As the spacing is increased the apparent resistivity is influenced by a greater depth of soil. If the resistivity increases with increasing electrode spacing it can be concluded that an underlying stratum of higher resistivity is beginning to influence the readings. If increased separation produces decreasing resistivity, on the other hand, a stratum of lower resistivity is beginning to influence the readings. The greater the thickness of a layer the greater the electrode spacing over which its influence will be observed, and vice versa.

Apparent resistivity is plotted against electrode spacing, preferably on log-log paper. Characteristic curves for a two-layer structure are illustrated in Fig. 10.9b. For curve A the resistivity of layer 1 is lower than that of layer 2: for curve B layer 1 has a higher resistivity than layer 2. The curves become asymptotic to lines representing the true resistivities R_1 and R_2 of the respective layers. Approximate layer thicknesses can be obtained by comparing the observed curve of resistivity versus electrode spacing with a set of standard curves. Other methods of interpretation have also been developed for two-layer and three-layer systems.

The procedure known as *profiling* is used in the investigation of lateral variation of soil types. A series of readings is taken, the four electrodes being moved laterally as a unit for each successive reading: the electrode spacing remains constant for each reading. Apparent resistivity is plotted against the centre position of the four electrodes, to natural scales: such plots can be used to locate the positions of soil of high or low resistivity. Contours of resistivity can be plotted over a given area.

The apparent resistivity for a particular soil or rock type can vary over a wide range of values: in addition, overlap occurs between the ranges for different types. This makes the identification of soil or rock type and the location of strata boundaries extremely uncertain. The presence of irregular features near the surface and of stray potentials can also cause difficulties in interpretation. It is essential, therefore, that the results obtained are correlated with borehole data. The method is not considered to be as reliable as the seismic method.

References

10.1　Bishop, A. W. (1948): 'A New Sampling Tool for use in Cohesionless Sands below Ground Water Level', *Geotechnique*, Vol. 1, No. 2.

10.2　British Standard 5930 (1981) *Code of Practice for Site Investigations*, British Standards Institution, London.

10.3　Dobrin, M. D. (1960): *Introduction to Geophysical Prospecting*, McGraw-Hill, New York.

10.4　Griffiths, D. H. and King, R. F. (1965): *Applied Geophysics for Engineers and Geologists*, Pergamon, London.

10.5　Hvorslev, M. J. (1949): *Subsurface Exploration and Sampling of Soils for Civil Engineering Purposes*, US Waterways Experimental Station, Vicksburg, Mississippi.

10.6　Kjellman, W., Kallstenius, T. and Wager, O. (1950): 'Soil Sampler with Metal Foils', *Proceedings Royal Swedish Geotechnical Institute*, No. 1.

10.7　Robertshaw, J. and Brown, P. D. (1955): 'Geophysical Methods of Exploration and their Application to Civil Engineering Problems', *Journal ICE*, Part I.

10.8　Rowe, P. W. (1972): 'The Relevance of Soil Fabric to Site Investigation Practice', *Geotechnique*, Vol. 22. No. 2.

10.9　Weltman, A. J. and Head, J. M. (1983): '*Site Investigation Manual*', CIRIA/PSA, London.

Principal Symbols

A, a	Area
A	Air content
$A, \bar{A}$	Pore pressure coefficients
a'	Modified shear strength parameter (effective stress)
a	Dial gauge reading in oedometer test
B	Width of footing
$B, \bar{B}$	Pore pressure coefficients
C_U	Coefficient of uniformity
C_C	Coefficient of curvature
C_N	Correction factor for overburden pressure
C_w	Correction factor for water table position
C_s	Isotropic compressibility of soil skeleton
C_{s0}	Uni-axial compressibility of soil skeleton
C_v	Compressibility of pore fluid
C_c	Compression index
C_α	Rate of secondary compression
c	Shear strength parameter
c_u	Undrained (total stress) shear strength parameter
c'	Drained (effective stress) shear strength parameter
c'_r	Drained residual shear strength parameter
c_w	Wall adhesion
c_v	Coefficient of consolidation (vertical drainage)
c_h	Coefficient of consolidation (horizontal drainage)
D	Depth of footing; depth of excavation
D	Depth factor
D	Particle size
D_b	Depth of embedment of pile in bearing stratum
D_r	Relative density
d	Length of drainage path
d	Diameter; depth
E	Young's modulus
e	Void ratio
e	Eccentricity
F	Factor of safety
f_s	Skin friction
G	Shear modulus
G_s	Specific gravity of solid particles
g	$9 \cdot 8 \text{ m/s}^2$
H, h	Height

H	Layer or specimen thickness
h	Total head
I	Influence factor
I_P	Plasticity index (or PI)
I_L	Liquidity index (or LI)
I_B	Brittleness index
i	Hydraulic gradient
i	Inclination factor
J	Seepage force
j	Seepage pressure
K	Lateral pressure coefficient
K_A	Active pressure coefficient
K_P	Passive pressure coefficient
K_0	Coefficient of earth pressure at rest
K	Absolute permeability
k	Coefficient of permeability
L, l	Length
LL	Liquid limit (or w_L)
LI	Liquidity index (or I_L)
M	Mass
m_v	Coefficient of volume compressibility
N	Normal force
N	Standard penetration resistance
N_d	Number of equipotential drops
N_f	Number of flow channels
N_γ	Bearing capacity factor
N_c	Bearing capacity factor
N_q	Bearing capacity factor
N_s	Stability coefficient
n	Porosity
n_d	Equipotential number
P_A	Total active thrust
P_P	Total passive resistance
PL	Plastic limit (or w_P)
PI	Plasticity index (or I_P)
p	Stress invariant
p	Pressure
p_A	Active pressure
p_P	Passive pressure
p_0	At-rest pressure
Q	Surface load
Q_f	Ultimate load
q	Flow per unit time
q	Stress invariant
q	Surface pressure; total foundation pressure

q_n	Net foundation pressure
q_a	Allowable bearing capacity
q_f	Ultimate bearing capacity
q_{nf}	Net ultimate bearing capacity
q_c	Cone penetration resistance
R, r	Radius
r	Compression ratio
r_u	Pore pressure ratio
S_r	Degree of saturation
s	Shape factor
s	Settlement
s_c	Consolidation settlement
s_i	Immediate settlement
T_v	Time factor (vertical drainage)
T_r	Time factor (radial drainage)
t	Time
U	Boundary water force
U	Degree of consolidation
u, u_w	Pore water pressure
u_a	Pore air pressure
V	Volume
v	Specific volume
v	Discharge velocity
v'	Seepage velocity
W	Weight
w	Water content
w_L	Liquid limit (or *LL*)
w_P	Plastic limit (or *PL*)
z	Depth coordinate
z	Elevation head
α'	Modified shear strength parameter (effective stress)
α	Angle of wall inclination
α	Skin friction coefficient
β	Skin friction coefficient
β	Slope angle
γ	Shear strain
γ	Unit weight
γ_d	Dry unit weight
γ_{sat}	Saturated unit weight
γ'	Buoyant unit weight
γ_w	Unit weight of water
δ	Angle of wall friction
ε	Normal strain
ε_v	Volumetric strain
η	Dynamic viscosity

κ	Slope of isotropic swelling/recompression line
λ	Slope of isotropic normal consolidation line
μ	Settlement coefficient
M	Slope of critical state line
v	Poisson's ratio
ρ	Bulk density
ρ_d	Dry density
ρ_{sat}	Saturated density
ρ_w	Density of water
σ	Total normal stress
σ'	Effective normal stress
$\sigma_1, \sigma_2, \sigma_3$	Total principal stresses
$\sigma'_1, \sigma'_2, \sigma'_3$	Effective principal stresses
τ	Shear stress
τ_f	Shear strength; peak shear strength
τ_r	Residual shear strength
ϕ	Potential function
ϕ	Shear strength parameter
ϕ_u	Undrained (total stress) shear strength parameter
ϕ'	Drained (effective stress) shear strength parameter
ϕ'_r	Drained residual shear strength parameter
ϕ'_{max}	Maximum (peak) angle of shearing resistance
ϕ'_{cv}	Angle of shearing resistance at constant volume
ϕ_μ	True angle of friction
χ	Parameter in effective stress equation for partially-saturated soil
ψ	Flow function
ψ	Angle of dilation

Answers to Problems

Chapter 1

1.1 SW, MS, ML, CV, (SW, SM, ML, CH)
1.2 0·55, 46·6%, 2·10 Mg/m³, 20·4%
1.3 15·7 kN/m³, 19·7 kN/m³, 9·9 kN/m³, 18·7 kN/m³, 19·3%
1.4 98%
1.5 1·92 Mg/m³, 0·38, 83·7%, 4·5%; no
1.6 15%, 1·83 Mg/m³, 3·5%

Chapter 2

2.1 $4·9 \times 10^{-8}$ m/s
2.2 $1·3 \times 10^{-6}$ m³/s (per m)
2.3 $5·8 \times 10^{-5}$ m³/s (per m), 316 kN/m
2.4 $2·0 \times 10^{-6}$ m³/s (per m)
2.5 $4·7 \times 10^{-6}$ m³/s (per m)
2.6 $1·1 \times 10^{-6}$ m³/s (per m)
2.7 $1·8 \times 10^{-5}$ m³/s (per m)
2.8 $1·0 \times 10^{-5}$ m³/s (per m)

Chapter 3

3.1 48·5 kN/m²
3.2 51·4 kN/m², 33·4 kN/m²
3.3 105·9 kN/m²
3.4 (a) 94·0 kN/m², 154·2 kN/m², (b) 94·0 kN/m², 133·8 kN/m²
3.5 9·9 kN, 73° below horizontal
3.6 30·2 kN/m², 10·6 kN/m²
3.7 1·5, 14 kN/m², 90 kN/m²
3.8 2·0, 0·65 m

Chapter 4

4.1 113 kN/m²
4.2 44°
4.3 110 kN/m², 0°
4.4 205 kN/m²
4.5 0, $25\frac{1}{2}°$, 170 kN/m²
4.6 15 kN/m², 28°
4.7 76 kN/m²
4.8 0·73
4.9 0·96, 0·23

Chapter 5

5.1	96 kN/m²
5.2	277 kN/m²
5.3	45 kN/m²
5.4	68 kN/m²
5.5	76 kN/m
5.6	7 mm

Chapter 6

6.1	76.5 kN/m, 122 kN/m
6.2	571 kN/m, 8·57 m
6.3	163 kN/m, 160 kN/m
6.4	175 kN/m², 69 kN/m², 1.9
6.5	3.95 m
6.6	228 kN/m², 35 kN/m², 1.2, 1.6
6.7	5·60 m, 226 kN
6.8	2·1, 314 kN
6.9	2·15
6.10	110 kN
6.11	1·7

Chapter 7

7.1	$c_v = 2·7$, $2·6\,\text{m}^2/\text{year}$, $m_v = 0·98\,\text{m}^2/\text{MN}$, $k = 8·1 \times 10^{-10}\,\text{m/s}$
7.2	318 mm, 38 mm (four sublayers)
7.3	2·6 years, 0·95 years
7.4	35·2 kN/m²
7.5	0·65
7.6	130 mm, 95 mm
7.7	124 mm, 38 mm, 72 mm, 65 mm
7.8	285 mm (six sublayers)
7.9	0·80
7.10	8·8 years, 0·7 years

Chapter 8

8.1	2·8, 2·9
8.2	4·8, 3·1
8.3	4100 kN
8.4	7·0, 5·3, 3·4
8.5	225 kN/m²
8.6	1·8
8.7	270 kN/m²
8.8	225 kN/m²
8.9	31 mm, 22 mm

8.10   9200 kN
8.11   1900 kN
8.12   230 kN

Chapter 9

9.1   1·43, 1·27
9.2   50°, 27°
9.3   0·91
9.4   1·01
9.5   1·22
9.6   13°, 3·1

Index